SIXTH EDITION

AUTOMOTIVE
AUTOMATIC
TRANSMISSIONS

WILLIAM H. CROUSE
DONALD L. ANGLIN

GREGG DIVISION/McGRAW-HILL BOOK COMPANY

New York ○ Atlanta ○ Dallas ○ St. Louis ○ San Francisco ○ Auckland ○ Bogotá ○ Guatemala
Hamburg ○ Lisbon ○ London ○ Madrid ○ Mexico ○ Montreal ○ New Delhi
Panama ○ Paris ○ San Juan ○ São Paulo ○ Singapore ○ Sydney ○ Tokyo ○ Toronto

ABOUT THE AUTHORS

William H. Crouse

Behind William H. Crouse's clear technical writing is a background of sound mechanical engineering training as well as a variety of practical industrial experience. After finishing high school, he spent a year working in a tinplate mill. Summers, while still in school, he worked in General Motors plants, and for three years he worked in the Delco-Remy division shops. Later he became director of field education in the Delco-Remy Division of General Motors Corporation, which gave him an opportunity to develop and use his writing talent in the preparation of service bulletins and educational literature.

During the war years, he wrote a number of technical manuals for the Armed Forces. After the war, he became editor of technical education books for the McGraw-Hill Book Company. He has contributed numerous articles to automotive and engineering magazines and has written many outstanding books. He was the first editor-in-chief of McGraw-Hill's *Encyclopedia of Science and Technology*.

William H. Crouse's outstanding work in the automotive field has earned for him membership in the Society of Automotive Engineers and in the American Society of Engineering Education.

Donald L. Anglin

Trained in the automotive and diesel service field, Donald L. Anglin has worked both as a mechanic and as a service manager. He has taught automotive courses and has also worked as curriculum supervisor and school administrator for an automotive trade school. Interested in all types of vehicle performance, he has served as a racing-car mechanic and as a consultant to truck fleets on maintenance problems.

Currently he devotes full time to technical writing, teaching, and visiting automotive instructors and service shops. Together with William H. Crouse he has co-authored magazine articles on automotive education and many automotive books published by McGraw-Hill.

Donald L. Anglin is a Certified General Automotive Mechanic, a Certified General Truck Mechanic, and holds many other licenses and certificates in automotive education, service, and related areas. His work in the automotive service field has earned for him membership in the American Society of Mechanical Engineers and the Society of Automotive Engineers. In addition, he is a member of the Board of Trustees of the National Automotive History Collection.

Library of Congress Cataloging in Publication Data

Crouse, William Harry (date)
  Automotive automatic transmissions.

  "Expanded and updated version of half of
Automotive transmissions and power trains, fifth
edition"—Pref.
  Includes index.
  1. Automobiles—Transmission devices, Automatic.
2. Automobiles—Transmission devices, Automatic—
Maintenance and repair.   I. Anglin, Donald L.
II. Crouse, William Harry (date). Automotive
transmissions and power trains. 5th ed. 1976.
III. Title.
TL263.C76   1983   629.2'446   81-14262
ISBN 0-07-014771-X   AACR2

AUTOMOTIVE AUTOMATIC TRANSMISSIONS
Sixth Edition

7 8 9 0   SMSM   8 9

Sponsoring Editor: D. Eugene Gilmore
Editing Supervisor: Iris Cohen
Design Supervisor: Caryl Valerie Spinka
Production Supervisor: Kathleen Morrissey

Text Designer: Linda Conway
Cover Designer: David Thurston
Technical Studio: J & R Services
Cover Illustration: Fine Line, Inc.

ISBN 0-07-014771-X

CONTENTS

The material contained in *Automotive Automatic Transmissions*, Sixth Edition, is an expanded and updated version of half of *Automotive Transmissions and Power Trains*, Fifth Edition. When the time came to revise and publish a new edition of *Automotive Transmissions and Power Trains*, the authors and editors were faced with a monumental task. Automotive manufacturers had brought out many new manual and automatic transmissions. There are now four- and five-speed manual transmissions with overdrive. There are now new automatic transmissions with torque-converter lockup in top gear, and also automatic transmissions with overdrive. Transaxles—manual and automatic, and also single- and dual-range—have appeared on the scene. Transfer cases have become more common with the increased production of four-wheel-drive vehicles.

All of these new developments demanded space in the new edition of *Automotive Transmissions and Power Trains*. However, if all of these new developments were included in a single book, the result would be a very large, expensive, and unwieldy volume. Therefore, the decision was made to split the book into two volumes: *Automotive Manual Transmissions and Power Trains* and *Automotive Automatic Transmissions*. *Automotive Automatic Transmissions* is one of nine books in the McGraw-Hill Automotive Technology Series. These books cover in detail the construction, operation, and maintenance of automotive vehicles. They are designed to give you all the information you need to become successful in the automotive service business.

The books satisfy the recommendations of the Motor Vehicle Manufacturers Association–American Vocational Association Industry Planning Council. They also meet the requirements for automotive mechanics certification and state vocational educational programs, as well as the recommendations for automotive trade apprenticeship training. Furthermore, the comprehensive coverage of the subject matter in these books makes them valuable additions to the library of anyone interested in any aspect of automotive engineering, manufacturing, sales, service, and operation.

Meeting the standards

The nine books in the McGraw-Hill Automotive Technology Series meet the standards set by the Motor Vehicle Manufacturers Association (MVMA) for an associate degree in automotive servicing and in automotive service management. The books also cover the subjects recommended by the American National Standards Institute in their detailed standard D18.1-1980, *American National Standard for Training of Automotive Mechanics for Passenger Cars, Recreational Vehicles, and Light Trucks*.

In addition, the books cover in depth the subject matter tested by the National Institute for Automotive Service Excellence (NIASE). The tests given by NIASE are used for certifying general automotive mechanics and automotive technicians working in specific areas of specialization under the NIASE voluntary mechanic testing and certification program.

Getting practical experience

At the same time that you study the books, you should be getting practical experience in the shop. You should handle automotive parts, automotive tools, and automotive servicing equipment, and you should perform actual servicing jobs. To assist you in your shop work, there is a workbook for each book in the Automotive Technology Series. For example, the *Workbook for Automotive Automatic Transmissions* includes the jobs which cover the basic procedures for servicing automotive automatic transmissions. If you do every job in the workbook, you will have hands-on experience with a variety of automotive automatic-transmission work.

While you are studying *Automotive Automatic Transmissions*, and demonstrating your ability in the shop, you and your instructor should be constantly aware of the need for congruence between your school curriculum and your future needs as you face and cope with real life in the automotive field. All of your learning about the automobile should be competency-based. Find out what the minimum standard of competence is in your chosen field. Then with the aid of your instructor develop your skills by mastering the necessary competencies and performance indicators outlined in the text.

If you are taking an automotive mechanics course in school, the instructor will guide you in your classroom and shop activities. But even if you are not taking a course, the workbook can act as an instructor. It tells you, step by step, how to do the various servicing jobs. Perhaps you can meet others who are taking a school course in automotive mechanics. You can talk over with them any problem you have. A local new-car dealership or independent garage is a good source of practical information. Get acquainted with the automotive mechanics there. Watch them at their work if you can. Make notes of important points for filing in a notebook.

Service publications

While you are in the service shop, study the various publications received at the shop. Automobile manufacturers, as well as suppliers of parts, accessories, and tools, publish shop manuals, service bulletins, and parts catalogs. All these help service technicians do a better job. In addition, numerous automotive magazines are published which deal with problems and methods of automotive service. All these publications will be of great value to you; study them carefully.

These activities will help you get practical experience in automotive mechanics. Sooner or later this experience, plus the knowledge that you have gained in studying the books in the McGraw-Hill Automotive Technology Series, will permit you to step into the automotive shop on a full-time basis. Or, if you are already in the shop, you will be equipped to step up to a more responsible job.

NIASE-type multiple-choice tests

The aim of anyone in automotive servicing should be to become certified. Certified mechanics have an edge over mechanics who are not certified. They have been tested by the NIASE (National Institute for Automotive Service Excellence) and have been found to possess the necessary knowledge and skills in the field to be qualified mechanics. To help prepare you to take the NIASE tests, NIASE-type multiple-choice tests are included at the end of each chapter. In these tests you will find a series of statements, followed by four words or phrases, labeled *a, b, c,* and *d*. You pick the *one* correct, best, or most probable answer to complete each statement. At the end of the book you will find the answers to the tests.

These multiple-choice tests do two things. First, they test your knowledge of the chapter you have just completed. If you have trouble with a test, you should review the chapter. Second, these tests prepare you for the NIASE tests, which you should be taking after you have had some practical experience in the shop. When you pass a test, you will become a *certified mechanic* in that area.

William H. Crouse
Donald L. Anglin

ACKNOWLEDGMENTS

During the preparation of this edition of *Automotive Automatic Transmissions*, the authors were given invaluable aid and inspiration by many people in the automotive industry and in the field of education. The authors gratefully acknowledge their indebtedness and offer their sincere thanks to these people. All cooperated with the aim of providing accurate and complete information that would be useful in training automotive mechanics.

Special thanks are owed to the following organizations for information and illustrations that they supplied: American Motors Corporation; Buick Motor Division of General Motors Corporation; Cadillac Motor Car Division of General Motors Corporation; Chevrolet Motor Division of General Motors Corporation; Chrysler Corporation; Ford Motor Com-

pany; General Motors Corporation; Oldsmobile Division of General Motors Corporation; Pontiac Motor Division of General Motors Corporation; Sun Electric Corporation; Toyota Motor Sales Company, Ltd.; Volkswagen of America, Inc.; and Weldun International. To all these organizations and the people who represent them, sincere thanks.

We would also like to acknowledge our debt to Paul H. Smith, Vale Technical Institute, for his help over the many years he has been a user and reviewer of this text. His helpful suggestions and technical expertise have been of tremendous value in the preparation of this edition.

William H. Crouse
Donald L. Anglin

POWER-TRAIN COMPONENTS

After studying this chapter, you should be able to:

1. Locate and identify power-train components on various vehicles.
2. Explain the purpose of each component.
3. Define the term *gear ratio* and discuss what it means in terms of gear size.
4. Explain the relationship between gear ratio and torque.
5. Discuss the purpose of the differential and explain how it works.

 1-1 Components of the automobile The automobile has five basic components, or parts:

1. The power plant, or engine, which is the source of power.
2. The chassis, which supports the engine and body and includes the brake, steering, and suspension systems.
3. The power train, or drive train, which is the power-transmission system that carries power from the engine to the drive wheels. It consists of the clutch (on vehicles with a manual transmission), transmission, transfer case (on vehicles with four-wheel drive), and drive-axle assembly. This unit includes the final drive, the differential, and the wheel axles.
4. The car body.
5. The car-body accessories, which include the heater and air conditioner, lights, radio and tape player, windshield wiper and washer, and electric windows and seat adjusters.

Many of these major components of the car can be seen in the illustration of the automotive chassis shown in Fig. 1-1.

The engine (Fig. 1-2) produces power by burning a mixture of gasoline vapor (or diesel fuel) and air in the engine combustion chambers. Combustion causes high pressure, which forces the engine pistons downward. The downward pushes on the pistons are carried through connecting rods to cranks on the engine crankshaft. These pushes on the cranks cause the crankshaft (Fig. 1-3) to rotate.

The engine *flywheel* (for a manual transmission) or drive plate (for an automatic transmission) is attached to the rear end of the crankshaft. The rear face of the flywheel is flat and smooth. It serves as the driving member of the clutch (on vehicles with manual transmissions). When the clutch is engaged, the rotary motion of the crankshaft is carried from the flywheel, through the clutch, to the transmission. With the transmission "in gear," the rotary motion is delivered by the transmission through a drive shaft to the differential. From there, the rotary motion is carried by the axle shafts and wheels to the tires, which push against the ground to move the car.

1-2 Power train The power train is a system of components that transmits power from the engine crankshaft to the drive wheels (Figs. 1-4 and 1-5). By shifting gears in the transmission, the power train can provide several different gear ratios between the engine crankshaft and the wheels.

For example, with a three-speed automatic transmission, the engine crankshaft will rotate about three, six, or nine times (in third, second, and first) to cause the wheels to rotate once. On cars with a three-speed manual transmission, about four, eight, or twelve revolutions of the crankshaft may be required to cause the wheels to rotate once. Cars with a manual transmission also must have a clutch (1-3).

1-3 Clutch (for manual transmission) Cars with a manual transmission (1-4) must have a means for temporarily disconnecting the transmission from the engine while the shifts are made. It is very difficult to shift gears while they are transmitting torque. Also, gear teeth would be moving at different speeds and they could be broken when the gears were brought together for meshing.

The clutch is located between the engine and the transmission (Fig. 1-6). Figure 1-7 shows a clutch disassembled. When the clutch is engaged (carrying

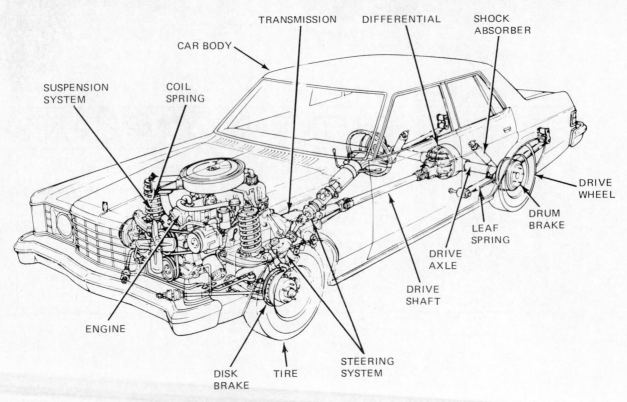

Fig. 1-1 Components of an automobile. The chassis contains the power plant, or engine; the frame, which supports the engine, wheels, and body; the power train, which carries the engine power to the drive wheels; and the brake, steering, and suspension systems. (*Ford Motor Company*)

torque throught it), the friction disk is clamped tightly between the engine flywheel (Fig. 1-8) and the pressure plate inside the clutch cover assembly. In this position, the friction disk must turn with the flywheel. However, the friction disk is splined to the transmission input

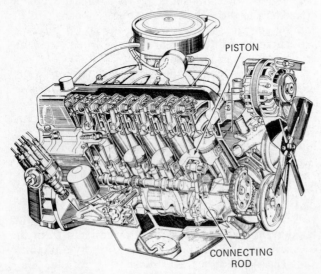

Fig. 1-2 Typical six-cylinder engine, partly cut away so that the internal construction can be seen. This engine is known as an *in-line* six because its six cylinders and pistons are located one behind the other. Note the cylindrical piston and its connecting rod shown in the cutaway section. (*Chrysler Corporation*)

shaft. When the disk turns, so does the input shaft. Torque passes from the engine crankshaft through the clutch into the transmission.

When the clutch pedal is pushed down by the driver, linkage moves the clutch fork. This pushes the throwout bearing into the pressure-plate-and-cover assembly. As the bearing moves in, it causes the spring force on the pressure plate to be released. This allows the friction disk to move away from the flywheel, which disengages the clutch (Fig. 1-8). Now, no torque can flow through the clutch. Gears may be shifted safely and easily because all torque between them has been relieved.

❂ **1-4 Manual transmission** A manual transmission (Fig. 1-6) is a transmission that the driver must shift by hand, or *manually*. It is an assembly of gears (❂ 1-6) and shafts that transmits power from the engine to the final drive or drive axle. In an automobile, the manual transmission provides three forward-gear ratios (a "three-speed" transmission) or more, and reverse. In addition, a neutral position is provided to permit disengaging the gears inside the transmission so that no power can flow through. Figure 1-9 shows the power flow through a four-speed manual transmission for each gear position.

Different gear ratios in the transmission are necessary because the internal-combustion engine develops relatively little power at low engine speeds. The engine must be turning at a fairly high speed before it can deliver enough power to get the car moving. To help

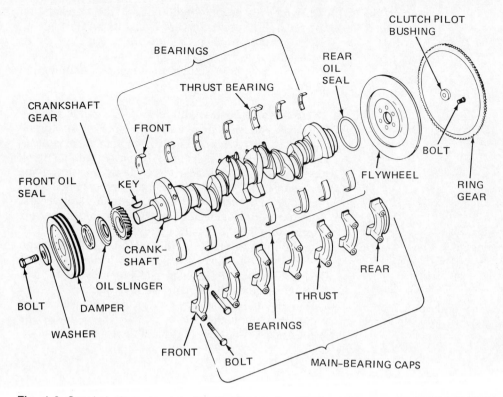

Fig. 1-3 Crankshaft and related parts for a six-cylinder in-line engine. *(Ford Motor Company)*

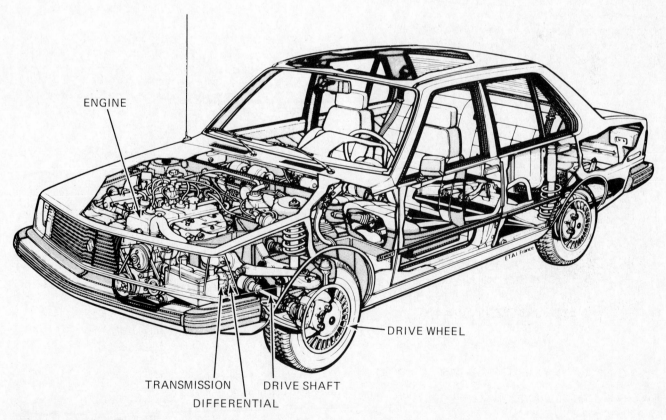

Fig. 1-4 Location of the major power-train components in a car with front-wheel drive. *(American Motors Corporation)*

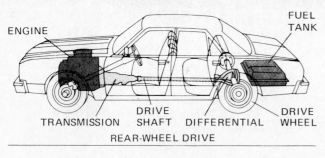

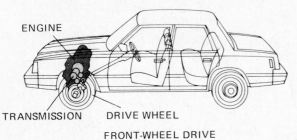

Fig. 1-5 Comparison of the power-train layout in a car with rear-wheel drive (top) and a car with front-wheel drive (bottom). (*Chrysler Corporation*)

overcome this problem with *inertia*, the transmission has several different forward-gear ratios. Through selection of the proper gear ratio, the engine torque or twisting force is increased. This permits putting the car into motion without stalling the engine or slipping the clutch excessively.

✿ 1-5 Automatic transmission The modern automatic transmission (Fig. 1-10) combines a fluid torque converter, a hydraulic control system, and a planetary-gear system in a single unit. In an automatic transmission, the various gear ratios between the engine crankshaft and wheels are selected and changed automatically. The driver does not manually shift gears. Automatic controls inside the transmission supply the proper ratio for the driving conditions.

In addition to the forward-gear ratios, neutral, and reverse, the automatic transmission has a *park* position. This locks the transmission to prevent the car from moving or rolling away while parked.

Both manual and automatic transmission use various types and combinations of gears (✿ 1-6) to change the speed and direction of rotation of the transmission out-

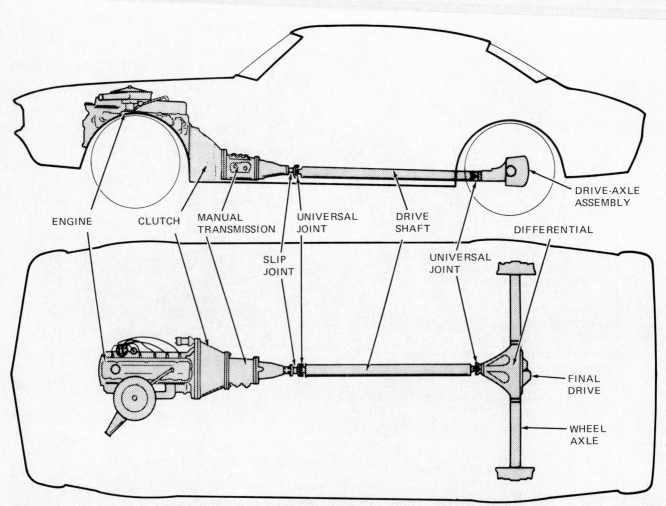

Fig. 1-6 Power train of an automobile with rear-wheel drive.

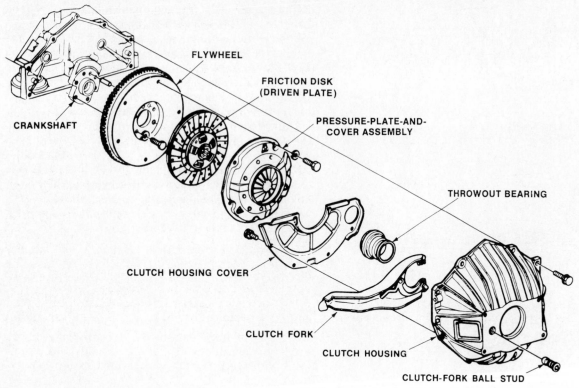

Fig. 1-7 Disassembled clutch assembly and flywheel. (*Chevrolet Motor Division of General Motors Corporation*)

put shaft. Figure 1-11 shows the power flow from the engine crankshaft to the wheels on a car with front-wheel drive. The torque converter and other major components of a typical three-speed automatic transmission are shown in Fig. 1-11.

⚙ 1-6 Gears Gears (Fig. 1-12) are wheels with teeth that transmit power between shafts (Fig. 1-13). The teeth may be on the inside or outside of the wheel, on the edge, or at an angle. Many types of gears are used in the automobile. The most basic type of gear is the *spur gear*. On the spur gear, the teeth are parallel to and align with the centerline or axis of the gear. The teeth are on the outside or external surface of the wheel. For this reason, the spur gear also is known as an *external gear*. Other types of gears differ from the spur gear mainly in the shape and alignment of the gear teeth.

For example, *helical gears* are like spur gears except that, in effect, the teeth have been twisted at an angle to the gear centerline. *Bevel gears* (Fig. 1-13) are shaped like imaginary cones with the tops cut off. The teeth point inward toward the apex or peak of the cone. Bevel gears are used to transmit motion through angles.

Other gears, called *ring gears* or *internal gears*, have their teeth pointing inward. These gears form part of the planetary gearset which is used in automotive automatic transmissions. Internal gears, a planetary gearset, and other types of gears are shown in Fig. 1-12.

Usually a gear is attached to a shaft and is meshed with another gear fastened to a different shaft. Gears are "meshed" when the teeth of the two gears are

interlaced (Fig. 1-13). In any set of gears, the smaller gear is the *pinion gear*.

When one gear of a two external-tooth gearset is turned, the teeth force the other gear to turn in the opposite direction. The relative speed between the two

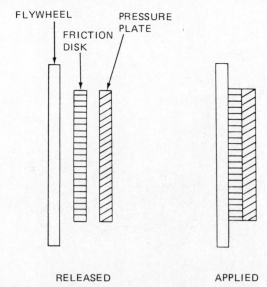

Fig. 1-8 Basic clutch elements, showing clutch action. Left, clutch released. The pressure plate and friction disk have moved away from the flywheel. Right, clutch applied. The pressure plate clamps the friction disk to the flywheel so that all parts must rotate together.

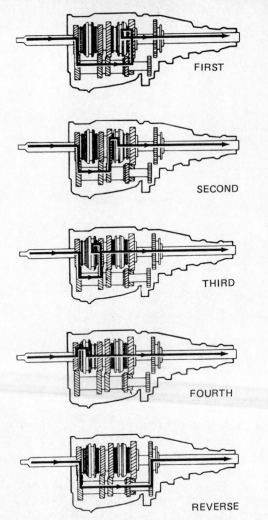

Fig. 1-9 Power flow through each gear position in a four-speed transmission.

FIRST

SECOND

THIRD

FOURTH

REVERSE

meshing gears is determined by the number of teeth on the gears. This is called the *gear ratio*.

For example, when two meshing gears have the same number of teeth, both gears turn at the same speed (Fig. 1-13). However, when the driven gear has more teeth than the driving gear, the smaller, driving gear will turn faster than the larger, driven gear. This means that a driving gear with 12 teeth (as shown in Fig. 1-14) will turn twice as fast as a driven gear with 24 teeth. The gear ratio between the gears is 2 to 1 (written as 2:1).

If a 12-tooth gear is meshed with a 36-tooth gear, the 12-tooth gear turns three times for every revolution of the larger gear. The gear ratio between these gears is 3:1. Gear ratio also may be determined by comparing the diameters of the two gears.

1. **Torque** The gear ratio changes as the number of teeth on each meshing gear changes. At the same time, *torque* also changes. Torque is twisting or turning effort. When you try to turn a shaft with your hand (Fig. 1-14), you are applying a twisting force or torque to the gear. Torque is measured in pound-feet (abbreviated lb-ft) or in newton-meters (N-m) in the metric system of measurement. Torque should not be confused with work, which is measured in foot-pounds (ft-lb) or joules (J).

To calculate torque, multiply the force (in pounds or newtons) times the distance (in feet or meters) from the center to the point where the force is exerted. For example, suppose you had a wrench 1 foot [0.31 m] long and that you used it to tighten a nut (Fig. 1-15). When you apply 10 pounds [44.5 N] of force to the wrench (as shown in Fig. 1-15), you are applying 10 lb-ft [13.6 N-m] of torque to the nut. If you apply 20 pounds [89 N] of force on the wrench, you are applying 20 lb-ft [27.1 N-m] of torque. If the wrench is 2 feet [0.61 m] long and you apply 10 pounds

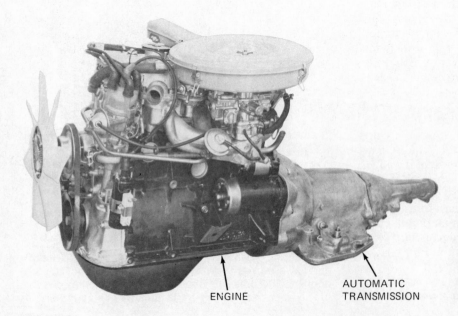

ENGINE

AUTOMATIC TRANSMISSION

Fig. 1-10 An automatic transmission connected to an engine. This engine-transmission assembly is for a car with the engine in the front and rear-wheel drive. (*Chrysler Corporation*)

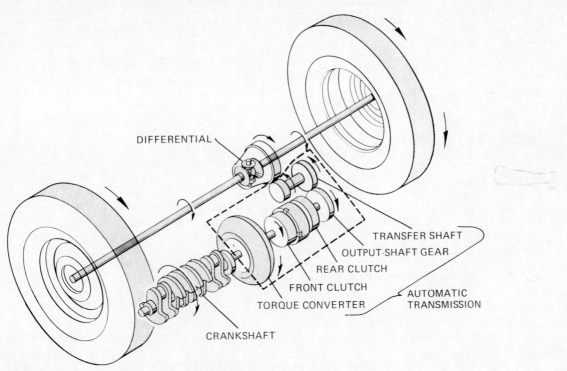

Fig. 1-11 Power flow from the engine crankshaft to the drive wheels and tires for a car with front-wheel drive. (*Chrysler Corporation*)

[44.5 N] of force, the torque on the nut is 20 lb-ft [27.1 N-m].

Any shaft or gear that turns has torque applied to it. The engine pistons and connecting rods push on the cranks on the crankshaft. This applies torque to the crankshaft and causes it to turn. The crankshaft applies torque through the flywheel and clutch to the gears in the manual transmission, and so the gears turn. This turning effort, or torque, is carried through the power train to the drive wheels.

2. **Torque in gears** Torque on shafts and gears is measured as a straight-line force at a distance from the center of the shaft or gear. For example, suppose we

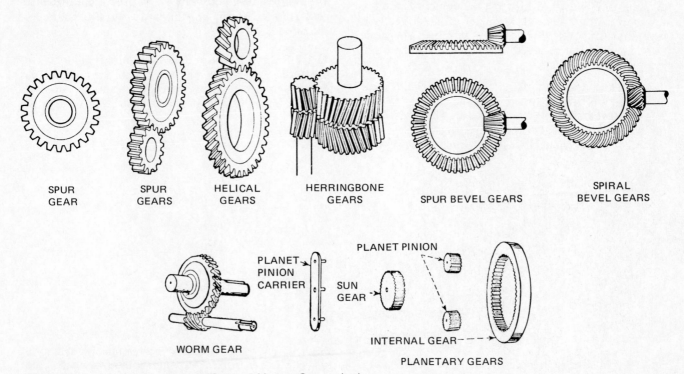

Fig. 1-12 Various types of gears. (*General Motors Cororation*)

Fig. 1-13 Meshed spiral-bevel gears.

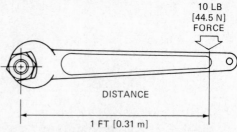

Fig. 1-15 Torque is found by multiplying the applied force times the distance through which the force acts.

want to measure the torque in the gears shown in Fig. 1-14. If we could hook a spring scale to the gear teeth and get a measurement of the pull on the scale, we could determine the torque. However, a spring scale actually is not used, because the teeth are moving. Other devices are used to measure the torque of rotating parts. Torque available at the engine crankshaft to do work is measured with an engine *dynamometer*. This is the torque that is applied to the transmission input shaft.

In a circle, the distance from the center to the outside edge is the *radius*. On a gear, the radius is the distance from the center to the point on the tooth where the force is applied. Now, suppose that a tooth on the driving gear is pushing against a tooth on the driven gear with a 25-pound [111-N] force (Fig. 1-16). When the force is applied at a distance of 1 foot [0.31 m], which is the radius of the driving gear, a torque of 25 lb-ft [33.9 N-m] is applied to the driven gear.

The 25-pound [111-N] force from the teeth of the smaller, driving gear is applied to the teeth of the larger, driven gear (Fig. 1-16). But the force is applied at a distance of 2 feet [0.61 m] from the center. Therefore the torque on the shaft at the center of the driven gear is 50 lb-ft (25 × 2) [67.8 N-m (33.9 × 2)]. The same force is acting at twice the distance from the shaft center.

3. Torque and gear ratio   When the smaller gear in Fig. 1-14 is driving the larger gear, the gear ratio is 2:1. However, the *torque ratio* is 1:2. The larger gear turns at half the speed of the smaller gear. As a result, the larger gear will have twice the torque of the smaller gear.

In gear systems, *speed reduction means torque increase*. For example, when a typical three-speed transmission is in first gear, there is a speed reduction (or gear reduction) of 9:1 from the engine to the drive wheels. The crankshaft turns nine times to turn the wheels once. Ignoring losses due to friction, this means that the torque increases nine times. If the engine produces a torque of 100 lb-ft [135.6 N-m], then 900 lb-ft [1220 N-m] of torque is applied to the drive wheels.

To see how the drive torque forces the car to move forward, refer to Fig. 1-17. In the example shown, the engine is delivering 100 lb-ft [135.6 N-m] of torque. The gear reduction from the engine to the drive wheels is 9:1, with a torque increase of 1:9. Wheel radius is assumed to be 1 foot [0.31 m], for ease of figuring.

With the torque acting on the ground at a distance of 1 foot [0.31 m], which is the radius of the wheel, the force of the tire pushing against the ground is 900 lb-ft [1220 N-m]. As a result, the car is pushed forward with a force of 900 pounds [4003 N].

NOTE Actually, the torque is split between the two drive wheels. Each tire pushes against the ground with a force of 450 pounds [2001.6 N]. It is both tires together that push the car forward with a total force of 900 pounds [4003 N].

⚙ **1-7 Engine torque**   The automotive engine has little torque at low speed (low engine revolutions per minute, or rpm). As its speed increases, so does its torque (Fig. 1-18). The engine must be turning above

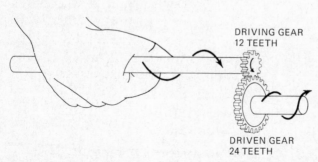

Fig.1-14 Two revolutions of the small gear are required to turn the large gear once. This is a gear ratio of 2:1.

DRIVING GEAR
12 TEETH

DRIVEN GEAR
24 TEETH

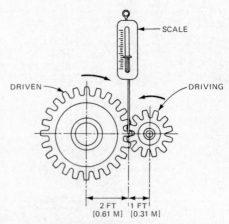

Fig. 1-16 The torque on a gear is the force on a gear tooth times the distance from the center of the shaft to the point on the tooth where the force is applied.

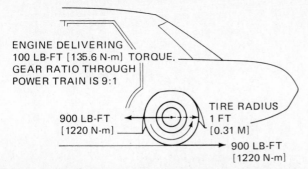

ENGINE DELIVERING
100 LB-FT [135.6 N-m] TORQUE.
GEAR RATIO THROUGH
POWER TRAIN IS 9:1

900 LB-FT
[1220 N-m]

TIRE RADIUS
1 FT
[0.31 M]

900 LB-FT
[1220 N-m]

Fig. 1-17 Engine torque is multiplied through the power train and delivered by the tires to the ground, forcing the car forward. The tire is turned with a torque of 900 lb-ft [1220 N-m]. Since the tire radius is 1 foot [0.31 m], the force of the tire against the ground will also be 900 lb-ft [1220 N-m]. As a result, the car is pushed forward with a force of 900 pounds [4003 N].

500 rpm to produce enough torque (and force at the wheels) to get the car moving and increase its speed.

When the engine is being cranked for starting, it must have no load on it. For this reason, both manual and automatic transmissions have neutral gear positions in which the power train is disconnected from the engine.

After the engine is started, its speed must be increased to a fairly high rpm so it will deliver enough torque to get the car moving. During this starting-out or "breakaway" period, the transmission is in first or low gear. The crankshaft is turning about nine times to turn the wheels once. Therefore engine torque is high, so the car starts moving and accelerates.

As a typical example, suppose the engine speed is increased to 1500 rpm. This means that the crankshaft is making 1500 complete revolutions per minute. At this engine speed, the car is moving at 15 miles per hour (mph) [24 km/h] with the transmission in low or first gear. The transmission gear ratio is about 3:1. Now a shift is made to second or intermediate gear. The transmission gear ratio is now about 2:1. An engine speed of 1500 rpm will move the car at a little under 25 mph [40 km/h]. When the shift is made to third or high,

there is direct drive through the transmission. The transmission output shaft turns at the same speed as the transmission input shaft. An engine speed of 1500 rpm will now move the car at more than 40 mph [64 km/h]. There is a further gear reduction of about 3:1 through the differential or final drive, as explained later.

Figure 1-19 shows the relationship between car speed and transmission output-shaft torque for each forward-gear position. Note that the numbers used in the examples are only approximate. They vary according to the design of the transmission. Also, how the shifts are made was not considered. In a mechanical transmission, the shifts are made by a mechanical movement of the shift lever. In an automatic transmission, the shifts are made automatically and are based on car and engine speed, throttle opening, and other factors. These are discussed later.

⚙ 1-8 Gear combinations Two gears in mesh will turn at different speeds if they are of different sizes. If one gear turns in one direction, the other gear will turn in the opposite direction, as shown by the arrows in Fig. 1-15. Also, when a gear with a larger diameter drives a smaller gear, the speed increases but the torque decreases. When a small gear drives a big gear, the speed decreases but the torque increases.

When two external gears are in mesh, they turn in opposite directions. But when a third gear is placed in the gear train, as shown in Fig. 1-20, the two outside gears will turn in the same direction. The middle gear is called an *idler gear*. It does not change the gear ratio.

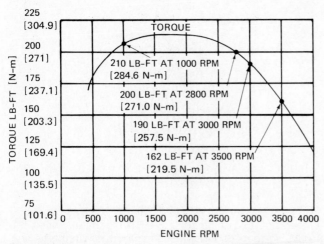

Fig. 1-18 Torque curve of an engine, showing how it increases and then falls off as engine speed increases.

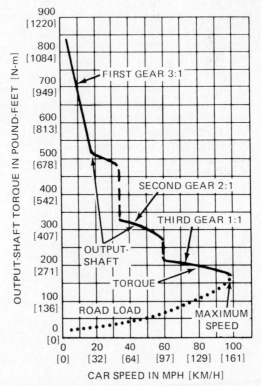

Fig. 1-19 Relationship between car speed and transmission output-shaft torque for each forward-gear position. (*Ford Motor Company*)

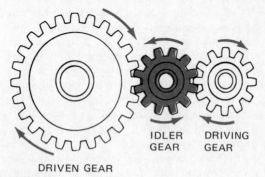

DRIVEN GEAR

Fig. 1-20 The idler gear causes the driven gear to turn in the same direction as the driving gear.

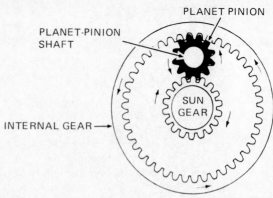

Fig. 1-22 The three gears that make up the basic planetary gearset.

It only changes the direction of rotation of the driven gear.

If the space is too small for an idler gear like that shown in Fig. 1-20, rotation of both gears in the same direction can be obtained by using an *internal gear* (Fig. 1-21). This gear is called an internal gear because the teeth are on the inside of the ring or wheel.

In Fig. 1-21 the small, driving gear is shown meshed with the teeth on the inside of the large, driven gear, or internal gear. When the small gear rotates, its teeth push on the internal teeth of the ring gear. This forces the ring gear to rotate in the same direction as the driving gear.

The addition of one more gear to the ring-gear system shown in Fig. 1-21 turns it into a planetary gearset. This last gear is centered in the ring gear (Fig. 1-22) and is called the *sun gear*. It is in the center of the planetary-gear system just as the sun is in the center of our solar system. The sun gear meshes with the spur gear, and the spur gear, in turn, meshes with the ring gear.

The intermediate spur gear is called the *planet pinion* or *planet gear*. This is because it revolves around the sun gear the way a planet revolves around the sun in our solar system. The planet pinion rotates around its own axis or shaft while revolving around the sun gear. This is the same motion that the planet Earth has as it revolves around the sun. Note the direction of rotation of each gear in the planetary-gear system shown in Fig. 1-22.

⚙ 1-9 Planetary-gear system operation Planetary gears are an essential component of automotive automatic transmissions. In its simplest form, a planetary gearset is made up of three gears (Fig. 1-22). In the center is the sun gear. All gears in the set revolve around the sun gear. Meshing with the sun gear are two or more planet-pinion gears, although only one is shown in Fig. 1-22.

A planetary gearset can provide any of five conditions:

1. A speed increase with a torque decrease (overdrive)
2. A speed decrease with a torque increase (reduction)
3. Direct drive (lockup)
4. Neutral
5. Reverse

All these variations (except neutral) are made possible by applying the input rotation to different planetary members while holding one of the other two members stationary.

To complete the actual planetary-gear system, another planet-pinion gear is added, as shown in Fig. 1-23. The two planet pinions are mounted on shafts that are part of a plate called the *planet-pinion carrier* or cage (Fig. 1-24). The carrier is also mounted on a shaft

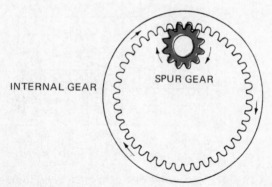

INTERNAL GEAR   SPUR GEAR

Fig. 1-21 If an internal gear is used with an external gear, both gears will turn in the same direction.

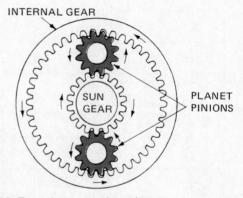

INTERNAL GEAR   PLANET PINIONS

Fig. 1-23 To make a complete planetary gearset, a second planet-pinion gear is added. This pinion balances the forces so that the system runs smoothly. Planetary-gear systems in automatic transmissions usually have three or four planet pinions.

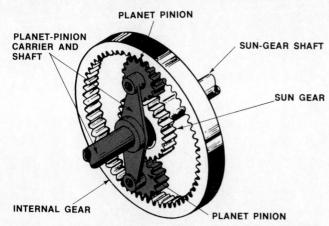

Fig. 1-24 Planet pinions rotate on shafts that are mounted on a planet-pinion carrier. The planet-pinion carrier is attached to a shaft that is aligned with the sun-gear shaft. These shafts are centered in the internal gear.

so that the shaft can rotate, thereby carrying the planet pinions around in a circle. When this happens, their teeth, meshed with both the sun gear and the ring gear, cause movement of the other gears. Each part of the planetary gearset is called a *member*. The three members are the sun gear, the planet-pinion carrier with the planet pinions, and the ring gear.

When any one member of a planetary gearset is held stationary and another is turned, the third member will produce either a speed increase, a speed reduction, or a direction reversal.

1. Speed increase—sun gear stationary If the planet-pinion carrier is turned while the sun gear is held stationary, the ring gear increases in speed (Fig. 1-25). This happens because as the planet-pinion-carrier shaft is rotated, it carries the pinion shafts around with it. As the planet pinions move in a circle around the sun gear, they also rotate on their shafts because the gears are meshed with the stationary sun gear.

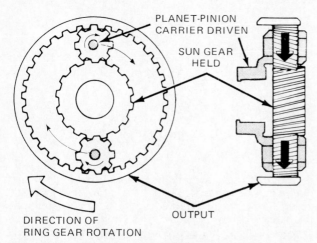

Fig. 1-25 End and side views of the planetary-gearset action when the sun gear is held and the pinion carrier is driven. The ring gear overdrives or turns faster than the pinion carrier.

The number of times a planet pinion rotates on its shaft while making one complete revolution around the sun gear depends on the gear ratio between the sun gear and the planet pinion. Suppose the sun gear has 36 teeth and the planet pinion has 6 teeth. According to the discussion of gear ratios in ⚙ 1-6, the ratio would be 36:6 or 6:1. However, this is not true in a planetary gearset. The planet pinion rotates *seven* times on its shaft as it completes one revolution around the stationary sun gear.

The extra rotation comes from the revolution that the planet pinion makes as it circles the stationary sun gear. Follow the planet pinion in Fig. 1-26 as it moves from the top around the sun gear and back to its starting position. Notice that the planet pinion must rotate seven times on its shaft while making one revolution around the sun gear. The seventh rotation comes from the change of position of the planet pinion as it moves around the sun gear.

Figure 1-27 shows that by the time the planet pinion has revolved only one-quarter of the way around the stationary sun gear, the planet pinion has already rotated one-half of a turn on its shaft. By the time the planet pinion has revolved halfway around the sun gear, the planet pinion has completed one full turn on its shaft.

You can try this yourself with two coins, such as nickels, dimes, or quarters. One coin represents the sun gear. Hold it stationary. The other coin represents the planet pinion. Place it next to the stationary coin. Then roll the "planet-pinion" coin around the stationary "sun-gear" coin. You will see how the moving coin gets its extra rotation from the change in its position as it revolves around the stationary coin.

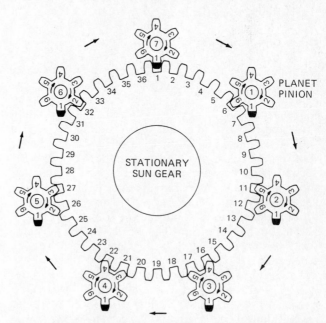

Fig. 1-26 The sun gear has 36 teeth. The planet pinion has 6 teeth. The positions marked 1 to 7 show the positions of the planet pinion as it rotates on its shaft and revolves around the sun gear. These positions prove that the planet pinion must rotate seven times as it makes one complete revolution around the sun gear.

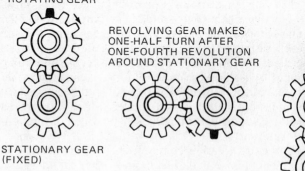

ROTATING GEAR

REVOLVING GEAR MAKES
ONE-HALF TURN AFTER
ONE-FOURTH REVOLUTION
AROUND STATIONARY GEAR

STATIONARY GEAR
(FIXED)

REVOLVING GEAR MAKES
ONE ROTATION AFTER
ONE-HALF REVOLUTION
AROUND STATIONARY GEAR

Fig. 1-27 Two external gears having the same number of teeth. One gear serves as the sun gear, and the other serves as the planet pinion. After the planet pinion makes one-half revolution around the sun gear, the planet pinion has made one complete rotation. After the planet pinion completes one revolution around the sun gear, the pinion will have made two complete rotations.

This extra rotation occurs regardless of the gear ratio of the sun gear and planet pinion. To take an extreme example, if the sun gear has 100 teeth and the planet pinion has 10 teeth, the planet pinion rotates 11 times on its shaft as it makes one revolution around the stationary sun gear. For another example, if the sun gear has 50 teeth and the planet pinion has 100 teeth, the planet pinion rotates 1½ turns as it revolves around the stationary sun gear. In this example, the sun gear is smaller than the planet pinion. However, the extra revolution is still added despite the actual gear ratio.

Holding the sun gear stationary and rotating the planet-pinion carrier causes the planet pinions to rotate on their shafts as they revolve around the sun gear. This action causes the ring gear to rotate. The ring gear rotates in the same direction that the planet-pinion carrier is turning and faster than the carrier.

Figure 1-28 shows how this speed increase is achieved with the sun gear stationary. The planet-pinion carrier rotates and forces the planet pinions to move in a circle around the sun gear. This action forces the planet pinions to rotate on their shafts. At any instant, the pinion tooth meshed with the sun gear is stationary. Therefore, the planet pinion pivots around the stationary tooth. If the pinion shaft moves at a speed of 1 foot per second [0.305 meter per second], then the outside tooth must move at twice this speed, or 2 feet per second [0.610 meter per second]. Now the ring-gear tooth meshed with this outside tooth must also move at 2 feet per second [0.610 meter per second]. Therefore, as shown in Fig. 1-28, the internal gear rotates faster than the planet-pinion carrier.

The ratio between the planet-pinion carrier and the internal gear can be altered by changing the sizes of the gears. In Fig. 1-29, the ring gear makes one complete revolution while the planet-pinion carrier turns only 0.7 revolution. This means that the ring gear turns faster than the planet-pinion carrier. The gear ratio between the two is 0.7:1, showing that the system is operating as a speed-increasing mechanism. The driven member (the ring gear) turns faster than the driving member (the

planet-pinion carrier). The ring gear rotates in the same direction as the planet-pinion carrier. Automatic transmissions can use this condition to provide an "overdrive" fourth gear.

Notice that the planetary-gear system in Fig. 1-29 has three planet pinions instead of the two shown in Fig. 1-24. Planet-pinion carriers in automatic transmissions usually have three or four planet pinions.

2. **Speed increase—ring gear stationary** Another combination is to hold the ring gear stationary and turn the planet-pinion carrier. Now, the sun gear is forced to rotate faster than the planet-pinion carrier and the system functions as a speed-increasing mechanism. The driven member (the sun gear) turns faster than the

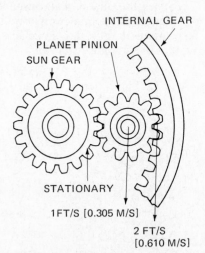

INTERNAL GEAR

PLANET PINION

SUN GEAR

STATIONARY

1 FT/S [0.305 M/S]

2 FT/S
[0.610 M/S]

Fig. 1-28 If the sun gear is stationary and the planet-pinion carrier turns, the ring gear will turn faster than the carrier. The planet pinion pivots about the stationary teeth. If the center of the pinion shaft moves at 1 foot per second [0.305 meter per second], the tooth opposite the stationary tooth will move at 2 feet per second [0.610 meter per second]. This is because the opposite tooth is twice as far away from the stationary tooth as the center of the shaft.

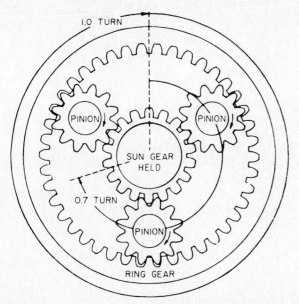

Fig. 1-29 The relative sizes of the gears in a planetary gearset that cause the ring gear to turn once while the planet-pinion carrier makes 0.7 turn. The sun gear is held stationary.

driving member (the planet-pinion carrier). This condition usually is not used in automatic transmissions.

3. Speed reduction—sun gear stationary   If the ring gear is turned while the sun gear is held stationary, the planet-pinion carrier will turn more slowly than the ring gear (Fig. 1-30). This situation is the opposite of the condition described in item 1, in which the planet-pinion carrier turns the internal gear. Now, with the internal gear turning the planet-pinion carrier, the planetary-gear system functions as a speed reducer. The drive member (the planet-pinion carrier) turns more slowly than the driving member (the ring gear). This is the way second gear is achieved in many automatic transmissions.

4. Speed reduction—ring gear stationary   When the ring gear is held stationary and the sun gear is turned, the planet pinions rotate on their shafts. At the same time, the planet pinions must "walk around" the ring gear, since they are in mesh with it. As they do this, the planet-pinion carrier is carried around. Therefore, the carrier rotates in the same direction as the sun gear, but at a slower speed. Now the system functions as a speed reducer. The driven member (the planet-pinion carrier) turns more slowly than the driving member (the sun gear). This condition provides the greatest increase in torque. It is used to obtain first gear in automatic transmissions.

5. Reverse 1   Another variation is to hold the planet-pinion carrier stationary and turn the ring gear (Fig. 1-31). In this case, the planet pinions act as idlers and cause the sun gear to turn in the direction opposite to that of ring-gear rotation. Now the system functions as a direction-changing mechanism, with the sun gear turning faster than the ring gear. However, there is no need for a "high-speed" reverse gear in an automatic transmission. Therefore, this condition is not used to obtain reverse in passenger cars.

6. Reverse 2   A second way to get reverse is to hold the planet-pinion carrier stationary and turn the sun gear. Then the ring gear turns in a reverse direction, but more slowly than the sun gear. This is the condition used to provide reverse gear in an automatic transmission.

7. Direct drive   If any two of the three members—sun gear, planet-pinion carrier, or ring gear—are locked together, then the entire planetary-gear system is locked (Fig. 1-32). The input and output shafts turn at the same speed. There is no change of speed or direction through the system, and the gear ratio is 1:1. This condition is used for third gear in automatic transmissions.

8. Neutral   When all of the members in a planetary gearset are free, no power can be transmitted through it. This condition provides the transmission with a neutral position for starting the engine without load.

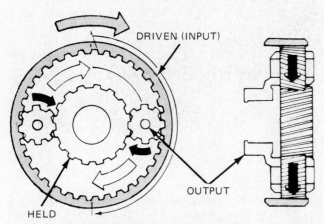

Fig. 1-30 Planetary gearset in reduction. The sun gear is stationary and the ring gear is driven. The planet carrier turns, but more slowly than the ring gear.

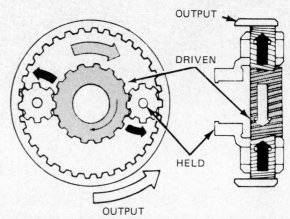

Fig. 1-31 Planetary gearset in reverse. The pinion carrier is held stationary. The sun gear is driven and this drives the ring gear. It turns in the direction opposite to the sun gear.

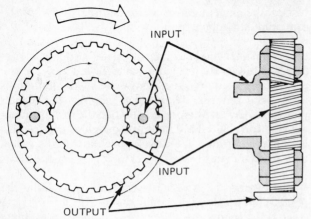

Fig. 1-32 Planetary gearset in direct drive. Locking any two members together puts the planetary-gear system in direct drive.

	1	2	3	4	5	6
Internal gear	D	H	T	H	T	D
Carrier	T	T	D	D	H	H
Sun gear	H	D	H	T	D	T
Speed	I	I	L	L	IR	LR

D—driven
H—held
I—increase speed
L—reduction in speed
R—reverse
T—turning or driving

Fig. 1-33 Various conditions that are possible in the planetary-gear system if one member is held and another is turned.

The planetary-gear system operating conditions are shown in Fig. 1-33. Consider condition 1. The letter T in this column indicates the driving member, which is the pinion carrier. H indicates that the sun gear is held stationary. D indicates that the ring gear is being driven. I designates an increase of speed between the pinion carrier and the ring gear.

The three conditions used most frequently in automatic transmissions are listed in columns 3, 4 and 6.

NOTE Different manufacturers give different names to the parts of the planetary gearset. For example, the internal gear is also called the *ring gear* or *annulus gear*. The planet-pinion carrier is also called the *planet-pinion cage*, and sometimes the *spider*. The planet pinions are also called *planet gears*. Planetary-gear systems are often called *planetary gearsets*. These different names are used throughout this book and in various manufacturers' service manuals.

✿ 1-10 Rear-wheel drive Many cars have power trains that drive the rear wheels. Others have power trains that drive the front wheels of the car. Front-wheel drive has become more common in recent years and is used extensively in the smaller, more compact automobiles (✿ 1-11). Rear-wheel drive uses a long drive shaft that connects the transmission to the final drive, the differential, and the rear-wheel axles (Figs. 1-1, 1-5, and 1-34). The drive shaft may be the single-piece type, or it may have two or more sections connected together. Slip joints and universal joints are used on the ends of the drive shaft tube to compensate for changes in driveline length and angularity.

✿ 1-11 Front-wheel drive In the front-wheel drive, the power train is short and carries the engine power to the front wheels. There are two basic arrangements. In one, the engine is mounted with its front end pointing to the front of the car (Fig. 1-4). In the other, the engine is mounted sideways, or transversely (Fig. 1-5). The engines in most of the smaller cars being built today have the transverse mounting.

These cars use a *transaxle*, which is a *trans*mission and drive *axle*, or differential, combined into a single unit. The transmission can be either the manual type or the automatic type. Differentials are described in ✿ 1-14. Automatic transaxles are described in following chapters.

The short drive shafts going to the two front wheels of a car with front-wheel drive also have universal joints and slip joints (Fig. 1-4). They permit the wheels to move up and down and to turn from side to side for steering the vehicle.

✿ 1-12 Four-wheel drive Some vehicles have power trains that can drive all four wheels (Fig. 1-35). These vehicles have better traction than vehicles with only two-wheel drive. Some vehicles in four-wheel drive can climb steep hills and travel over rough, muddy, slippery ground that would stop cars with two-wheel drive. In four-wheel drive, the power from the engine passes through the transmission and enters a *transfer case*. The transfer case has gears that can be engaged to send power to both front and rear wheels.

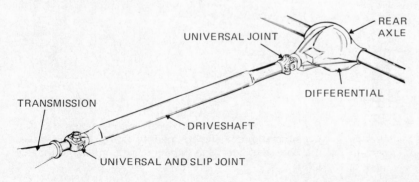

Fig. 1-34 The drive shaft connects the transmission with the differential (in a car with front engine and rear-wheel drive). This is a one-piece drive shaft with two universal joints and one slip joint.

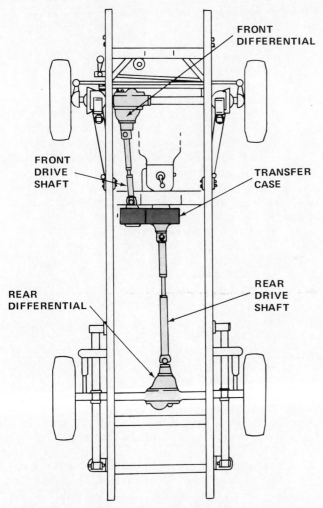

Fig. 1-35 Suspension and drive-train parts for a typical four-wheel-drive vehicle. The transfer case allows the driver to select rear-wheel drive or four-wheel drive. (*Ford Motor Company*)

NOTE: Four-wheel-drive vehicles are often referred to as 4WD vehicles. Front-wheel drive cars are often called FWD cars. So there are two similar abbreviations: 4WD, meaning four-wheel drive, and FWD, meaning front-wheel drive.

⚙ 1-13 Final drive The final drive is the final speed-reduction gearing in the power train. It multiplies the torque from the transmission output shaft to the value required for use by the drive axles (⚙ 1-2). For example, typical automatic-transmission gear ratios are 3:1, 2:1, and 1:1 in first, second, and third gears. However, overall gear ratios between the engine crankshaft and the tires are about 9:1, 6:1, and 3:1 (Fig. 1-19). This is because of the multiplication of the torque from the transmission output shaft provided by the final-drive gearing of 3:1. (Final-drive gearing for a car with manual transmission is usually about 4:1.)

In the power train, the final drive is located between the drive shaft (in rear-wheel-drive vehicles) and the differential (Fig. 1-36). Reduction gearing, usually called a *ring gear* and a *pinion gear* or *drive pinion*

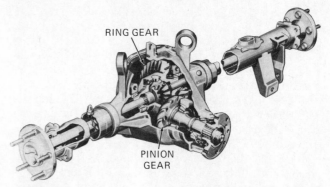

Fig. 1-36 Cutaway view of a rear drive-axle assembly showing the differential gears. (*Ford Motor Company*)

provides the necessary torque multiplication. In most power trains, the ring gear is attached to the differential case (⚙ 1-14). The pinion gear is connected through the rear universal joint to one end of the drive shaft (in rear-wheel-drive cars), as shown in Fig. 1-34.

⚙ 1-14 Differential If a car were driven in a straight line only, no differential would be necessary. However, when a car rounds a turn, the outer wheel must travel farther than the inner wheel. Look at the action in a rear-wheel-drive car (Fig. 1-37). If a right turn is made with the inner wheel turning on a 20-foot [6.1-m] radius, this wheel travels about 31 feet (9.5 m). The outer wheel, being nearly 5 feet [1.5 m], or 56 inches, from the inner wheel, turns on a 24⅔-foot [7.5-m] radius and travels nearly 39 feet [11.9 m].

If the drive shaft were geared rigidly to both rear wheels so that they both had to rotate together, then each wheel would have to skid an average of 4 feet [1.2 m] in making the turn. As a result, tires would not last long. In addition, the skidding would make the car hard to control around turns. The differential eliminates these troubles because it allows the wheels to rotate by different amounts when turns are made.

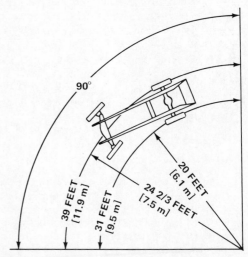

Fig. 1-37 The difference in wheel travel as the car makes a 90° turn with the inner wheel turning on a 20-foot [6.1 m] radius.

The construction and action of the differential are shown in Fig. 1-38. Two rear wheels are attached, through the axles, to two small bevel gears called *differential side gears* (Fig. 1-38*a*). There is a differential case assembled around the left axle (Fig. 1-38*b*). The case has a bearing that permits it to turn independently of the left axle. Inside the case is a shaft that supports a third bevel gear (Fig. 1-38*c*). This third bevel gear, called the *differential pinion* gear, is meshed with the differential side gears.

When the differential case is rotated, both differential side gears rotate. Therefore both wheels turn. However, suppose that one wheel is held stationary. Then, when the differential case is rotated, the differential pinion gear will also rotate as it "runs around" on the stationary differential side gear. As the differential pinion gear rotates in this manner, it carries rotary motion to the other differential side gear, causing it and the wheel to rotate.

When one rear wheel turns faster than the other, the differential pinion gear spins on its shaft (Fig. 1-38*d*). This transmits more rotary motion to one drive wheel than to the other. When both drive wheels turn at the same speed, the differential pinion gear does not rotate on its shaft.

The differential case is rotated by a ring gear attached to it. This ring gear is meshed with a drive pinion on the end of the drive shaft (Fig. 1-34). When the car is on a straight road, the ring gear, differential case, dif-ferential pinion gear, and two differential side gears all turn as a unit without any relative motion. However, when the car begins to round a curve, the differential pinion gear rotates on its shaft to permit the outer wheel to turn faster than the inner wheel.

An actual automotive differential, partly cut away to show the parts, is shown in Fig. 1-36. The driving power enters the differential through the drive pinion on the end of the drive shaft. The drive pinion is meshed with a large ring gear so that the ring gear revolves with the pinion. Attached to the ring gear (through the differential case) is a differential pinion shaft on which are assembled two differential pinion gears. Each drive wheel has a separate axle, and there are two side gears splined to the inner ends of the two wheel axles. The two differential pinion gears mesh with these two side gears. When the car is on a straight road, the two differential pinion gears do not rotate on the pinion shaft. However, they do exert force on the two side gears so that they turn at the same speed as the ring gear. This causes both drive wheels to turn at the same speed also.

When the car rounds a curve, the outer wheel must turn faster than the inner wheel. To permit this, the two pinion gears rotate on their pinion shaft, transmitting more turning movement to the outer side gear than to the inner side gear. Therefore, the side gear on the outer-wheel axle turns faster than the side gear on the inner-wheel axle. This permits the outer wheel to turn faster while the car is rounding the curve.

✿1-15 Dynamometer tests of the power train

Some automotive shops have chassis dynamometers that can check the power train (Fig. 1-39). The chassis dynamometer can duplicate almost any kind of road test at any load or speed. The visible part

Fig. 1-38 Drive axles and differential. (*a*) The drive axles are attached to the wheels. Bevel gears are located on the inner ends of the drive axles. (*b*) The differential case is assembled on the left axle but can rotate on a bearing independently of the axle. (*c*) The differential case supports the differential pinion gear on a shaft. This gear meshes with the two bevel gears. (*d*) The ring gear is attached to the differential case so that the case rotates with the ring gear when the ring gear is driven by the drive pinion.

Fig. 1-39 Automobile in place on a chassis dynamometer. The drive wheels turn the dynamometer rollers. At the same time, instruments on the control console measure roller speed and power or torque available at the drive wheels. (*Sun Electric Corporation*)

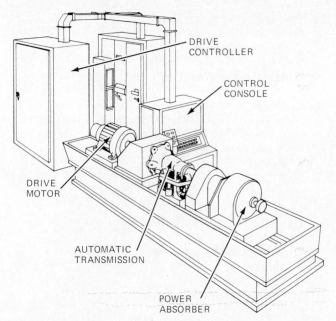

DRIVE
CONTROLLER

CONTROL
CONSOLE

DRIVE
MOTOR

AUTOMATIC
TRANSMISSION

POWER
ABSORBER

Fig. 1-40 Testing an automatic transmission before installing it in the car. (*Weldun International*)

tion, and idle. Next, the engine is started and the transmission put into gear so that the drive wheels of the car can spin the rollers. Then the car is operated as though it were actually on the road.

Under the floor is the part of the dynamometer called a *power absorber*. It can place varying loads on the engine and power train. By changing the load placed on the drive wheels by the power absorber, the technician can check engine performance and power-train operation. On some dynamometers, the shift points of automatic transmissions can be determined under varying load conditions. Also, faulty conditions of the clutch, transmission, and other power-train components can be detected.

✿ 1-16 Automatic-transmission test stands After an automatic transmission is assembled at the factory by the new-car manufacturer, the transmission is placed in a test stand (Fig. 1-40). This device conducts a performance test of the transmission to ensure that it is operating properly and does not leak prior to installation in the new car.

In the service field, some shops specializing in automatic-transmission repair, and many rebuilders and remanufacturers, have a similar test stand. By making pressure tests and checking input and output torque in the various gears, any defect in the transmission can be detected. This prevents shop "comebacks" and helps to assure a quality repair job.

of the dynamometer in Fig. 1-39 consists of two rollers mounted at about floor level. The car is driven onto these rollers. Instruments are connected to measure engine action during acceleration, cruising, decelera-

Chapter 1 review questions Quizes

Select the *one* correct, best, or most probable answer to each question. Then check your answers against the correct answers given at the end of the book.

1. Two meshed gears have a gear ratio of 3:1. Every time the larger gear turns once, the smaller gear turns:
   a. one-third revolution
   b. two-thirds revolution
   c. one revolution
   d. three revolutions

2. If two meshing gears have a 4:1 gear ratio and the smaller gear has 12 teeth, the larger gear will have:
   a. 12 teeth
   b. 24 teeth
   c. 36 teeth
   d. 48 teeth

3. The device that produces different gear ratios in the power train is called the:
   a. differential
   b. transmission
   c. drive shaft
   d. output shaft

4. In gear systems, speed reduction means:
   a. torque reduction
   b. gear ratio
   c. torque increase
   d. lost power

5. In the example shown in Fig. 1-17, if the gear reduction through the power train is 6:1, the car will be pushed forward by a torque of:
   a. 400 lb-ft [542 N-m]
   b. 600 lb-ft [810 N-m]
   c. 800 lb-ft [1085 N-m]
   d. 1200 lb-ft [1627 N-m]

6. In the example shown in Fig. 1-17, if the final drive or rear-axle ratio is 4:1, the car will be pushed forward by a torque of:
   a. 400 lb-ft [542 N-m]
   b. 600 lb-ft [810 N-m]
   c. 800 lb-ft [1085 N-m]
   d. 1200 lb-ft [1627 N-m]

7. To take care of the difference in drive angle as the rear axle moves up and down, the drive shaft has one or more:
   a. slip joints
   b. elbow joints
   c. release joints
   d. universal joints

8. To take care of the lengthening and shortening of the drive shaft with rear-axle movement, the drive shaft has a:
   a. slip joint
   b. elbow joint
   c. release joint
   d. universal joint

9. In the differential, the drive pinion meshes with the:
   a. bevel gear
   b. pinion gear
   c. ring gear
   d. axle gear

10. The power train includes the clutch (on some cars), transmission, differential, and:
   a. carburetor
   b. drive shaft
   c. axle housing
   d. engine

11. In all cars, the power train transmits power from the engine to the:
   a. crankshaft
   b. steering wheels
   c. drive wheels
   d. idler wheels

12. In a front-engine, rear-wheel-drive car, the drive shaft usually is:
   a. a solid bar
   b. hollow
   c. not used
   d. none of the above

13. In the standard differential described in this chapter, the total number of gears and pinions is:
   a. 2
   b. 3
   c. 6
   d. 11

14. In the differential, the ring gear is attached to the:
   a. bevel gear
   b. drive gear
   c. differential case
   d. drive shaft

15. The gear reduction through the drive axle is called the final-drive ratio, or:
   a. the transmission ratio
   b. the torque-reduction ratio
   c. the drive-axle ratio
   d. none of the above

16. Mechanic A says a vehicle with four-wheel drive always has a transfer case. Mechanic B says a vehicle with four-wheel drive always has a transaxle. Who is right?
   a. A only
   b. B only
   c. both A and B
   d. neither A nor B

17. The transaxle combines the transmission and the:
   a. drive shafts
   b. differential
   c. transfer case
   d. wheel axles

18. The planetary-gear system can act as:
   a. a speed reducer
   b. a direct-drive unit
   c. a speed increaser
   d. all of the above

19. If the ring gear is held stationary and the sun gear is turned, the planetary-gear system acts as a:
   a. speed reducer
   b. speed increaser
   c. direct-drive unit
   d. torque reducer

20. If the ring gear and the planet-pinion carrier are locked together, the planetary-gear system acts as a:
   a. speed reducer
   b. speed increaser
   c. direct-drive unit
   d. torque reducer

HYDRAULICS AND TORQUE CONVERTERS

After studying this chapter, you should be able to:

1. Explain how liquids can be used to send pressure and movement from one place to another.
2. Describe how a pressure-regulator valve works.
3. Discuss the construction and operation of a balanced valve.
4. Explain how bands can be used to control planetary gears.
5. Discuss the construction and operation of hydraulically operated clutches.
6. Explain how a fluid coupling works.
7. Describe the construction and operation of a torque converter.
8. Explain the purpose and operation of the stator.

2-1 Introduction to the torque converter Manual transmissions require a clutch (❖ 1-3) to uncouple the transmission from the engine so that gears can be shifted. In an automatic transmission (Fig. 2-1), hydraulic pressures are used to control the shifts. Instead of flowing through the clutch to the gears, the power in an automatic-transmission engine flows from the crankshaft to the *torque converter*.

The torque converter is filled with oil and acts as a *fluid coupling* or *fluid clutch*. It allows gear shifting while power is flowing through the transmission. The way in which shifts are made (through planetary gearsets) also allows the shifts to be made smoothly.

❖ 2-2 Principles of hydraulics Early in the seventeenth century, the French scientist Blaise Pascal proved that force and motion can be transmitted by means of a confined liquid. He discovered that the force multiplication in the hydraulic-pressure system was the same as when the mechanical advantage was provided by gears and shafts (❖ 1-6). The relationships between force and distance in multiplying torque were the same in both systems.

To understand the operation of the torque converter and the hydraulic-control systems in an automatic transmission, a basic knowledge of *hydraulics* is required. Hydraulics is the science of liquids, such as water and oil. A hydraulic device uses the force that can be exerted by a liquid to transmit motion and to multiply force. One type of device is *hydrodynamic*, such as the torque converter. In it, the force of the fluid (*hydro*) in motion (*dynamic*) is used to multiply torque. A *hydrostatic* device, such as a valve in the hydraulic control system, operates because of the incompressibility of the fluid working on it. The valve moves as the oil pressure increases.

❖ 2-3 Incompressibility of liquids If a gas, such as air, is put under pressure, the gas can be compressed into a smaller volume (Fig. 2-2). However, applying pressure to a liquid cannot change its volume. This is because liquids are incompressible.

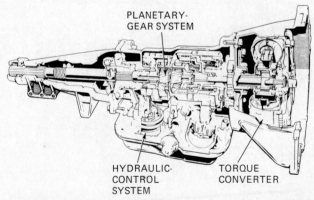

Fig. 2-1 Cutaway view of an automatic transmission, showing the torque converter, the planetary-gear system, and the hydraulic-control system. (*Ford Motor Company*)

PLANETARY-GEAR SYSTEM

HYDRAULIC-CONTROL SYSTEM

TORQUE CONVERTER

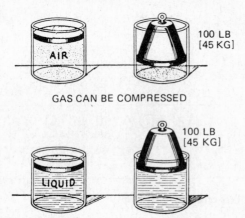

GAS CAN BE COMPRESSED

LIQUID CANNOT BE COMPRESSED

Fig. 2-2 Gas can be compressed when pressure is applied. However, liquid cannot be compressed when pressure is applied. (*Pontiac Motor Division of General Motors Corporation*)

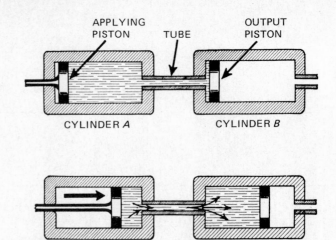

Fig. 2-4 Motion may be transmitted through a tube from one cylinder to another by hydraulic pressure. (*Pontiac Motor Division of General Motors Corporation*)

☼ 2-4 Transmission of motion by liquid Since liquid is not compressible, it can be used to transmit motion. For example, Fig. 2-3 shows two pistons in a cylinder with a liquid between them. When the applying piston is moved 8 inches [203.2 mm] into the cylinder, as shown in the lower illustration, then the output piston will be pushed the same distance along the cylinder.

In Fig. 2-3, a solid connecting rod between pistons A and B could be substituted to get the same result. But the advantage of a hydraulic system is that motion can be transmitted between cylinders by a tube (Fig. 2-4). As the applying piston is moved (in Fig. 2-4), liquid is forced out of cylinder A, through the tube, and into cylinder B. This causes the output piston to move in its cylinder.

☼ 2-5 Transmission of pressure by liquid The pressure applied to a liquid is transmitted by the liquid in all directions to every part of the system. An example is shown in Fig. 2-5. When a piston with an area of 1 square inch [6.45 cm²] applies a force of 100 pounds [445 N] on a liquid, the pressure on the liquid is 100 pounds per square inch (psi) [690 kPa]. This pressure will show up throughout the entire hydraulic system. If the area of the piston is 2 square inches [12.9 cm²] and the piston applies a force of 100 pounds [445 N], then the pressure is only 50 psi [345 kPa] (Fig. 2-6).

Pressure and force are closely related. A *force* is simply the push or pull on an object. The force produced by a spring when it is compressed or stretched is called *spring force*. Pressure is defined as the force divided by the surface area upon which the force acts, or *force per unit area*. In the U.S. Customary System of measurement, a force, such as spring force, is given in pounds or ounces. Pressure is measured in pounds per square inch (psi).

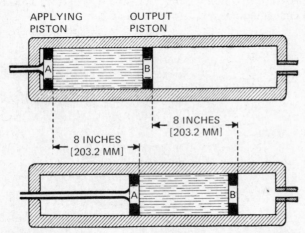

Fig. 2-3 Motion can be transmitted by liquids. When the applying piston A is moved 8 inches [203.2 mm], then the output piston B is also moved 8 inches [203.2 mm]. (*Pontiac Motor Division of General Motors Corporation*)

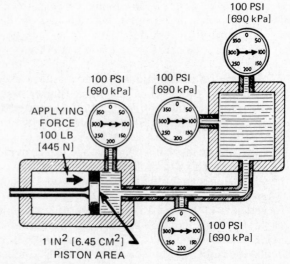

Fig. 2-5 Pressure applied to a liquid is transmitted equally in all directions. (*Pontiac Motor Division of General Motors Corporation*)

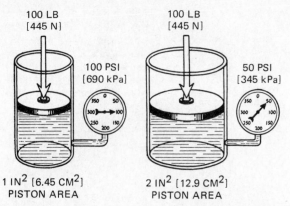

Fig. 2-6 Hydraulic pressure in the system is determined by dividing the applying force by the area of the applying piston. (*Pontiac Motor Division of General Motors Corporation*)

In an input-output system (Fig. 2-7), the force of the output piston can be determined. Output force is given by the equation

$$F = P \times A$$

where F = output force
P = pressure applied to piston
A = area of piston upon which pressure acts

To find the output force, multiply the applied pressure (in psi) times the area of the output piston (in square inches). For example, the hydraulic pressure shown in Fig. 2-7 is 100 psi [690 kPa]. The output piston to the left has an area of 0.5 square inch [3.23 cm²]. Therefore, the output force of this piston is 100 times 0.5, or 50 pounds [222 N].

The center piston in Fig. 2-7 has an area of 1 square inch [6.45 cm²]. Therefore its output force is 100

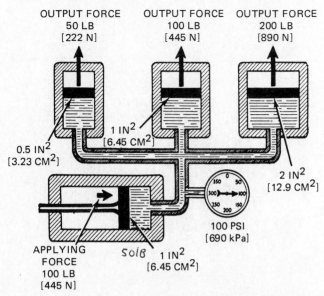

Fig. 2-7 The force applied by the output piston is the pressure in the system times the area of the output piston. (*Pontiac Motor Division of General Motors Corporation*)

pounds [445 N]. The right piston has an area of 2 square inches [12.9 cm²]. Its output force is 200 pounds [100 × 2] [890 N]. This shows that the bigger the output piston, the greater the output force. For example, if the area of the piston were 100 square inches [645 cm²], then the output force would be 10,000 pounds [44,480 N].

Another way to increase the output force is to raise the applied pressure. The higher the hydraulic pressure, the greater the output force. If the hydraulic pressure on the 2-square-inch [12.9-cm²] piston were to go up to 1000 psi [6900 kPa], then the output force on the piston would be 2000 pounds [8900 N].

In Fig. 2-7 and previous illustrations, a piston-cylinder arrangement has been shown as the means of producing the hydraulic pressure. However, any type of pump can be used. In automatic transmissions, the three most common types are gear, rotor, and vane.

⚙ 2-6 Hydraulic valves The two basic hydraulic valves used in automatic transmissions are the *pressure-regulator valve* and the *balanced valve*. All of the valves in automatic transmissions are of one of these types.

The pressure-regulator valve (Fig. 2-8) uses hydraulic action to prevent excessive pressure in a hydraulic system. The valve is spring-loaded and has a small piston that can move back and forth in a cylinder bore. Valve movement produces a constant pressure from a variable pressure source. For example, suppose the pressure source is an oil pump driven by the engine. When the engine runs at high speed, the pump also runs at high speed. Therefore it delivers a large volume of oil. This would produce excessively high oil pressure if it were not for the action of the pressure-regulator valve.

Figure 2-8a shows the position of the spring-loaded valve when the pump pressure is low. Spring force has pushed the valve to the left (in Fig. 2-8a), closing off the return line to the reservoir.

As the pressure goes up, there is an increasing force on the valve. When the preset value (determined by the force of the spring) is reached, the oil pressure overcomes the spring force. The valve moves back in its cylinder bore (to the right in Fig. 2-8b). As the valve moves back, it uncovers an opening, or *port*, which is connected to a return line to the oil reservoir. Now part of the oil from the pump can flow through the return line (Fig. 2-8b). This reduces the pressure so that the valve starts forward again, moved by the force of the spring. However, as the valve moves forward, it partly shuts off the port to the return line. Since less oil can now escape, the oil pressure goes up and the valve is again moved back.

Actually, the valve does not normally move back and forth as described above. Instead, it seeks and finds the position at which the oil pressure just balances the spring force. Then, if the incoming oil volume changes because of a change in pump speed, the valve position will change. In operation, the valve maintains a constant output pressure by bleeding off more or less oil from the pump. For example, as the pump speed

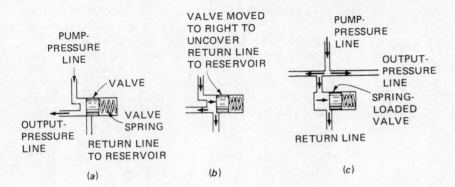

Fig. 2-8 (a) A pressure-regulator valve. As pump pressure increases, the spring-loaded valve moves back against spring force (b), dumping more of the oil from the pump into the return line. This action maintains a constant output pressure. (c) Arrangement in which neither the output-pressure line nor the pump-pressure line is directly connected to the pressure-regulator valve.

goes up—which means the pump delivers more oil and the pressure goes up—the valve is forced to open the port wider. This permits more oil to flow into the return line, preventing excessively high pressure.

The output-pressure and pump-pressure lines do not actually have to run to the piston as shown in Fig. 2-8a. The arrangement can be as shown in Fig. 2-8c. Pump pressure entering the output-pressure line is reduced because the valve has positioned itself to allow excess oil to escape through the line. Just enough oil escapes to maintain the pressure in the output line at the value determined by the valve spring. In the actual transmission, the pressure-regulator valve shown in Fig. 2-8c is used with one or more *boost valves* in line with it. Under certain conditions, a boost valve will have oil pressure applied to it. This causes the boost valve to add to the force of the spring in the pressure-regulator valve. The result is that the output pressure is increased, or *boosted*. Boost-valve action is covered later.

⚙ 2-7 Balanced valves Balanced valves may be used to produce pressure changes that are proportional to the movement of mechanical linkage, or to variations in spring force or control pressure. A balanced valve is shown in Fig. 2-9. It contains a spool valve, which is basically a solid bar with an undercut section. The spool valve moves back and forth in a cylinder. Oil pressure works against one end of the spool valve and spring force works against the other end.

In the centered position shown in Fig. 2-9, input oil under constant pressure enters the cylinder and passes around the undercut section of the spool valve. Then the oil flows through the bypass line and out of the cylinder through the return line and the output-pressure line. Input oil pressure is held constant by a pressure-regulator valve. In operation, variations in spring force against one end of the spool valve cause variations in output pressure.

In Fig. 2-10a, the mechanical linkage has been moved to increase the spring force against the right end of the spool valve. This causes the spool valve to move to the left (in Fig. 2-10a). Now the valve moves to close the return line. As the return line is closed, pressure increases on the output (left) end of the valve. The increasing pressure acts against the output end of the valve, which is pushed to the right against the spring

force. When output pressure equals spring force, a balanced condition is reached. Then the valve stops moving.

If the spring force is reduced by movement of the linkage (Fig. 2-10b), then the pressure on the output end of the valve can move it to the right in Fig. 2-10b. This occurs when the return line cannot release enough oil to control the pressure. As the valve moves, it tends to close off the input-pressure line. Therefore no oil can enter and a lower pressure results. The pressure drops until a balanced condition again results, with the output pressure balancing the spring force. Then the valve stops moving.

To show exactly how the valve might work, here are two examples. Assume the output end of the valve has an area of 1 square inch [6.45 cm²] and that the input pressure is 100 psi [690 kPa]. Now a spring force of 10 pounds [44.5 N] is applied to the spool valve (Fig. 2-11). For the valve to balance, there must be only 10 pounds [44.5 N] of force (from the oil) on the output end. As the oil enters and goes through the bypass line to the output end, the oil pressure forces the valve to the right (in Fig. 2-11). This movement tends to shut off the input pressure. The pressure increases until a

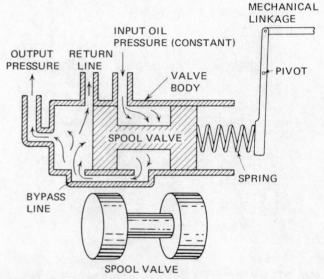

Fig. 2-9 Schematic layout of a balanced valve. The spool valve is shown removed from the valve body at the bottom.

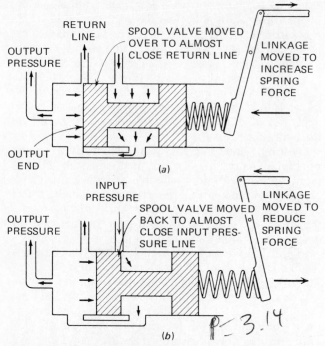

Fig. 2-10 If the linkage is moved to increase the spring force (a), then the valve spool will be moved toward the output end to partly close off the return line and permit a hydraulic-pressure increase to balance the spring-pressure increase. However, if the linkage is moved to reduce the spring force (b), then the hydraulic pressure forces the valve spool away from the output end to partly close off the input-pressure lines. Hydraulic pressure is reduced to balance the spring-force reduction.

balance is attained with 10 pounds [44.5 N] of force acting on each end of the valve.

If the force on the spring reaches 100 pounds [445 N], the full input pressure of 100 psi [690 kPa] can pass unhampered through the valve. This pressure then becomes the output pressure. However, a spring force above 100 pounds [445 N] cannot raise the pressure above 100 psi [690 kPa], since that is all the input pressure available. The input-pressure line will always be completely open.

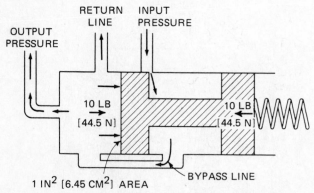

Fig. 2-11 If a spring force of 10 pounds [44.5 N] is applied, the valve will be moved to the left to allow oil to enter until a force (from the oil) of 10 pounds [44.5 N] is acting against the output end of the valve.

✿ 2-8 Control by varying hydraulic pressure In the typical hydraulic system of an automatic transmission, there are valves that produce a varying output pressure. This pressure depends on spring force, and on a varying hydraulic control pressure. A typical arrangement is shown in Fig. 2-12. The input or control pressure enters the valve in a space, or *annular groove*, between two lands of different sizes. In the example shown in Fig. 2-12, one land has an area of ½ square inch [3.2 cm²] while the other has an area of only ¼ square inch [1.6 cm²]. The effect is exactly as though the control pressure were working on the end of a valve with an area of ¼ square inch [1.6 cm²].

Now if the control pressure were 20 psi [138 kPa], there would be an effective hydraulic force on the valve of 5 (20 × ¼) pounds [22.2 N] pushing it to the left (in Fig. 2-12). Suppose the spring produced a force of 10 pounds [44.5 N]. Then the valve would be pushed to the left with a total force of 15 pounds [66.7 N]. Input pressure would be reduced by the valve action. Now the output pressure would be 15 psi [103 kPa]. If the control pressure went up to 60 psi [413 kPa], then its effect on the valve would amount to 15 (60 × ¼) pounds [66.7 N] of force. With the added force of 10 pounds [44.5 N] from the spring, the output pressure would be 25 psi [172 kPa].

By varying the control pressure, the same effect is achieved as though the spring force were varied. This is because the output pressure varies in proportion to the control pressure. In the hydraulic systems of automatic transmissions, several valves use this principle. They have a difference in face areas between lands, and a varying oil pressure is introduced into this space.

✿ 2-9 Drums, bands, and clutches In automatic transmissions, drums, bands, and clutches are used to control planetary gears and cause a change of gear ratio through the transmission. These control devices are hydraulically operated. When the ring gear of a planetary gearset is held stationary, a speed increase or reduction occurs. If the sun gear is driving when the internal or ring gear is held stationary, the planet-pinion carrier will turn more slowly than the sun gear and in the same direction. One way to hold the ring gear is by clamping a brake band around the outside surface of the ring gear, which now acts as a brake drum.

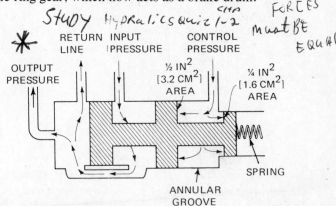

Fig. 2-12 Balanced valve using a varying hydraulic pressure to control output pressure.

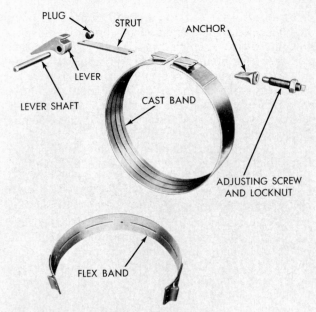

Fig. 2-13 Band with related linkages for an automatic transmission. (*Chrysler Corporation*)

A typical band is shown in Fig. 2-13. It is made of steel and has a lining of tough asbestos friction material bonded to it. In a typical arrangement used in automatic transmissions, the band fits loosely around the drum (Fig. 2-14). The band is tightened or *applied* by hydraulic pressure under certain conditions to bring the ring gear to a halt. This produces a speed change through the planetary gearset.

✿ 2-10 Servo action The device that hydraulically applies the band is called a *servo* (Fig. 2-14). It converts hydraulic pressure to mechanical movement. The typical servo consists of a piston that moves in a cylinder when hydraulic pressure acts on the piston. The transmission servo shown in Fig. 2-14 illustrates the operation of the servo as it applies and releases the band. This, in turn, allows or stops rotation of the drum. There are usually one or two servos in an automatic transmission.

In a servo, hydraulic pressure can be admitted to one or the other end (or both ends) of the cylinder. This pressure causes the piston to move. Then mechanical linkage from the piston causes movement of some mechanism. Therefore, hydraulic pressure is used to produce a mechanical action. In Fig. 2-14, when hydraulic pressure is applied to the piston, the piston will move to the right. This action causes the piston rod to apply the band. Drum rotation is stopped by increasing the hydraulic pressure.

Hydraulic brakes on cars work on exactly this same principle. Depressing the foot pedal increases the hydraulic pressure in the brake master cylinder. This increasing hydraulic pressure causes pistons in the wheel cylinder to move. As the wheel-cylinder pistons move, they force the brake shoes against the brake drums or disks to slow or stop their rotation.

Figure 2-15 is a sectional view of a servo that controls the band in an automatic transmission. The *apply pressure* for applying the band comes from the control valves in the hydraulic-control system. Transmission fluid flows from the hydraulic system through a hole in the piston stem and fills the chamber on the apply side of the piston. This pressure forces the piston to move to the left (in Fig. 2-15). As the piston moves, it tightens the band around the drum, stopping it.

When car speed and throttle opening signal the need for an upshift, hydraulic pressure is applied to the release side of the piston (which has a larger area than the apply side). This release-side pressure plus the spring force pushes the piston back to the right (in Fig. 2-15). Now the band loosens on the drum. The drum and the member of the planetary gearset it is connected to are free to revolve. This prepares the transmission for an upshift.

✿ 2-11 Accumulators Sudden application of a band or engagement of a clutch would produce a rough shift. To prevent this, an accumulator is used in automatic transmissions to cushion the shock of clutch and servo

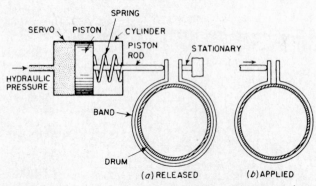

Fig. 2-14 The servo consists of a piston that moves in a cylinder in accordance with changing hydraulic pressure. This movement may be used to perform work, such as applying the band.

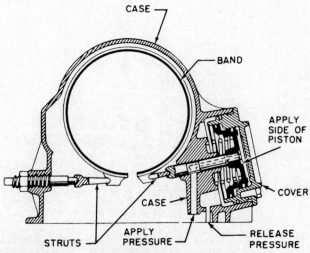

Fig. 2-15 Band with servo positioned in the transmission case. (*Ford Motor Company*)

actions (Fig. 2-16). This device produces a rapid but smooth clutch engagement or band application.

When oil is directed to the servo by the hydraulic control system, the oil enters in back of the apply piston. At first, the apply piston is briefly held in place by the heavy spring. But the oil passes through the check valve and applies pressure against the inside of the accumulator piston. This forces the accumulator to move down, operating the actuating lever so that the band contacts the drum. Then the hydraulic pressure moves the apply piston so that it adds its force against the actuating lever for maximum clamping action of the band. This combination two-step application gives a smooth band application. In many automatic transmissions, the accumulator is a separate valve in the hydraulic circuit between the pump and the servo or clutch piston.

An accumulator may also be used to cushion the shock of clutch engagement. For example, the General Motors 350 Turbo Hydra-Matic transmission has two accumulators with two clutches to produce the upshifts from first to second and from second to third.

✿ 2-12 Multiple-disk clutches Hydraulically operated clutches provide another means for locking or releasing a rotating member of the planetary-gear system. The hydraulic clutches used in automatic transmissions are the multiple-disk or -plate type (Fig. 2-17). Figure 2-18 shows a disassembled clutch. The plates are alternately attached to an outer housing, or drum, and to an inner hub. When the clutch is disengaged, the two members can rotate independently of each other. However when the clutch is engaged, the plates are forced together. Then the friction between them locks the two members so they rotate together. With the clutch engaged, there is direct drive through the planetary gearset.

Engagement of the clutch is produced by oil pressure, which forces an annular piston to push the plates together. The annular piston is a ring that fits snugly into the *clutch-drum assembly*, also called the *piston-retainer assembly*. This piston is ring-shaped, and is shown in sectional view in Fig. 2-17. An actual piston

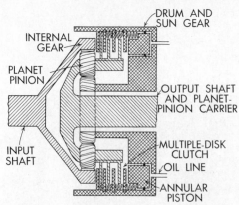

Fig. 2-17 Sectional view of a multiple-disk clutch used in automatic transmissions.

can be seen in the disassembled view of a clutch drum shown in Fig. 2-19. The piston fits snugly into the drum, sealed by a sliding seal to the drum's inner surface. A spring or a series of springs that rest against the piston return seat normally hold the piston slightly away from the plates. To engage the clutch, oil pressure must overcome the spring force and move the piston. This causes it to force the plates together.

In various transmissions, when the band is applied it may hold the sun gear, the ring gear, or the planet-pinion carrier stationary. Different transmissions may lock different members of the planetary gearset together when the clutch is engaged. However, the principle is the same in all transmissions. There is direct drive, reduction, reverse, or overdrive through the planetary gearset when the clutch is engaged. There is gear reduction or overdrive when the band is applied. A clutch can also be used, like a band, to hold one member stationary.

✿ 2-13 Fluid couplings The clutch used with manual transmissions is a mechanical coupling (✿ 1-3). When engaged, it couples the engine to the transmission. In automatic transmissions, instead of a mechan-

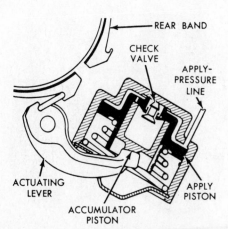

Fig. 2-16 Sectional view of an accumulator. (*Ford Motor Company*)

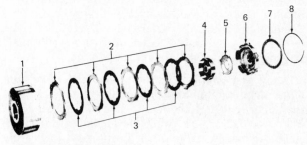

1. Clutch-drum assembly
2. Clutch-driven plates
3. Clutch-drive plates
4. Clutch hub
5. Clutch-hub thrust washer
6. Low-sun-gear-and-thrust-flange assembly
7. Clutch-flange retainer
8. Retainer snap ring

Fig. 2-18 Exploded view of a clutch. (*Chevrolet Motor Division of General Motors Corporation*)

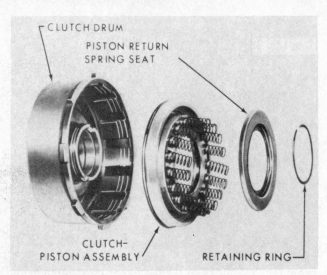

Fig. 2-19 Disassembled view of a clutch drum, showing the piston, springs, and return seat. (*Oldsmobile Division of General Motors Corporation*)

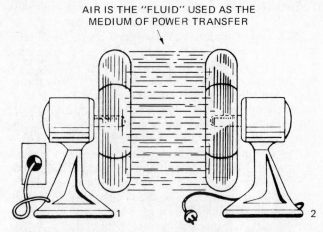

Fig. 2-20 Rotation of fan 1 (on the left) causes fan 2 (on the right) to rotate. This is a simple fluid coupling, with air serving as the fluid.

ical connection, a *fluid*, or hydraulic, coupling is used. It is always "engaged." But since the coupling is produced by a fluid, the driven member can slip and turn more slowly than the driving member. In any type of fluid coupling, a *member* is a functional component such as a *pump*, a *turbine*, or a *stator*. The action of each of these members is described later.

Essentially, the fluid coupling is a special form of clutch which provides a smooth, vibrationless coupling between the engine and the transmission. The fluid coupling operates at maximum efficiency when the driving member approaches the speed of the driven member. If there is a big difference in the speeds of the two members, power is lost and efficiency is low. The reason for this loss of efficiency is discussed later.

The fluid coupling was used in older cars with manual transmissions and clutches, and with early-model automatic transmissions. With the fluid coupling, some type of gearbox was always required to vary the gear ratios between the engine and the drive wheels. Automobiles no longer are equipped with the simple fluid coupling described in detail below. However, the *torque converter* now used in almost all automatic transmissions is a type of fluid coupling.

1. Fluid-coupling operation   A simple fluid coupling can be illustrated with two electric fans placed a short distance apart and facing each other (Fig. 2-20). When one fan is turned on, the air flow from it will cause the blades of the other fan to spin. In this example, the air acts as the fluid to transmit power from one fan to the other. But since the two fans are not enclosed or closely coupled, this kind of fluid coupling is very inefficient.

To make a more efficient fluid coupling, oil is used as the fluid. The two halves or members of the coupling are mounted very close together and enclosed in a housing. Figure 2-21 shows the two members of a fluid coupling. They resemble a hollowed-out doughnut

sliced in half, with blades or vanes set radially into the hollow halves.

Figure 2-22 is a cross section of a fluid coupling. The driving member, called the *pump* or *impeller,* is attached to a drive plate or *flexplate* which takes the place of a flywheel on the engine crankshaft. The driven member, called the *turbine,* is attached to the transmission input shaft. This shaft is, in turn, connected through other shafts and gears to the differential and drive wheels.

With the engine off, the hollow spaces in the pump and turbine are filled with oil. When the engine starts running, the pump begins to rotate and the oil is set into motion. The vanes in the pump start to carry the oil around with them. As the oil is spun around, it is thrown outward and away from the pump vanes by centrifugal force. In the pump, the oil moves outward in the circular path shown by the dotted arrows in Fig. 2-22.

Since the oil is being carried around with the rotating driving member, or pump, the oil is thrown into the driven member, or turbine, at an angle (Fig. 2-23). As the oil leaves the pump and strikes the vanes in the

Fig. 2-21 Sectional view of a fluid coupling, showing the two members. (*Chevrolet Motor Division of General Motors Corporation*)

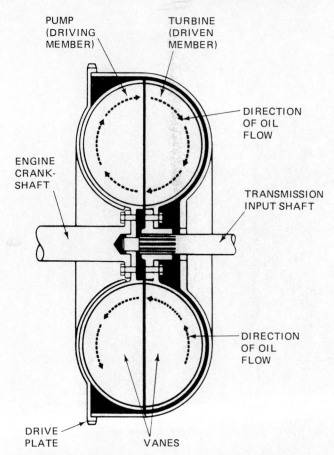

Fig. 2-22 Cross-sectional view of a fluid coupling.

turbine, this applies a torque or turning force to it. The faster the pump turns with the turbine stationary, the harder the oil strikes the turbine vanes. The harder the oil strikes the turbine vanes, the greater the torque applied to the turbine and to the transmission input shaft which is attached to it. However, in a fluid coupling, the torque applied to the turbine can never exceed the torque applied by the pump. This means that a fluid coupling can transmit power but cannot multiply torque.

When sufficient torque is applied to the turbine, it begins to turn and the car begins to move. As the

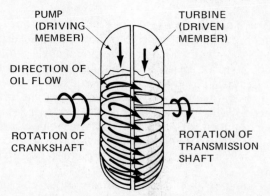

Fig. 2-23 Fluid coupling in action. Oil is thrown from the pump or driving member into the turbine or driven member. The outer housings have been cut away so that the vanes can be seen.

turbine approaches the speed of the pump, the effective force of the oil on the turbine vanes is reduced. If the two members turn at the same speed, no oil passes from one member to the other. When no oil passes from one member to the other, then no power is transmitted through the coupling. However, the pump and turbine cannot turn at the same speed when the engine is driving the car. The pump always turns slightly faster than the turbine.

If the engine speed is reduced so that the car begins to drive the engine, there will be a point at which both members turn at the same speed. As the engine slows further, the driven member or pump will temporarily become the driving member, since the car will be back-driving it. As this happens, the turbine begins to pass oil into the pump. Now the engine will exert a braking effect on the car. This is the same condition that results in a clutch-equipped car when the accelerator is released. Then the engine helps slow the car through *compression braking*.

2. Guide-ring oil-flow control   There are three different types of flow patterns for the oil in a fluid coupling: rotary flow, vortex flow, and turbulent flow. (Fig. 2-24).

Rotary flow is the path a drop of oil takes as the oil rotates with the coupling between the members. Vortex flow is the spiraling or whirlpool path within a fluid coupling that a drop of oil takes due to centrifugal force and the curvature of the coupling. The actual vortex flow occurs as the oil flows from the pump and on through the turbine (and stator in a torque converter; see ✿ 2-15).

However, the simple fluid coupling described earlier is not very efficient under many conditions, because of the turbulence that is set up in the oil. Turbulence is violent random motion or agitation. When there is a

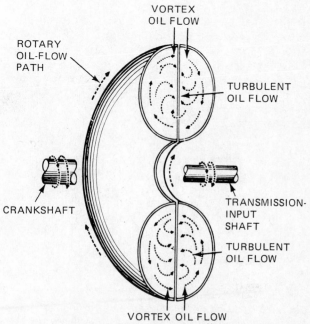

Fig. 2-24 Oil-flow patterns in a fluid coupling.

large difference in speed between the pump and the turbine, there is a great amount of vortex flow. The oil strikes the turbine vanes with great force. This makes the oil swirl about in all directions, especially in the center sections of the members (Fig. 2-24).

To reduce this turbulent flow, and thereby make the fluid coupling more efficient, a split-guide ring is centered in the members (Fig. 2-25). The guide ring also looks like a hollowed-out doughnut cut in half. Each half is set into the vanes of one of the coupling members. This arrangement prevents the oil from creating the turbulence shown in Fig. 2-24.

INNER -outer GEARS

✿ 2-14 Torque converter The torque converter (Figs. 2-1 and 2-26) is a special form of fluid coupling. It has both a driving member and a driven member. Each is equipped with vanes between which oil passes. The torque converter has curved vanes and like the fluid coupling is essentially a special form of clutch. It transmits torque at maximum efficiency when both members are turning at close to the same speed. When the driving member turns much faster than the driven member, then the efficiency of any fluid coupling is reduced.

Reduced efficiency occurs because, as the pump rotates, the oil is thrown onto the turbine vanes with considerable force. When the oil strikes the turbine vanes, the oil splashes and bounces back toward the pump vanes, striking the oil being thrown out of the pump vanes. When there is a big difference in pump and turbine speeds, much of the driving torque is used up in overcoming this "bounce-back" effect. As a result, torque is lost and there is a torque reduction through the fluid coupling.

A torque converter is designed to prevent, or reduce to a minimum, the bounce-back effect. The result is that torque is not reduced when there is a large difference between the pump and turbine speeds. Instead,

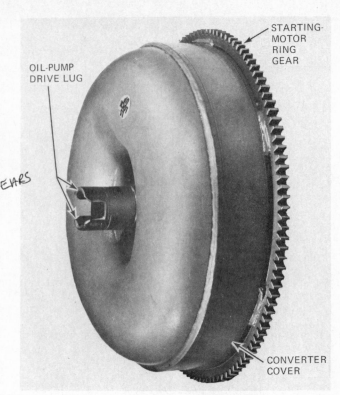

Fig. 2-26 A torque converter, showing the starting-motor ring gear and the drive lugs for the oil pump in the front of the transmission. (*Chrysler Corporation*)

torque is increased or multiplied in the torque converter.

The torque converter acts in a sense like a gear transmission with many usable gear ratios. For example, it can transmit torque at a 1:1 ratio. Under certain conditions, it can increase or multiply this torque so that more torque is delivered than applied. This compares to a manual transmission in first gear. In first gear, the speed through the manual transmission is reduced, but speed reduction means torque increase (✿ 1-6). In a like manner, speed reduction through the torque converter means a torque increase. When torque output is greater than the torque input, there is torque multiplication.

✿ 2-15 Torque-converter action The torque converter provides varying drive ratios between the driving and driven members, with corresponding variations in torque increase. This is accomplished by means of curved vanes in the pump and turbine, and by the use of an extra member between the driving and driven members. This extra member acts to reduce the splashing or bounce-back of the oil. This effect causes torque loss, since the bouncing oil strikes the forward faces of the pump vanes. As a result, the pump tends to slow down.

To improve the efficiency of the torque converter, curved vanes are used in the pump and turbine. The curvature of the vanes is shown in Fig. 2-27. They allow the oil to change directions rather gradually as it passes from the driving member to the driven member. The

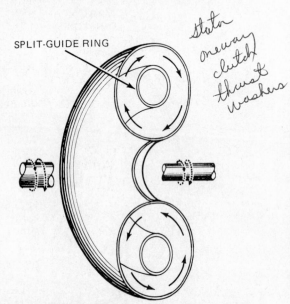

SPLIT-GUIDE RING

stator oneway clutch thrust washers

Fig. 2-25 Split-guide ring designed to reduce oil turbulence. (*Chevrolet Motor Division of General Motors Corporation*)

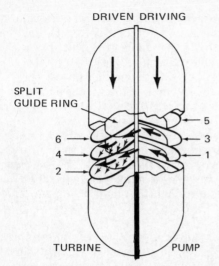

Fig. 2-27 Simplified cutaway view of two members of a torque converter. The heavy arrows show how oil circulates between driving- and driven-member vanes. In operation, the oil is forced by vane 1 downward toward vane 2. Therefore, the oil pushes downward against vane 2, as shown by the small arrows. Oil then passes around behind the split-guide ring and into the driving member again, or between vanes 1 and 3. Then it is thrown against vane 4 and continues this circulatory or vortex flow, passing continuously from one member to the other.

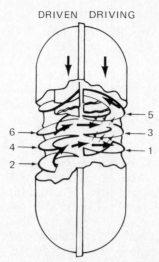

Fig. 2-28 This illustration shows what would happen if the vanes in Fig. 2-27 were continuous. Actually, the inner ends of the vanes are not as shown here but are as pictured in other illustrations. Here, the split-guide ring and outer ends of the vanes have been cut away. If the vanes were as shown here, the oil leaving the trailing edges of the turbine would be thrown upward against the front faces of the pump vanes. Therefore, this oil would oppose the driving force. The effect, shown by the small arrows, would cause wasted energy and loss of torque.

heavy arrows in Fig. 2-27 show the oil paths. The small arrows indicate the driving force with which the oil strikes the turbine vanes.

The oil, which is moving around with the pump, is thrown with a forward motion into the turbine. As the oil passes into the turbine, the oil presses forward all along the vanes, as shown by the small arrows in Fig. 2-27. This is the force which causes the turbine to rotate, turning the attached transmission input shaft.

However, when the pump is revolving much faster than the driven member, the oil still has great forward motion as it bounces back and leaves the turbine. The bouncing oil hits the forward faces of the pump vanes with considerable force. This action is shown by the small arrows in Fig. 2-28. Notice that the oil is opposing the rotation of the pump. With a big difference in speed between the pump and turbine, much of the driving force is wasted in overcoming this action of the oil. As a result, torque is lost.

Figure 2-29 shows how to overcome this effect and at the same time increase the force or torque on the turbine. In Fig. 2-29a, a jet of oil is striking a hemispherical cup attached to the rim of a wheel. This compares to the oil being thrown from the pump into the curved vanes of the turbine in the fluid coupling. The jet of oil swirls around the curved surface of the cup and leaves it with almost the same velocity the oil had when entering the cup. The curved vanes of the driven member react against the oil, change its direction, and throw it back to the driving member. Less torque is transmitted to the turbine than is applied by the pump.

In Fig. 2-29a, the oil gives up very little of its energy to the cup. Therefore, only a small torque is applied to

the turbine. It is the returning oil which bounces back into the pump that causes this loss of torque. However, the effect can be prevented by installing an additional stationary curved vane or *stator* in the path of the

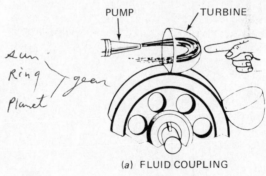

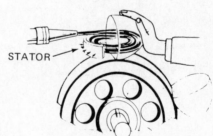

Fig. 2-29 Effect of a jet of oil on a bucket attached to a wheel. If the oil enters and leaves as at (a), the push imparted to the bucket and wheel is small. But if the oil jet is redirected into the bucket by a curved vane or stator, as at (b), the push is increased. (*Chrysler Corporation*)

bouncing oil, as shown in Fig. 2-29b. Now the oil leaves the cup and enters the stationary curved vane, which redirects the oil once more back into the cup.

Theoretically, the oil can make many complete circuits between the cup and the stator. Each time the oil reenters the cup, the oil imparts additional force or added torque to the cup. This effect is known as *torque multiplication* and is a characteristic of a torque converter.

In an actual torque converter, there are stationary curved vanes which reverse the direction of the oil as it leaves the turbine. An additional member is placed between the trailing edges of the turbine blades. The trailing edges are the edges of the vanes that the oil passes last as it leaves a member. The leading edges are the edges onto which the oil first flows.

The oil enters the pump in a "helping" direction, passes through, and reenters the turbine, where the oil gives the turbine vanes another "push." The oil is repeatedly redirected into the turbine, adding torque to the driven member during each pass through. The result is an increase in torque, or torque multiplication, through the torque converter. The greater the vortex flow (in a forward direction) the greater the torque multiplication. Engine braking lowers vortex flow in the reverse direction—from the turbine to the pump to the stator. The less the vortex flow, the greater the rotary flow (✿ 2-13).

Today most torque converters have three members (although some have more). These are the driving member, or pump; the driven member, or turbine; and the reaction member, or stator. It is the stator that changes the direction of oil flow into a helping direction, as shown in Fig. 2-29.

✿ **2-16 Stator action** When a third member or stator is added to a fluid coupling, it become a torque converter. Figure 2-30 is a partial cutaway view of an automatic transmission using a three-member torque converter. Figure 2-31 shows an assembled and disas-

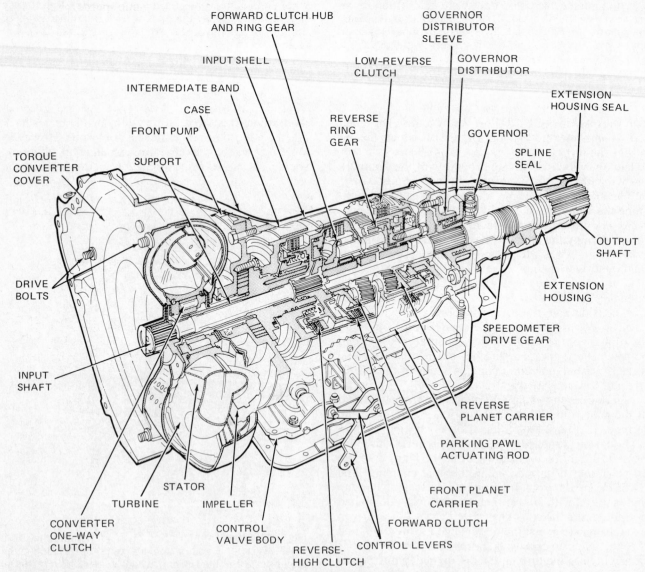

Fig. 2-30 Partial cutaway view of an automatic transmission which has a three-member torque converter.

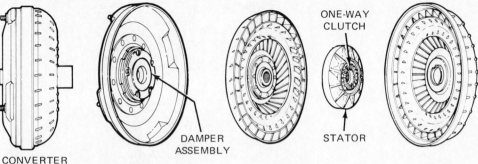

ONE-WAY CLUTCH

DAMPER ASSEMBLY

STATOR

Fig. 2-31 Assembled and disassembled torque converter. (*Ford Motor Company*)

CONVERTER ASSEMBLY   CONVERTER COVER   TURBINE   IMPELLER

sembled torque converter. Figure 2-32 is a simplified drawing of a torque converter.

The stator is mounted on a one-way clutch called a *freewheeling clutch, overrunning clutch,* or *sprag clutch.* This device permits the stator to run free when both the pump and turbine are turning at about the same speed. However, when torque increase and speed reduction take place, the stator stops and acts as a reaction member, or reactor. This action is possible because the stator is mounted, through the overrunning clutch, on a tube or hollow stationary shaft called the *stator shaft* (Fig. 2-33). The shaft is stationary because it is connected to the oil pump, which is attached to the transmission case.

Figure 2-33 shows how the transmission input shaft fits through the stator shaft. Splines on the end of the transmission input shaft or turbine shaft (Fig. 2-32) mesh with splines in the center of the turbine in the torque converter. Splines on the end of the stator shaft mesh with splines in the inner race of the one-way clutch in the stator.

The one-way clutch allows the stator to rotate only in the same direction as the pump in the torque converter. When the stator tries to turn backward, it locks to the stator shaft and torque multiplication takes place.

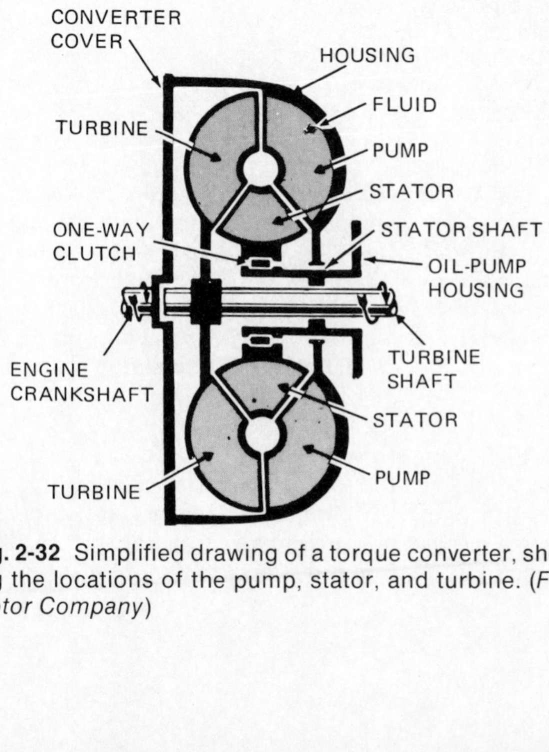

Now the stator vanes turn the oil from the trailing edges of the turbine blades into a helping direction before the oil reenters the pump. This produces a maximum possible torque increase of more than 2:1 through the torque converter when the turbine is stopped or stalled.

As turbine speed approaches pump speed, the torque increase gradually drops off until it becomes 1:1. This occurs when the turbine and pump speeds reach a ratio of approximately 9:10. At this ratio, which is called the *coupling point,* the oil begins to strike the back faces of the stator vanes so that the stator begins to turn. In effect, the oil pushes the stator ahead so that it no longer enters into the torque-converter action. The torque converter acts simply as a fluid coupling under these conditions. Figure 2-34 shows the operation of the torque converter.

✿2-17 One-way-clutch operation A one-way clutch is completely mechanical in operation and requires no type of control mechanism. It locks and unlocks or freewheels automatically as the result of the forces acting on it. The construction of a typical one-way clutch is shown in Fig. 2-35. It contains a series of

CONVERTER COVER

HOUSING

FLUID

TURBINE

PUMP

STATOR

ONE-WAY CLUTCH

STATOR SHAFT

OIL-PUMP HOUSING

ENGINE CRANKSHAFT

TURBINE SHAFT

STATOR

TURBINE

PUMP

Fig. 2-32 Simplified drawing of a torque converter, showing the locations of the pump, stator, and turbine. (*Ford Motor Company*)

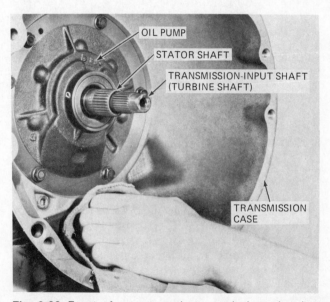

OIL PUMP

STATOR SHAFT

TRANSMISSION-INPUT SHAFT (TURBINE SHAFT)

TRANSMISSION CASE

Fig. 2-33 Front of an automatic transmission, showing how the transmission input shaft fits through the hollow stator shaft, which is attached to the oil pump. (*Chrysler Corporation*)

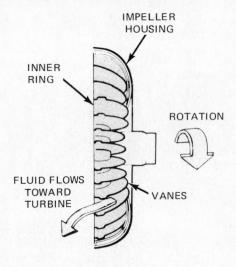

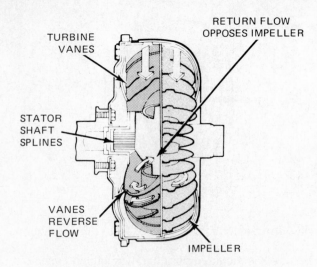

(a) IMPELLER OPERATION

(b) DRIVING THE TURBINE

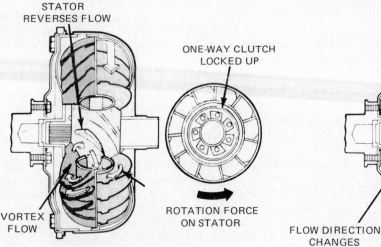

(c) TORQUE MULTIPLICATION

(d) COUPLING PHASE

Fig. 2-34 Torque-converter action. (a) The impeller, or pump, sends a flow of oil or fluid toward the turbine. (b) The turbine vanes receive the flow and this spins the turbine. The vanes reverse the direction of fluid flow and send it back toward the impeller. (c) The stator reverses the flow of the fluid into a helping direction and this multiplies the torque. (d) When the turbine speed nears the impeller speed, the fluid strikes the backs of the stator vanes, causing the stator to spin forward. This prevents the stator vanes from interfering with the fluid flow. (*Ford Motor Company*)

rollers placed between the stator and the hub. The rollers permit the stator to overrun or freewheel when the oil strikes the back faces of the stator vanes.

However, when the oil strikes the front faces of the vanes, the oil attempts to turn the stator in the opposite direction. This action locks the one-way clutch, so the stator is held stationary. Now the stator acts as a reaction member, changing the direction of the oil flow as it passes between the turbine and pump vanes.

Some one-way clutches use sprags instead of rollers. The sprags are shaped like slightly flattened rollers. A series of sprags are placed between the inner and outer races (Fig. 2-36). The sprags are held in place by two cages and small springs. During overrunning, when stator action is not needed, the outer race is unlocked

(Fig. 2-36a). It is attached to the stator, so the stator can spin freely. The inner case is attached to the stator shaft (Figs. 2-32 and 2-33). When stator action is needed, the oil is directed into the stator vanes. This attempts to spin the stator backward (Fig. 2-36b). When this happens, the sprags jam between the outer and inner races to lock the stator to the stator shaft (Fig. 2-34c).

✿ 2-18 Torque-converter lockup There is always some slippage in the torque converter. The turbine can never turn at exactly pump speed, as long as the turbine is being driven by oil thrown from the pump. This slippage through the fluid represents a loss of engine power.

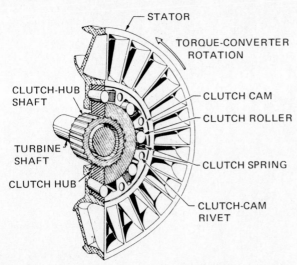

Fig. 2-35 Details of a one-way clutch used to support and control the stator in the torque converter. (*Chrysler Corporation*)

quired, the clutch will partially disengage. This permits the engine to reach a higher speed where it can produce more power for acceleration. However, all lockup mechanisms do not work this way. In many torque converters, lockup occurs at about 30 mph [48 km/h], and only in third gear (direct drive) or fourth gear (overdrive).

By locking the torque converter, the engine is then mechanically driving the transmission. Power flows through the torque converter as if it were a solid piece. This increases fuel economy by up to 8 percent in highway driving and 2 percent in city driving. Construction and operation of the hydraulically operated converter clutch is covered in ✿ 4-23 and 4-24, and in other chapters.

To eliminate this loss, many torque converters now have a clutch, similar to the clutch used with a manual transmission (✿ 1-3). The clutch engages automatically, locking the pump and turbine together (Fig. 2-37). Two types of locking mechanism are used, centrifugal (Fig. 2-37) and hydraulic (Fig. 4-21). One centrifugal type (Fig. 2-37) has 10 shoes made of a friction material similar to that used in transmission bands. The shoes are thrown outward against the converter cover by centrifugal force. This forms a direct mechanical connection between the pump and the turbine.

With the converter clutch shown in Fig. 2-37, clutch engagement begins in any gear between 400 and 900 rpm. It increases gradually with speed until lockup is fully achieved. When additional acceleration is re-

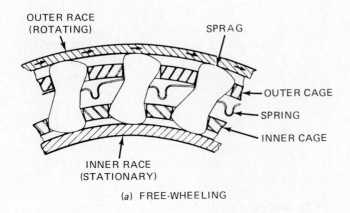

(a) FREE-WHEELING

(c) ASSEMBLED

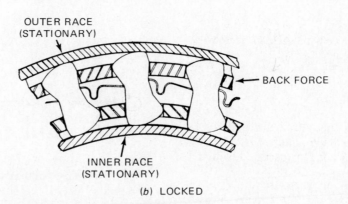

(b) LOCKED

Fig. 2-36 Operation of a sprag type of one-way clutch. The complete sprag clutch is shown at the bottom.

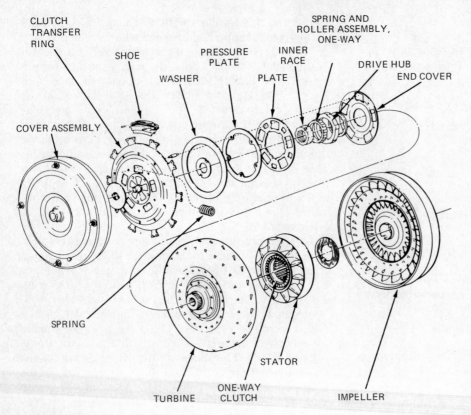

CLUTCH TRANSFER RING

SHOE

WASHER

PRESSURE PLATE

PLATE

INNER RACE

SPRING AND ROLLER ASSEMBLY, ONE-WAY

DRIVE HUB

END COVER

COVER ASSEMBLY

SPRING

TURBINE

ONE-WAY CLUTCH

STATOR

IMPELLER

Fig. 2-37 A centrifugal type of torque-converter locking clutch. (*Ford Motor Company*)

Chapter 2 review questions

Select the *one* correct, best, or most probable answer to each question. Then check your answers against the correct answers given at the end of the book.

1. If a liquid is put under pressure, it will:
   a. compress
   b. not compress
   c. increase in volume
   d. decrease in volume

2. A piston with an area of 5 square inches [32.26 cm²] is pushed into a cylinder filled with a liquid with a force of 200 pounds [890 N]. The pressure developed is:
   a. 40 psi [1379 kPa]
   b. 200 psi [276 kPa]
   c. 1000 psi [6895 kPa]
   d. 10,000 psi [68,948 kPa]

3. Since a liquid is incompressible, it can be used to transmit:
   a. force and motion
   b. vacuum and heat
   c. compression and release
   d. none of these

4. A piston with an area of 0.25 square inch [1.55 cm²] is pushed into a cylinder filled with liquid with a force of 200 lb [890 N]. The pressure developed is
   a. 50 psi [345 kPa]
   b. 200 psi [1379 kPa]
   c. 500 psi [3447 kPa]
   d. 800 psi [5516 kPa]

5. An output piston against which hydraulic pressure is working has an area of 12 square inches [77.4 cm²]. The hydraulic pressure is 240 psi [1655 kPa]. The force on the piston is:
   a. 20 pounds [89 N]
   b. 240 pounds [1068 N]
   c. 2400 pounds [10,675 N]
   d. 2880 pounds [12,810 N]

6. If the pressure-regulator valve has an area of 2 square inches [12.9 cm²] and the spring applies a force of 25 pounds [111 N] on it, the regulated pressure will be:
   a. 12.5 psi [86 kPa]
   b. 25 psi [172 kPa]
   c. 50 psi [345 kPa]
   d. 250 psi [1724 kPa]

7. If the spring force on the balanced valve is 25 pounds [111 N], the input pressure is 100 psi [690 kPa], and the area of the output end of the valve is 2 square inches [12.9 cm²], then the output pressure will be:
   a. 12.5 psi [86 kPa]
   b. 25 psi [172 kPa]
   c. 50 psi [345 kPa]
   d. 125 psi [861 kPa]

8. If the control pressure on the valve shown in Fig. 2-12 is 100 psi [690 kPa] and the spring force is 20 pounds [89 N], the output pressure will be:
   a. 120 psi [827 kPa]
   b. 60 psi [413 kPa]

c. 45 psi [301 kPa]
d. 450 psi [3013 kPa]

9. The band is operated by:
   a. the clutch piston
   b. the output valve
   c. a servo
   d. the master cylinder

10. The clutch disks are forced together by movement of the:
    a. clutch springs
    b. clutch piston
    c. servo piston
    d. clutch fork

11. The purpose of the fluid coupling is to act as a:
    a. synchronizing device
    b. automatic gear changer
    c. flexible power-transmitting coupling
    d. direct drive

12. The fluid coupling consists essentially of two:
    a. doughnuts
    b. vaned members
    c. guide rings
    d. driving shafts

13. In the fluid coupling, oil passes from the driving member to the:
    a. coupling
    b. vanes
    c. driven member
    d. gear

14. The purpose of the guide ring in the fluid coupling is to reduce oil:
    a. movement between members
    b. turbulence
    c. level in coupling
    d. overheating

15. The fluid coupling has maximum efficiency when the driving and driven members are turning at:
    a. high speed
    b. low speed
    c. different speeds
    d. about the same speed

16. The oil flow that is used to multiply torque in a torque converter is:
    a. the vortex flow
    b. the rotary flow
    c. the turbulent flow
    d. none of the above

17. In the torque converter, the stator is placed between the pump and the:
    a. clutch
    b. rotor
    c. turbine
    d. flywheel

18. The typical torque converter has:
    a. three members
    b. four members
    c. five members
    d. two members

19. Two types of one-way clutches are:
    a. ball and roller
    b. sprag and roller
    c. needle bearing and friction bearing
    d. taper bearing and antifriction bearing

20. Two controlling devices in the transmission operated by hydraulic pressure are the bands and:
    a. pistons
    b. clutches
    c. gears
    d. planetary sets

3 to 4
NEXT
WEEK

AUTOMATIC-TRANSMISSION FUNDAMENTALS

After studying this chapter, you should be able to:

1. Explain how the planetary gears are controlled in automatic transmissions.
2. Describe the action of the band and what happens when it is applied.
3. Describe the action of the clutch and what happens when it is engaged.
4. Explain how the shift control works to produce gear reduction or an upshift.
5. Discuss the effects of intake-manifold vacuum on the modulator valve and what changing vacuum does to the shift point.

 3-1 Introduction to automatic transmissions For many years, engineers and inventors have searched for a way to make gear shifting easier. The introduction of synchronizing devices such as the synchromesh unit for manual transmissions is one result of their efforts. This device makes it almost impossible to clash gears when shifting a manual transmission. However, the driver must operate a clutch to interrupt the flow of power while moving a lever to shift the gears from one power-transmitting position to another.

Over the years, automatic devices which eliminate manual gear shifting have been developed. They change the gear ratio through the transmission semiautomatically or automatically in accordance with car and engine speed and the driver's selection.

With an automatic transmission, the power flow is not interrupted even momentarily during a gear-ratio change. Ratio changes, or *shifts,* are achieved by use of hydraulically actuated clutches and bands. If the gear-ratio change is made too abruptly, the passengers in the car will be jarred and the transmission will wear quickly. If the shift is made too softly, the friction faces in the clutches and bands will be destroyed by heat. To eliminate these problems, control of clutches and bands must be precise.

Instead of using a foot-operated clutch, a fluid coupling connects the engine with the power train. Engine power is transmitted through a fluid or oil, and it allows the transmission to be in gear at engine idle. The velocity of the fluid in the fluid coupling is too low at idle speed to move the car.

☼ 3-2 Fundamentals of automatic transmissions Automatic transmissions do the job of shifting gears without assistance from the driver. Most start out in first gear as the car begins to move forward. Then they shift from first gear into second and then third or direct drive as the car picks up speed. The actions are produced hydraulically by oil pressure.

There are three basic parts to the automatic transmission. These are the torque converter, the gear system, and the hydraulic control system (Fig. 2-1). The torque converter passes the engine power from the crankshaft to the planetary-gear system (Figs. 3-1 and 3-2).

Although there are several variations of automatic transmissions, all work in about the same way. All have

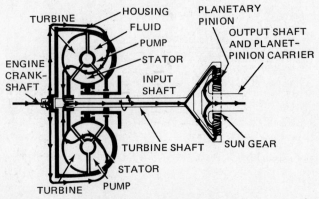

Fig. 3-1 Power flow from the crankshaft through the torque converter to the planetary gearset.

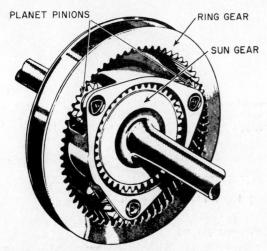

Fig. 3-2 Construction of a typical planetary gearset used in automatic transmissions.

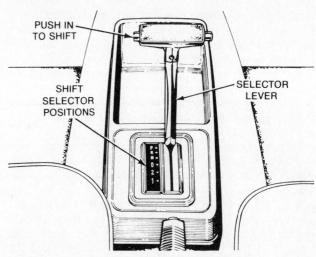

Fig. 3-4 Selector-lever positions for a shift quadrant located in the center console mounted on the floor of the car. (*Ford Motor Company*)

a selector lever on the steering column (Fig. 3-3), or in the center console or on the floor (Fig. 3-4). In some automatic transmissions there are five selector-lever positions. These are P (park), R (reverse), N (neutral), D (drive), and L (low). Others have six selector-lever positions. In addition to D, there are a 2 and a 1 position to the right of D or below it (Fig. 3-4).

In P, the transmission is locked so the car cannot move. In N, no power flows through the transmission but the locking effect is off. In 1, there is maximum gear reduction through the transmission. In 2, there is some gear reduction through the transmission, which provides extra torque at the wheels for climbing steep grades or for engine-compression braking when descending a long hill. In drive, the transmission automatically shifts up or down according to car speed and throttle position. All these positions are covered later.

Some automatic transmissions have a select-shift feature which enables the driver to select the desired gear. Then the transmission will stay in this gear until the driver manually shifts out of it. For example, in Fig. 3-4, when the shift lever is placed in 1, the transmission remains in first gear. When the shift-lever is placed in 2, the transmission starts out and remains in

second gear, regardless of car speed or throttle position.

⚙ 3-3 Planetary gears in automatic transmissions Many planetary gearsets used in automatic transmissions are simple planetary gearsets (Fig. 3-2). They consist of a ring gear, a sun gear, and planet pinions held in place by a planet-pinion carrier. Other automatic-transmission planetary gearsets are *compound* planetary gears. They consist of a sun gear, two sets of planetary gears with carriers, and two ring gears (Fig. 3-5).

Holding any one of the three members of the simple planetary gearset stationary turns the planetary into a speed-reducing or speed-increasing gearset, according to which member is held stationary. Allowing all of the members to spin freely provides neutral. Locking any

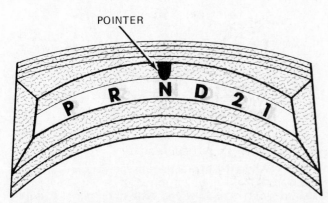

Fig. 3-3 Selector-lever positions for a shift quadrant located on the steering column. (*Chrysler Corporation*)

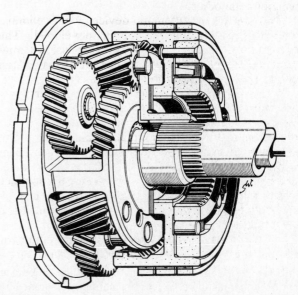

Fig. 3-5 Cutaway view of a compound planetary gearset. (*Volkswagon of America, Inc.*)

two of the members together produces direct drive through the gearset. In addition, if the planet-pinion carrier is held stationary and the ring gear or sun gear is turned, the planet pinions act as reverse-idler gears. Then the direction of rotation is reversed through the gearset. Briefly, these are the actions of the planetary gears in the automatic transmission. They provide either neutral, speed reduction, direct-drive, speed increase, or reverse.

In some automatic transmissions, the output shaft turns faster than, or overdrives, the input shaft. Transmissions with this ability are called *automatic overdrive transmissions* (AOT).

Figure 3-1 shows one arrangement for carrying the power flow from the torque converter into the planetary gears. This is only one of three possible paths. The power can flow to the ring gear, the carrier, or the sun gear.

⚙ **3-4 Planetary-gear system control** Now let's discuss how to hold any one of the members stationary. Also, we will cover how to lock two members together to get a 1:1 gear ratio through the planetary gearset. When this happens, both the input and the output shafts of the planetary gearset are turning at the same speed. The three devices that provide control of the planetary gearset are the multiple-plate clutch, the band and drum, and the overrunning clutch.

Figure 3-6 shows how a multiple-plate clutch and a band and drum can be used to obtain the various conditions through the planetary gearset. Note in Fig. 3-6 that the input shaft has the ring or internal gear on it. The planet-pinion carrier is attached to the output shaft. The sun gear is separately mounted, and a drum is part of the sun-gear assembly. The sun gear and drum turn together as a unit.

The band is operated by a servo, as shown in Fig. 3-7. The servo consists of a piston on the end of a stem or rod and is mounted in a cylinder called the *servo body*. The other end of the stem acts against the band positioned around the drum. Without oil pressure acting against the end of the servo piston, the release spring pushes the piston to the left (in

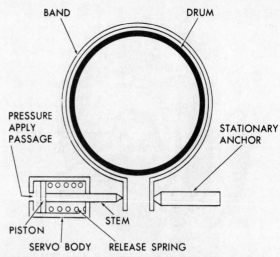

Fig. 3-7 Arrangement of the servo, band, and drum.

Fig. 3-7). This allows the band to move away from around the drum. Now the sun gear is free to turn.

However, under the proper operating conditions, oil under pressure is admitted to the left side of the servo cylinder (Fig. 3-8), forcing the piston to the right. This action compresses the spring and tightens the band around the drum, which brings the drum and sun gear to a halt. With the ring gear being turned and the sun gear held, there is speed reduction through the planetary gearset.

NOTE: The arrangement described above of a band holding the sun gear stationary is only one of several arrangements used in automatic transmissions. In many transmissions, when the band is applied, it holds the ring gear or the planet-pinion carrier stationary. Different transmissions may lock different members together when the clutch is engaged. However, the principle is the same in all two- and three-speed automatic

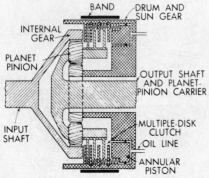

Fig. 3-6 Sectional view of the two hydraulically operated controlling mechanisms for the planetary gearset. One mechanism is the band and drum. The other is the multiple-disk clutch.

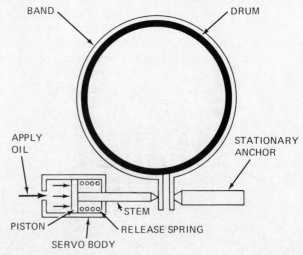

Fig. 3-8 When the hydraulic pressure is applied behind the piston, the piston and rod move to the right (in the illustration). This causes the band to tighten around the drum and hold it. With the band applied, the drum is held stationary.

transmissions. There is speed reduction when a band is applied, and there usually is direct drive when a multiple-disk clutch is engaged. In some transmissions, the band function is taken over by an overrunning clutch to provide speed reduction through a planetary gearset.

To get direct drive, the band must be released. This actions takes place when the hydraulic control system dumps the oil from the left chamber of the servo cylinder. Then the spring forces the piston to move again to the left, releasing the band (Fig. 3-7). At the same time, the clutch is engaged to lock the sun gear and planet-pinion carrier together. The clutch consists of a series of clutch plates alternately splined to the carrier and to the inner face of the drum (Fig. 2-17, 3-6, and 3-9).

When the clutch is disengaged, these plates are held apart (Fig. 3-6). However, simultaneously with the releasing of the band, oil is directed into the chamber in back of the annular, or ringlike, piston in the drum–sun-gear assembly. This action forces the piston to the left (in Fig. 3-9) so that the clutch plates are forced together. Then, friction between the plates locks the sun gear and planet-pinion carrier together. When this happens, the planetary-gear systems acts as a direct-drive coupling. Both the input and output shafts turn at the same speed with a 1:1 ratio.

To achieve reverse in the planetary-gear system shown in Fig. 3-9, the planet-pinion carrier must be held while both the ring gear, which is on the input shaft, and the sun-gear-and-drum rotate. With this arrangement, the sun gear turns backward. Actually,

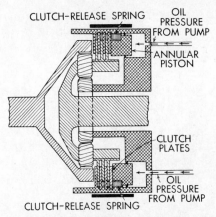

Fig. 3-9 One set of clutch plates is splined to the sun-gear drum. The other set is splined to the planet-pinion carrier. When the hydraulic control system directs oil behind the annular piston, as shown by the arrows, the clutch plates are forced together so that the sun gear and planet-pinion carrier are locked together.

more complex arrangements are used in automatic transmissions.

☼ 3-5 Shift control   Shifting from one gear to another must be controlled so that it takes place only under the desired operating conditions. Figure 3-10 is a simplified diagram of a hydraulic control system for a single planetary gearset in an automatic transmission. The band is controlled by a servo. In hydraulic systems, a servo is a device that converts hydraulic pressure to mechanical movement as the oil acts on a piston inside the

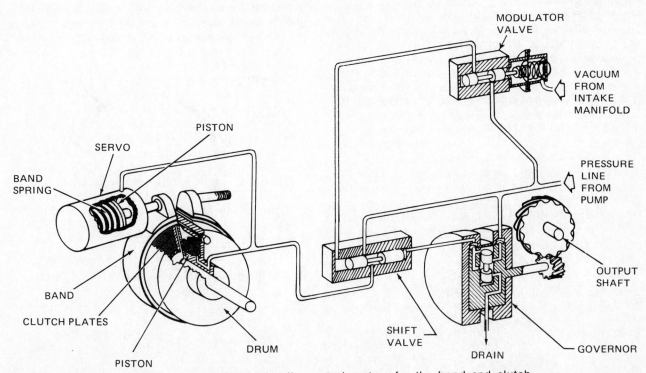

Fig. 3-10 Schematic diagram showing the hydraulic control system for the band and clutch.

servo. Later we will look at the circuits for automatic transmissions that use two or more planetary gearsets. Automatic transmissions can use more than one planetary gearset.

The major purpose of the hydraulic control system is to apply and release a band or to engage and disengage a clutch. By this action, the hydraulic system controls the shifting of the planetary gearset from speed reduction to direct drive. The shift must take place at the right time. This depends on car speed and throttle opening, or engine load. These two factors produce varying oil pressures that work against the two ends of the *shift valve*.

The shift valve is a spool-type valve inside a bore or hole in the valve body. Figure 2-9 shows the spool valve. Pressure at one end of the shift valve (Fig. 3-10) comes from the governor and is known as *governor pressure*. Pressure at the other end of the shift valve changes as the engine intake-manifold vacuum changes and is known as *throttle pressure*. Throttle pressure, which is discussed later, and governor pressure are the two signals or forces that determine when the transmission will shift.

☼ 3-6 Governor action The automatic-transmission governor is a device that controls or governs gear shifting in relation to car speed. In the hydraulic system, the governor controls pressure on one end of the shift valve. Governor pressure changes with car speed, because the governor is driven by the transmission output shaft. As output-shaft speed and car speed go up, governor pressure increases proportionally.

Governor pressure is actually a modified line pressure. An oil pump in the transmission produces the line pressure, which then passes into the governor, as shown in Fig. 3-10. As car speed increases, the governor spins faster. This allows more pressure to pass through, resulting in greater pressure against the right end of the shift valve. The higher the car speed, the higher the pressure released by the governor.

☼ 3-7 Throttle pressure Working on the left end of the shift valve is a pressure called *throttle pressure* that changes as intake-manifold vacuum changes. Line pressure enters the modulator valve at the upper right in Fig. 3-10. The modulator valve contains a spool valve attached to a spring-loaded diaphragm. Vacuum increases in the engine intake manifold when the throttle valve is partly closed, or when engine load decreases. This vacuum pulls the diaphragm in (to the right in Fig. 3-10), which then causes the modulator valve to also move to the right. Movement of the modulator valve reduces line pressure to the left end of the shift valve.

As the modulator valve moves, a land on the valve closes off the line-pressure passage from the pump. When this happens, it reduces the pressure on the left end of the shift valve. Then the shift valve is pushed to the left by governor pressure. As a result, line pressure can now pass through the shift valve to the band servo and clutch for the planetary gearset. The line pressure releases the band and engages the clutch. This puts the

planetary gearset into direct drive, as the shift from speed reduction to direct drive is made.

☼ 3-8 Shift action When the planetary gearset changes from a lower-speed ratio to a higher-speed ratio, such as from speed reduction to direct drive, this is called an *upshift*. When the speed ratio through the planetary gearset is changed from a higher-speed to a lower-speed ratio, a *downshift* has taken place. In an automatic transmission, the hydraulic control system must control both the upshifts and the downshifts. Now let's review how upshifts take place. Then we will cover shift action during the downshift.

With the clutch disengaged and the band applied, the planetary gearset is in speed reduction or low gear. As car speed increases, the governor releases more and more pressure to one end of the shift valve. The pressure on the other end of the shift valve depends on intake-manifold vacuum. As long as the throttle is held open, there is little manifold vacuum. Therefore, the throttle pressure on the left end of the shift valve is high. This high pressure holds the shift valve to the right. The planetary gearset is in low or reduction for good acceleration.

However, as car speed continues to increase, governor pressure becomes great enough to push the shift valve to the left (in Fig. 3-10). This allows line pressure through the annular groove in the shift valve. The line pressure releases the band and engages the clutch. This action shifts the planetary gearset from reduction or low into direct drive.

The upshift will also take place if the throttle is partly closed after the car reaches intermediate speed. Closing the throttle increases the intake-manifold vacuum. The vacuum cuts down modulator pressure to the left end of the shift valve. Then the shift valve moves to the left (in Fig. 3-10), pushed by governor pressure. This applies line pressure to the planetary gearset. The clutch engages and the band releases, shifting the planetary gearset into direct drive.

The reason for varying the pressure at the left end of the shift valve is to change the upshift according to driving conditions. When the car is accelerating, high torque is needed. The gears should stay in low. Then, when cruising speed is reached, less torque is needed. The driver eases up on the throttle. This increases intake-manifold vacuum so that the upshift occurs.

If the driver again wants fast acceleration, the throttle is pushed to the floor. This reduces intake-manifold vacuum, and the planetary gearset downshifts into reduction or low gear. This increases torque to the drive wheels.

☼ 3-9 Hydraulic valves The preceding discussion of shift action is a simplified description of how upshifting and downshifting are accomplished. The actual valves are more complicated, having springs for initial loading of the valves. There are other valves in the hydraulic control system, including valves to ease the shifts, regulate pressures, and time downshifts. There is also a manual-shift valve.

3-10 Manual valve The manual valve is a special spool valve in the automatic-transmission valve body that is manually positioned by the driver through linkage. This occurs when the driver moves the shift lever to the selected transmission operating range (3-4). As the driver moves the selector lever on the steering column or on the floor, the manual valve moves in the valve body inside the transmission. The manual valve opens and closes passages that direct oil to the different circuits in the transmission as needed for the selected transmission range.

Figure 3-11 shows the manual valve and the section of the valve body in which the manual valve operates. It is the same as the spool valve (Fig. 2-9) except that the manual valve has another major undercut section. Sometimes this type of valve is called a *multiple-groove spool valve*. On the left end (in Fig. 3-11) is an additional groove for the *manual lever*, which moves the manual valve to the right or left as the shift lever moves.

The different positions of the manual valve with the shift lever in P (park), R (reverse), N (neutral), D (drive), 2 (second or intermediate), and 1 (low or first) are shown in Fig. 3-11. When the selector lever is moved from neutral to drive, the manual valve is moved to the right (in Fig. 3-11). Now line pressure is directed to drive, intermediate, and low. The low hydraulic circuit puts the transmission in first gear, or low.

As car speed increases, shifts will be made from low to intermediate, and then from intermediate to direct drive. Under the proper conditions, downshifts will also be made. Drive is the normal range used for most driving conditions, both in the city and on the highway, in most automatic transmissions.

3-11 Selector lever Automatic transmissions have a selector lever which corresponds to the shift lever used in manual transmissions. However, the selector lever selects a driving *range*. The automatic transmission does the rest—upshifting or downshifting automatically as car speed and throttle position require.

The selector lever is located either on the steering column (Fig. 3-3) or on the console or floor (Fig. 3-4). With either arrangement, movement of the selector lever causes the manual valve inside the transmission to move. This action selects the driving range.

The driving range selected is indicated by a pointer on the selector quadrant (Fig. 3-12). If the driver selects D or drive when first starting out, the transmission will shift to low or first gear. Then as car speed increases or the accelerator pedal position changes, the transmission will upshift through intermediate or second gear to high or direct gear. The transmission will remain in direct gear until the driving situation changes. If the driver slows or stops the car, the transmission will downshift through intermediate to low. Also, if the driver wants an increase of speed to pass another car, pressing the accelerator pedal to the floor will cause the transmission to downshift to second gear.

If the selector lever is moved to 2, the transmission will shift to second. It will not upshift to direct. If the selector lever is moved to 1, or low, the transmission will stay in low gear. This gear position is used for pulling heavy loads or for going down a steep grade. In the low-gear position, the engine helps to slow the car through engine-compression braking. This saves the car brakes from long periods of use when going down a long hill.

Some automatic transmissions have overdrive. In overdrive, the output shaft turns faster than, or overdrives, the input shaft. On these, the D on the quadrant has a circle, or O, around it—Ⓓ—which means *overdrive*. At highway speed, the transmission automatically shifts from direct drive to overdrive. This type of transmission is described later.

In R, or reverse, the car is backed up. In N, or neutral, the transmission is unlocked and no power can flow through it. N can be used for pushing or pulling the car if it must be moved. However, most cars with automatic transmissions cannot be "push-started."

P is park. In park, the transmission is locked (Fig. 3-13). A lever with a parking pawl on it moves into teeth in a gear splined to the transmission output shaft. This locks the output shaft to the transmission case so the car cannot be moved.

3-12 Starting the engine When a vehicle is equipped with an automatic transmission, the engine cannot be started unless the selector lever is in N or P. The reason for this is to prevent accidents. If the engine could be started with the selector lever in a driving range, the car would suddenly start to move. This could result in an accident.

There are two different starting-control systems. One is for the steering-column-mounted selector lever and quadrant. The other may be used with either floor shift or column shift.

Figure 3-14 shows an ignition switch which is also a steering-wheel lock and a starting switch. When the

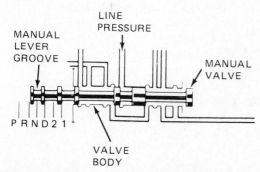

Fig. 3-11 A manual valve, and the section of the valve body in which the manual valve operates. (*Ford Motor Company*)

Fig. 3-12 Selector quadrant with the pointer at D. In this position, the transmission will start out in first gear and shift through second to direct drive.

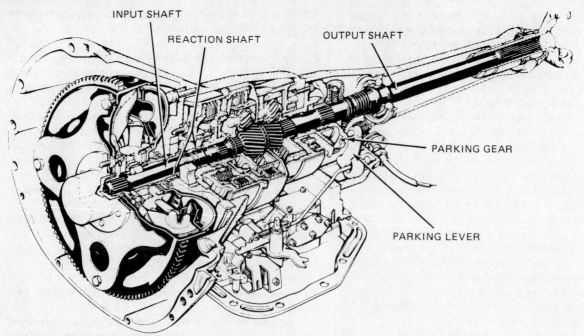

Fig. 3-13 In park, the transmission output shaft is locked to the transmission case by a pawl on the end of the parking lever. (*Chrysler Corporation*)

ignition key and switch are turned to the start position, the starting motor operates to crank the engine. At the same time, the steering wheel is unlocked. When the engine starts, the driver releases the ignition key and the ignition switch moves back to the ON, or RUN, position. The ignition key and switch cannot be started if the transmission is in any driving range.

The floor-mounted selector lever has a different arrangement. This includes a neutral safety switch which is open in all positions except P and N. If the selector lever is in any other position, the safety switch is open so the starting motor cannot operate. But if the selector lever is moved to N or P, the safety switch points are closed. Then, turning the ignition switch to start will allow the starting motor to operate.

⚙ **3-13 Automatic transmission fluid (ATF)** The fluid used in an automatic transmission is a special oil.

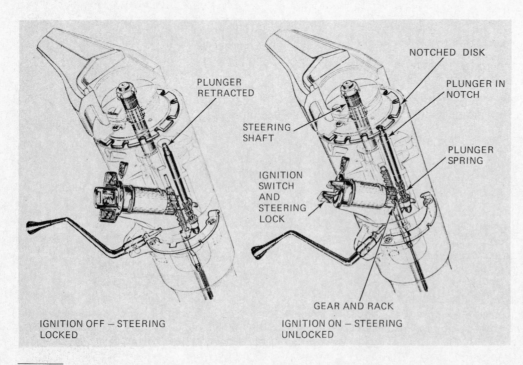

Fig. 3-14 Combination ignition switch and steering lock in phantom view, showing the two positions of the lock. (*General Motors Corporation*)

The oil has had added to it several additives such as viscosity-index improvers, oxidation and corrosion inhibitors, extreme-pressure and antifoam agents, detergents, dispersants, friction modifiers, pour-point depressants, and fluidity modifiers. The oil usually is dyed red. If leakage occurs, you can easily tell whether it is engine oil or transmission fluid.

There are several types of automatic-transmission fluid. This is because various models of automatic transmission have different requirements for the lubricating oil. Most automotive automatic transmissions require either type F or Dexron II automatic-transmission fluid.

In the automatic transmission, the fluid performs the following tasks:

1. It serves as the medium for transmitting power from the pump to the turbine in the torque converter.
2. It transmits hydraulic force as required to operate servos, clutches, and other devices.
3. It transfers heat from the inside of the transmission to the outside air.
4. It serves as a lubricant for bearings, bushings, clutches, and gears.

The four most important jobs of the automatic-transmission fluid are listed above. In addition, the fluid must have the following properties.

1. It must not react chemically with any other material in the transmission, such as metals, plastics, friction materials, or seals.
2. It must have high resistance to oxidation, and must not catch fire, break down, or form varnish at high operating temperatures.
3. It must act as a rust inhibitor to prevent the moisture that enters the transmission in the ventilating air from allowing rust to form.
4. It must be a multiviscosity oil, thin at low temperature to pump easily and thick enough at high temperatures to permit the valves and orifices to properly control the flow.

✿ 3-14 Types of automatic-transmission fluid

Type A automatic-transmission fluid met these basic requirements and was widely used in the past. However, type A fluid was replaced by Dexron. Dexron, in turn, was replaced by Dexron II. Today, type F and Dexron II are the most common types of automatic-transmission fluid. Most Ford transmissions should have type F. Other transmissions use Dexron II. To identify automatic-transmission fluid, the type is stamped on the can.

Type F and Dexron II are not the same. One difference is in the friction characteristics of the fluid. Theoretically, Dexron II is more slippery than type F. Therefore, a transmission using Dexron II will have a softer shift. However, during transmission design, the material chosen for the friction elements, such as band facings and clutch-plate linings, is selected in combination with the type of fluid to be used. For this reason, filling an automatic transmission with the wrong type of fluid may cause the transmission to fail.

Dexron II and type F fluid have similar flow characteristics and heat stability. However, each has different friction characteristics. The automatic transmissions manufactured by Ford require a fast lockup with little slippage as the clutches engage. Therefore, the type F fluid is specified for use in these Ford transmissions. It has a higher coefficient of friction than Dexron II.

To provide a smooth shift, transmissions built by General Motors and Chrysler allow some slippage of the plates as the clutch engages. Therefore, Dexron II contains a friction modifier to reduce the coefficient of friction. This means that, compared to type F, Dexron is more slippery. If Dexron II is used in Ford transmissions, rapid clutch-plate wear and slippage might result. The transmission and torque converter should be drained and refilled with type F fluid.

✿ 3-15 Transmission-fluid coolers

Transmission fluid may become very hot, especially under severe operating conditions. Some of the heat comes from the engine because it is closely coupled with the transmission. Much of the heat develops in the transmission itself. As the fluid is tossed from the torque-converter pump to the turbine and back again, the fluid gets hot. Rapid stirring of any liquid will heat it. The fluid is also heated by the friction effect of the shafts and bearings rotating with the fluid serving as lubricating oil.

To prevent overheating of the fluid and the transmission, most automatic transmissions have a fluid-cooling system (Fig. 3-15). This system includes a fluid-cooler tube in the outlet tank of the engine cooling-system radiator (Fig. 3-16). The tube is immersed in the coolant of the engine-cooling system. This arrangement allows the engine-cooling system to keep the transmission fluid and transmission from overheating.

Some vehicles are equipped with an air-cooled auxiliary fluid cooler which is mounted in front of the radiator or air-conditioner condenser (Fig. 3-17). It is connected in series with the transmission fluid-cooler tube to provide additional cooling. Figure 3-17 shows the direction in which the fluid flows.

The fluid-cooler tube in the outlet tank of the engine radiator is not shown. However, the fluid from the transmission goes through this tube first, as shown by the arrows in Fig. 3-17. The fluid then passes through the auxiliary fluid cooler, where it loses additional heat.

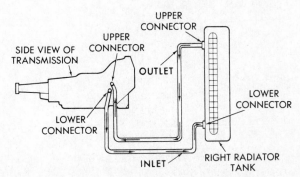

Fig. 3-15 Transmission oil-cooling system. (*Cadillac Motor Car Division of General Motors Corporation*)

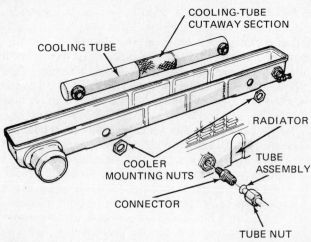

Fig. 3-16 Transmission oil-cooler tube location in outlet tank of the engine cooling-system radiator. (*Chrysler Corporation*)

Then it flows back to the transmission. This continuous circulation of the fluid between the transmission and the cooler removes excess heat from the transmission and keeps it, and the fluid, from overheating.

✸ 3-16 Automatic-transmission variations A basic design of automatic transmission may be used with small variations with a variety of engines and vehicles.

The operation of the clutches and bands in a transmission is due to hydraulic pressures. The pressures and the timing of the shifts can be changed by changing valve springs, hydraulic connections, and servos. In this way, one basic model of automatic transmission can be adapted to suit a particular engine and vehicle.

The planetary-gear system, the clutches, and the bands can remain essentially the same. But the valves and other components of the hydraulic system can be altered to suit engine requirements. For example, when the transmission is used with a four-cylinder engine, the shift points selected might require fairly high car speed. But with a six-cylinder engine having a higher torque, best car operation and drivability might require that the transmission shift at a lower speed.

Also, the clutches may differ with different engines. For example, with a larger engine, the transmission clutches may have more plates for greater holding force to handle the higher engine power.

✸ 3-17 Electronic gear shifting A few models of cars are now equipped with computers that control the shifting of automatic transmissions. The system, first introduced on production cars by Renault and Toyota, uses most of the automatic-transmission components discussed earlier. However, the hydraulic-control system (✸ 3-5 to 3-11) is replaced by electric solenoids. The solenoids are controlled by a microprocessor, which is a small on-board computer. Sensors measure

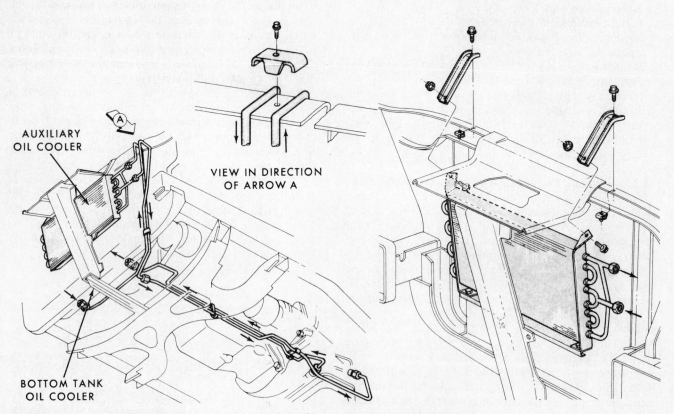

Fig. 3-17 Location of auxiliary transmission-oil cooler. This is a small radiator, similar to the engine cooling-system radiator, mounted in front of the engine radiator. The arrows show the flow of the transmission oil to and from the radiator. (*Chrysler Corporation*)

engine speed, throttle opening, engine temperature, and other conditions. Then this information is fed to the microprocessor, which signals the solenoids when to shift gears.

Computer-controlled shifting is reported to provide up to 7 percent improvement in fuel economy. The improvement comes from the ability of the microprocessor to calculate the ideal time to shift gears. Although automatic transmissions control the operation of the converter clutch (✿ 2-18) electronically, gear-shifting is still done by the hydraulic-control system. In some advanced automatic-transmission designs, the computer decides when to shift gears, and then selects the best gear ratio.

One advantage to electronically controlled transmissions is that they may allow the manufacturer to build one basic model of automatic transmission and then easily tailor it to various car models. This could be done by simply changing the program in the microprocessor. With electronic engine control systems now widely used, predictions are being made that eventually both the engine and transmission will be under computer control.

KNOW CHP 3 to 4

Chapter 3 review questions

Select the *one* correct, best, or most probable answer to each question. Then check your answers against the correct answers given at the end of the book.

1. The selector levers for most automatic transmissions have:
   a. four positions
   b. five positions
   c. six positions
   d. seven positions
2. The selector-lever position in which there is no power flow through the transmission but the locking effect is off is:
   a. P
   b. D
   c. N
   d. R
3. The typical planetary gearset consists of planet pinions, a ring or internal gear, and:
   a. a sun gear
   b. pinions
   c. clutch
   d. a hand brake
4. Three actions of the planetary gears in an automatic transmission are to provide reverse, direct drive, and:
   a. high gear
   b. gear reduction
   c. gear locking
   d. braking
5. The two control mechanisms used with a planetary-gear system are:
   a. valve and pump
   b. torque converter and sun gear
   c. bands and clutches
   d. driving gear and sun gear

6. The band is actuated by a:
   a. servo
   b. clutch piston
   c. manual valve
   d. linkage to foot pedal
7. The ring-shaped part in the clutch which puts the pressure on the clutch disks is called the: *Study*
   a. spring
   b. piston
   c. ring gear
   d. planet-pinion carrier
8. The two controlling factors that cooperate to produce gear shifting are:
   a. hydraulic pressure and governor pressure
   b. car speed and governor pressure
   c. intake-manifold vacuum and car speed
   d. intake-manifold vacuum and throttle opening
9. The bellows in the modulator valve is actuated by:
   a. intake-manifold vacuum
   b. car speed   *Study*
   c. engine speed
   d. governor pressure
10. The valve which is controlled by linkage to the selector lever is called the:
    a. shift valve
    b. control valve
    c. pressure valve
    d. manual valve

GENERAL MOTORS 350C TURBO HYDRA-MATIC TRANSMISSION

After studying this chapter, you should be able to:

1. Explain the construction of planetary gearsets and how they produce three forward and one reverse speed.
2. Name the holding members in action in all gear positions and describe their effect on the gearset.
3. List six of the major valves in the transmission and explain their purpose and how they work.
4. Explain the purpose of the accumulators.
5. Discuss the torque-converter clutch and explain its purpose and how it works.
6. Describe the actions in the transmission during the shifts in the drive range.
7. Explain what happens when the selector lever is moved to L; also, when it is moved to 2.
8. Describe what happens when the selector lever is moved to R.

☼ 4-1 General Motors applications The General Motors type 350C Hydra-Matic automatic transmission is widely used in the General Motors line of cars, including Buick, Chevrolet, Oldsmobile, and Pontiac. It includes planetary gearsets, multiple-disk clutches, roller or one-way clutches, and a band to provide three forward speeds and reverse. The 350C automatic transmission uses a fluid cooler as explained in ☼ 3-15. Servicing the 350C transmission is described in Chap. 10.

☼ 4-2 Type 350C automatic transmission The 350C transmission is often referred to as the 350C Turbo Hydra-Matic. It is essentially the same automatic transmission as the earlier 350 transmission. However, the 350C transmission has, as of 1980, an added feature. This feature is a converter-clutch assembly. When the car reaches cruising speed on the highway, the clutch locks up the torque converter by locking the pump and turbine together. This prevents any slippage through the converter.

There is always some slippage in an operating torque converter (☼ 2-14 to 2-18). This is because the turbine can never turn as fast as the pump when the engine is driving the wheels. Slippage represents a power loss. The converter clutch, as it locks up the converter,

prevents slippage and the resulting power loss. It also improves fuel mileage. The converter clutch is described in ☼ 4-23.

NOTE The C after the 350 stands for *clutch* and means the transmission has a torque-converter clutch.

Figure 4-1 is a cutaway view of the 350C automatic transmission. In addition to the torque converter and converter clutch, it has two planetary gearsets, four multiple-disk clutches, two roller or one-way clutches, and a band. The hydraulic system uses valves and servos to operate the clutches and band. By operating them in different combinations, the transmission provides three forward speeds and reverse.

The 350C transmission is used on many different cars with several types of engines. The transmission is basically the same, regardless of what car or engine it is used with. However, there are different models of the type 350C automatic transmission, and they vary in some details. For example, some models may have more plates in the multiple-plate clutches than others. Also, valves and valve springs may be different to produce different shifting patterns.

When the transmission is used with one engine, the upshift points might be on the high side. But on another engine with a different torque curve, best car operation

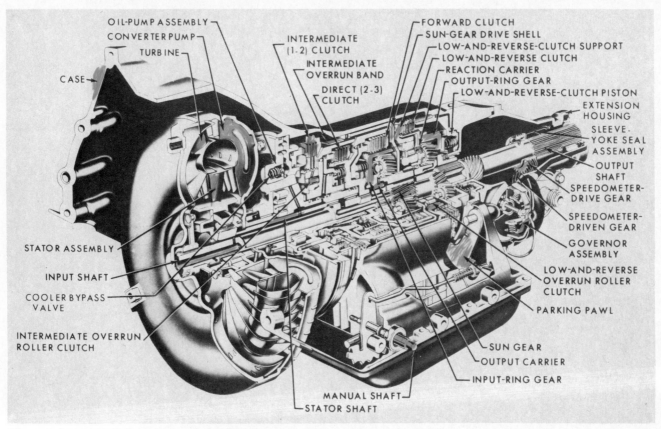

Fig. 4-1 Cutaway view of the type 350 Turbo Hydra-Matic automatic transmission. (*Oldsmobile Division of General Motors Corporation*)

might require the transmission to upshift at somewhat lower speeds. However, all versions of the type 350C transmission have the same general servicing procedure.

Figure 4-1 is a cutaway view of the complete transmission showing the locations of the various components. Figure 4-2 shows the components in the upper half of the transmission in sectional view. The components, from left to right in Fig. 4-2, are named as follows. First is the intermediate overrun roller clutch. This clutch is a one-way clutch that works with the intermediate clutch, as explained later. The intermediate clutch (B in Fig. 4-2) is a multiple-disk clutch. How one-way and multiple-disk clutches work was explained in Chap. 3. The one-way clutch permits rotation in one direction but not in the other. Also, it permits one member to run faster than, or overrun, another member. The multiple-disk clutch can lock up two members if the disks are pressed together by hydraulic pressure.

The intermediate overrun band (C in Fig. 4-2) is positioned around the extension from the front-gearset sun gear. When the band is applied, it holds the sun gear stationary. We explain the effect of this action shortly.

The direct clutch (D in Fig. 4-2) is a multiple-disk clutch. Then there are the front planetary gearset, forward clutch (E in Fig. 4-2), the low-and-reverse mul-

tiple-disk clutch (F in Fig. 4-2), the low-and-reverse roller (G in Fig. 4-2), and the rear planetary gearset.

⚙ 4-3 Power flow in D, low gear When the driver moves the selector lever to D (drive), linkage to the valve body in the transmission moves the manual valve to the D or driving position (Fig. 3-11). Hydraulic pressure can then flow from the valve body to other valves in the hydraulic system and from there to the forward clutch. Now let us see what happens when the selector lever is moved to D and the car pulls away from the curb.

When first starting out in D, or in the drive range, the transmission is in low gear. Fluid pressure is applied to the forward clutch (Fig. 4-3). The other disk clutches and band are off. With this condition, the power flow is as shown in Fig. 4-3. Power input is at the torque-converter pump (1). Power flows to the turbine (2) and to the input shaft (3). The turbine hub is splined to the input shaft. The forward-clutch drum assembly (5) is splined to the input shaft. Therefore, this drum assembly must rotate with the input shaft. The forward clutch is on, which causes the ring gear of the front gearset to rotate as shown to the upper left in Fig. 4-3. The ring gear drives the planet gears clockwise, which causes the sun gear to turn counterclockwise. The sun gear of the front gearset (7) and the sun gear of the rear gearset (8) make up a single piece. The two sun gears are at the

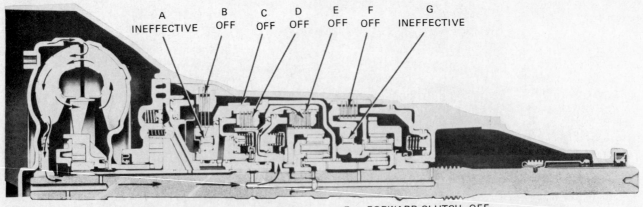

| | A
INEFFECTIVE | B
OFF | C
OFF | D
OFF | E
OFF | F
OFF | G
INEFFECTIVE |

A   INTERMEDIATE-OVERRUN ROLLER CLUTCH—INEFFECTIVE

B   INTERMEDIATE CLUTCH—OFF

C   INTERMEDIATE OVERRUN BAND—OFF

D   DIRECT CLUTCH—OFF

E   FORWARD CLUTCH—OFF

F   LOW-AND-REVERSE CLUTCH—OFF

G   LOW-AND-REVERSE ROLLER CLUTCH—INEFFECTIVE

Fig. 4-2 Sectional view of the upper half of the type 350 Turbo Hydra-Matic automatic transmission. Transmission is in neutral with engine running. (*Oldsmobile Division of General Motors Corporation*)

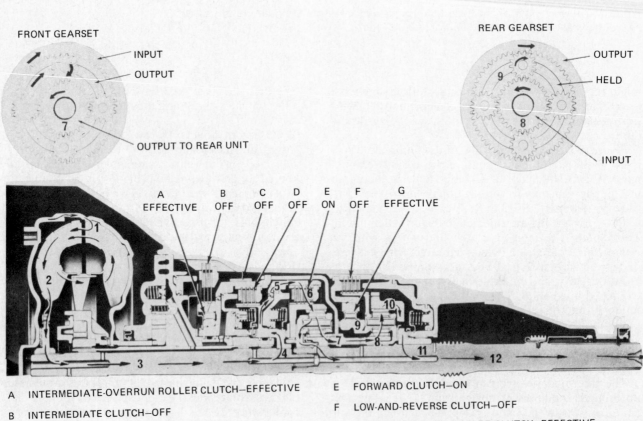

FRONT GEARSET — INPUT — OUTPUT — 7 — OUTPUT TO REAR UNIT

REAR GEARSET — OUTPUT — HELD — 9 — 8 — INPUT

A   INTERMEDIATE-OVERRUN ROLLER CLUTCH—EFFECTIVE

B   INTERMEDIATE CLUTCH—OFF

C   INTERMEDIATE-OVERRUN BAND—OFF

D   DIRECT CLUTCH—OFF

E   FORWARD CLUTCH—ON

F   LOW-AND-REVERSE CLUTCH—OFF

G   LOW-AND-REVERSE ROLLER CLUTCH—EFFECTIVE

Fig. 4-3 Power flow, shown by arrows, through the transmission in drive range and low gear. (*Oldsmobile Division of General Motors Corporation*)

two ends of a hollow shaft. This hollow shaft rides on the output shaft.

When the front sun gear turns, the rear sun gear must turn also because they are a single piece. This action causes the planet gears of the rear set to rotate in a clockwise direction, as shown in Fig. 4-3. The planet-gear cage is held and prevented from rotating by the low-and-reverse roller clutch (9). Note that the outer race of the roller clutch is attached to the transmission case so that the outer race cannot turn. The inner race is part of the planet-pinion-cage assembly. Therefore, when the sun gear rotates counterclockwise, as shown to the upper right in Fig. 4-3, it attempts to cause the planet-pinion cage to turn counterclockwise also. However, the roller clutch (9) locks to prevent this counterclockwise rotation. As a result, the planet pinions act as idlers and drive the ring gear (10) of the rear gearset in a clockwise direction.

The ring gear (10) of the rear gearset is splined to the output shaft (12) at 11. Therefore, the output shaft turns in a clockwise direction to give the car forward motion. Note that there is gear reduction in both the front and rear gearsets. Their combined gear reduction is 2.52:1. The input shaft must turn 2.52 times to turn the output shaft once. However, there can be additional "gear" reduction in the torque converter. The pump can turn several times faster than the turbine, producing a speed reduction and a torque increase.

Note that the intermediate overrun roller clutch is effective, or locked in D, low gear. This means that the sun gear, sun-gear drive shell, direct-clutch housing, intermediate roller clutch, and intermediate-clutch plates are all turning in a counterclockwise direction.

⚙ **4-4 Power flow in D, second gear** As engine and car speed increase, an upshift from low gear to second gear occurs. The manner in which the upshift occurs was explained in ⚙ 3-5. How the hydraulic system controls the upshift is discussed in ⚙ 4-9.

When conditions are right for the upshift, the hydraulic system applies pressure to the intermediate clutch (4 in Fig. 4-4). This action changes the power flow through the transmission. The power enters through the torque converter and input shaft as before (1, 2, and 3 in Fig. 4-4). Engagement of the intermediate clutch locks the outer race of the intermediate overrun roller clutch (5) to the transmission case—through the clutch disks (4). The inner race of the intermediate overrun roller clutch (6) is assembled to the sun-gear shell (7) and sun gear (8). When the outer race is locked stationary and the roller clutch is effective (also called *locked* or *engaged*), then the sun gear is locked and cannot turn. It tries to turn in a counterclockwise direction just as it did in first gear. But the intermediate overrun roller clutch locks up to prevent this movement, and so the sun gear is held stationary.

The ring gear of the front gearset (9) is locked to the input shaft (3) through the forward clutch (9), which is

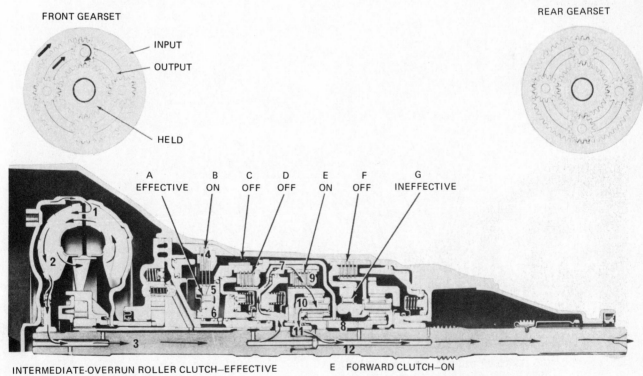

A INTERMEDIATE-OVERRUN ROLLER CLUTCH—EFFECTIVE

B INTERMEDIATE CLUTCH—ON

C INTERMEDIATE-OVERRUN BAND—OFF

D DIRECT CLUTCH—OFF

E FORWARD CLUTCH—ON

F LOW-AND-REVERSE CLUTCH—OFF

G LOW-AND-REVERSE ROLLER CLUTCH—INEFFECTIVE

Fig. 4-4 Power flow, shown by arrows, through the transmission in drive range and second gear. (*Oldsmobile Division of General Motors Corporation*)

still engaged. With the hydraulic system signaling for the upshift from low to second gear, both the intermediate clutch (4) and the forward clutch (9) are effective. When this happens, the ring gear of the front gearset must turn with the input shaft. The sun gear (8) is held as the intermediate clutch (4) engages. This means that the clockwise rotation of the ring gear rotates the planet pinions against the stationary sun gear, forcing the planet pinions to walk around the sun gear. As a result, the planet-pinion carrier (10) is carried around in a clockwise direction. The planet-pinion carrier is splined to the output shaft (12) at 11. Therefore, when the planet-pinion carrier rotates, it causes the output shaft to rotate.

Note that the rear gearset is ineffective. There is gear reduction through the transmission of 1.52:1. This means that the input shaft must turn 1.52 times to cause the output shaft to turn once. Additional speed reduction and torque increase are available through the torque converter.

The reason that the rear gearset is ineffective is that the output shaft turns the ring gear of the rear gearset at the same speed as the output shaft is turning. The two are splined together. The sun gear is held stationary. This combination causes the planet gears to walk around the stationary sun gear, carrying the planet-pinion cage with them. This effect is the same as in the front gearset. The planet-pinion cage of the rear gearset is assembled to the inner race of the low-and-reverse

roller clutch. This clutch now permits the inner race to freewheel so the planet-pinion cage can revolve without restraint.

✿ 4-5 Power flow in D, direct drive
In D (direct drive, or third gear) the conditions are as shown in Fig. 4-5. The intermediate clutch (4) and forward clutch (5) remain engaged, but now the direct clutch (6) becomes engaged. The direct clutch locks the ring gear (7) of the front gearset to the sun gear. Note that the sun gear (8) is locked to the input shaft (3) through the direct clutch (6) and the sun-gear shell (9). At the same time, the ring gear (7) is locked to the input shaft (3) through the forward clutch (5). With both the sun gear and the ring gear locked to the input shaft, both gears must turn with the input shaft. This means that the planet gearset is locked. The planet-pinion carrier turns at the same speed as the input shaft. Since the carrier is splined to the output shaft, the output shaft also turns at the same speed as the input shaft. The transmission is now in direct drive, with a 1:1 gear ratio.

✿ 4-6 Power flow in L, first gear
If the driver moves the selector lever to L—the low range—the transmission will shift to low gear, and remain in that gear. The internal conditions are shown in Fig. 4-6. The manual valve is positioned by the shift to L so as to produce the following: The intermediate clutch is off; the intermediate overrun band is off; the direct clutch is off; the

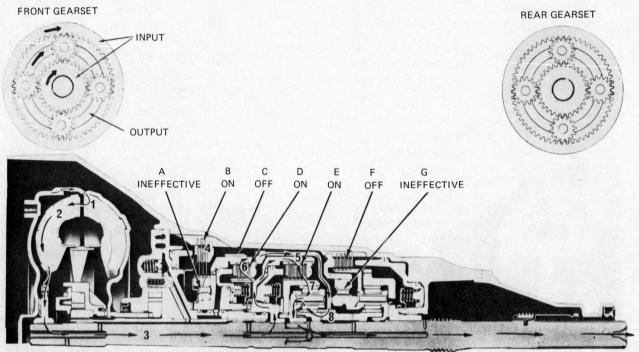

A	B	C	D	E	F	G
INEFFECTIVE	ON	OFF	ON	ON	OFF	INEFFECTIVE

A   INTERMEDIATE-OVERRUN ROLLER CLUTCH—INEFFECTIVE

B   INTERMEDIATE CLUTCH—ON

C   INTERMEDIATE-OVERRUN BAND—OFF

D   DIRECT CLUTCH—ON

E   FORWARD CLUTCH—ON

F   LOW-AND-REVERSE CLUTCH—OFF

G   LOW-AND-REVERSE ROLLER CLUTCH—INEFFECTIVE

Fig. 4-5 Power flow, shown by arrows, through the transmission in drive range and direct, or third, gear. (*Oldsmobile Division of General Motors Corporation*)

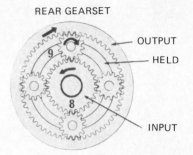

FRONT GEARSET
INPUT
OUTPUT
OUTPUT TO REAR UNIT

REAR GEARSET
OUTPUT
HELD
INPUT

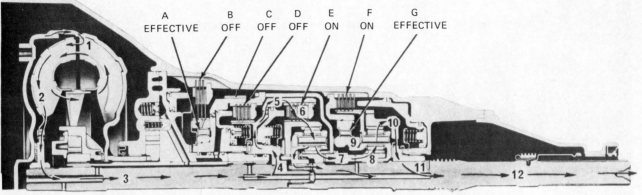

A	B	C	D	E	F	G
EFFECTIVE	OFF	OFF	OFF	ON	ON	EFFECTIVE

A INTERMEDIATE-OVERRUN ROLLER CLUTCH—EFFECTIVE

B INTERMEDIATE CLUTCH—OFF

C INTERMEDIATE-OVERRUN BAND—OFF

D DIRECT CLUTCH—OFF

E FORWARD CLUTCH—ON

F LOW-AND-REVERSE CLUTCH—ON

G LOW-AND-REVERSE ROLLER CLUTCH—EFFECTIVE

Fig. 4-6 Power flow, shown by arrows, through the transmission in low range and low, or first, gear. (*Oldsmobile Division of General Motors Corporation*)

forward clutch is on; and the low-and-reverse clutch is on. Now, the power flow is as shown in Fig. 4-6: from the input shaft through the forward clutch to the front-gearset ring gear. The ring gear is turned in a clockwise direction. This causes the planet pinions of the front gearset to turn in a clockwise direction, driving the sun gear in a counterclockwise direction. As the sun gear turns counterclockwise, it causes the planet pinions of the rear gearset to turn clockwise, driving the ring gear of the rear gearset clockwise. (The rear carrier is being held by the low-and-reverse clutch.) This ring gear is splined to the output shaft, and so the output shaft turns. With gear reduction in both the front and rear gearsets, the total gear reduction through the transmission is 2.52:1.

4-7 Power flow in D2 or S, second gear If the driver selects D2 or S, the hydraulic system will shift the transmission into second gear and hold it there, except at very low car speed, which will cause the transmission to downshift into low. The conditions will be as shown in Fig. 4-7. The intermediate clutch is engaged to allow the intermediate overrun roller clutch to hold the shell and sun gear stationary (against counterclockwise rotation). Power flows through the engaged forward clutch to the ring gear of the front gearset. The sun gear is stationary, and so the planet gears of the front gearset walk around the sun gear. This movement turns the planet-pinion cage of the front

gearset and output shaft in a clockwise direction. The reaction of the planet pinions against the sun gear is taken either by the intermediate overrun roller clutch or by the intermediate overrun band. The intermediate overrun band provides overrun braking as it holds the sun gear stationary. The gear ratio is 1.52:1.

4-8 Power flow in reverse When the driver moves the selector lever to R (reverse), the hydraulic system will shift the transmission into reverse. The conditions will be as shown in Fig. 4-8. Only the direct clutch and the low-and-reverse clutch will be engaged. Power flows from the input shaft through the direct clutch to the sun-gear shell and sun gear. The sun gear rotates clockwise. This causes the planet pinions of the rear gearset to rotate counterclockwise. The planet-pinion carrier of the rear gearset is held by the low-and-reverse clutch. Therefore, the planet pinions act as idlers and cause the ring gear of the rear gearset to rotate in a counterclockwise direction. Since this ring gear is splined to the output shaft, the output shaft will also turn counterclockwise. This causes the car to back up. The gear ratio through the transmission is 1.93:1.

4-9 Hydraulic system The hydraulic system includes a pump, driven by the engine, which produces hydraulic pressure. Also, there are several valves that control the pressure and direct it to the proper clutch or band servo to produce the proper gearshift. In

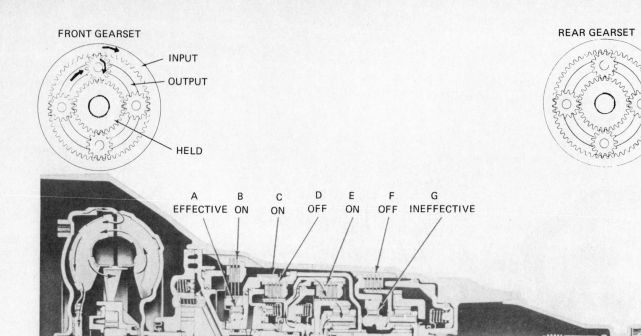

A INTERMEDIATE-OVERRUN ROLLER CLUTCH—EFFECTIVE   E FORWARD CLUTCH—ON

B INTERMEDIATE CLUTCH—ON   F LOW-AND-REVERSE CLUTCH—OFF

C INTERMEDIATE-OVERRUN BAND—ON   G LOW-AND-REVERSE ROLLER CLUTCH—INEFFECTIVE

D DIRECT CLUTCH—OFF

Fig. 4-7 Power flow, shown by arrows, through the transmission in super range (D2) and second gear. (*Oldsmobile Division of General Motors Corporation*)

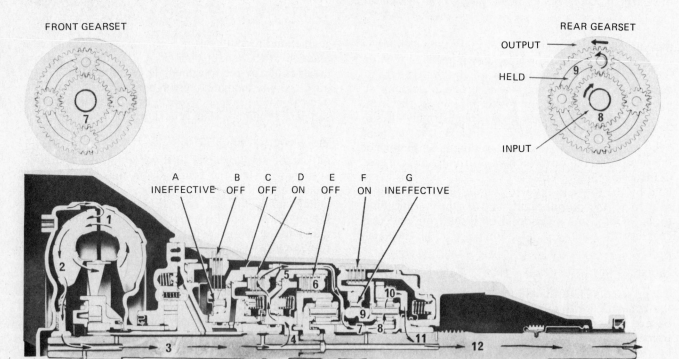

A INTERMEDIATE-OVERRUN ROLLER CLUTCH—INEFFECTIVE   E FORWARD CLUTCH—OFF

B INTERMEDIATE CLUTCH—OFF   F LOW-AND-REVERSE CLUTCH—ON

C INTERMEDIATE-OVERRUN BAND—OFF   G LOW-AND-REVERSE ROLLER CLUTCH—INEFFECTIVE

D DIRECT CLUTCH—ON

Fig. 4-8 Power flow, shown by arrows, through the transmission in reverse. (*Oldsmobile Division of General Motors Corporation*)

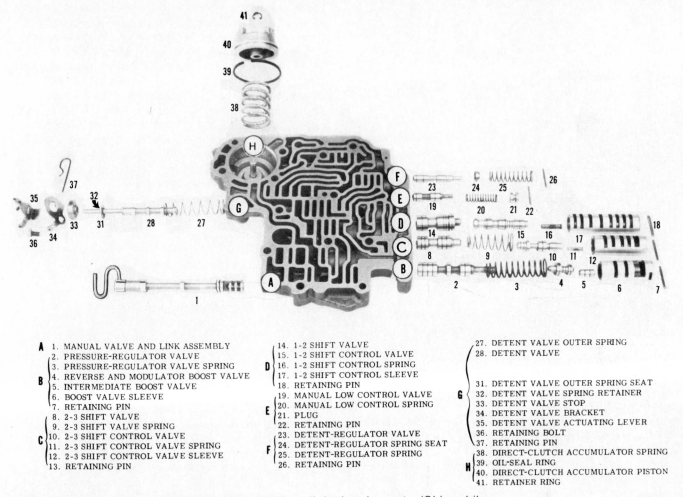

Fig. 4-9 Exploded view of the valve body showing all the interior parts. (*Oldsmobile Division of General Motors Corporation*)

A   1. MANUAL VALVE AND LINK ASSEMBLY
    2. PRESSURE-REGULATOR VALVE
    3. PRESSURE-REGULATOR VALVE SPRING
B   4. REVERSE AND MODULATOR BOOST VALVE
    5. INTERMEDIATE BOOST VALVE
    6. BOOST VALVE SLEEVE
    7. RETAINING PIN
    8. 2-3 SHIFT VALVE
    9. 2-3 SHIFT VALVE SPRING
C 10. 2-3 SHIFT CONTROL VALVE
  11. 2-3 SHIFT CONTROL VALVE SPRING
  12. 2-3 SHIFT CONTROL VALVE SLEEVE
  13. RETAINING PIN

  14. 1-2 SHIFT VALVE
  15. 1-2 SHIFT CONTROL VALVE
D 16. 1-2 SHIFT CONTROL SPRING
  17. 1-2 SHIFT CONTROL SLEEVE
  18. RETAINING PIN
  19. MANUAL LOW CONTROL VALVE
E 20. MANUAL LOW CONTROL SPRING
  21. PLUG
  22. RETAINING PIN
  23. DETENT-REGULATOR VALVE
F 24. DETENT-REGULATOR SPRING SEAT
  25. DETENT-REGULATOR SPRING
  26. RETAINING PIN

  27. DETENT VALVE OUTER SPRING
  28. DETENT VALVE
  31. DETENT VALVE OUTER SPRING SEAT
G 32. DETENT VALVE SPRING RETAINER
  33. DETENT VALVE STOP
  34. DETENT VALVE BRACKET
  35. DETENT VALVE ACTUATING LEVER
  36. RETAINING BOLT
  37. RETAINING PIN
  38. DIRECT-CLUTCH ACCUMULATOR SPRING
  39. OIL-SEAL RING
H 40. DIRECT-CLUTCH ACCUMULATOR PISTON
  41. RETAINER RING

addition, there are devices for smoothing shifts and modifying them to suit engine and highway operating conditions.

Almost all the valves in the transmission are located in the valve body. The valve body is a die casting, a little less than 1 foot [0.305 m] square, which is located at the bottom of the transmission in the oil pan. The valve body contains most of the valves, as shown in Fig. 4-9. Each valve is described in following sections. After the function of each valve is discussed, then how the complete system works is covered. Color Plates 1 to 8 show the complete hydraulic system during different operating conditions.

NOTE See the ''Note'' at the end of ❀ 4-25 for an explanation of the plates, where to find them, and how to use them.

❀ **4-10 Oil pump** The oil pump is located at the front of the transmission just behind the torque converter, as shown in Fig. 4-1. The oil pump is driven by the engine and is the gear type, as shown in Fig. 4-10. As the gears rotate, the spaces between the teeth first become larger so that oil is drawn into the spaces. Actually, atmospheric pressure pushes the oil into the

increasing spaces between the teeth. Then, when the teeth approach each other and mesh, the oil between them is forced out into the line. Now this pump pressure varies greatly according to engine speed. At high engine speed, the pump pressure would be very high if it were not for the pressure-regulating valve train. This valve reduces the pressure as necessary to meet the operating conditions (❀ 4-13).

❀ **4-11 Vacuum modulator** The vacuum modulator contains a metal bellows, a diaphragm, and two springs

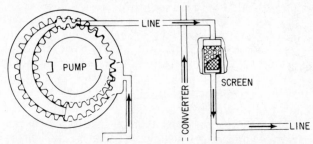

Fig. 4-10 Schematic view of the oil pump. (*Oldsmobile Division of General Motors Corporation*)

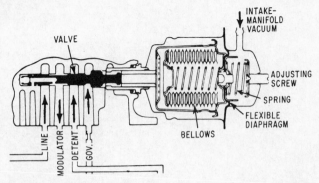

Fig. 4-11 Sectional view of the vacuum-modulator assembly. (*Oldsmobile Division of General Motors Corporation*)

(Fig. 4-11). Its purpose is to modulate, or modify, the line pressure with changing engine output and also with changes in altitude. It does this by changing the valve position in its cylinder. This cylinder has several openings, or ports, to admit oil and allow it to flow out. As the spool valve moves it can partly shut off or increase the valve-port openings. If the port that admits pump pressure is partly shut off, the effect is to lower the pressure that exits from the valve assembly. Note, in Fig. 4-11, that there are four ports: line, modulator, detent, and governor. The line and modulator ports are discussed first.

The metal bellows is evacuated. It contains no air and therefore has a vacuum inside it. Atmospheric pressure acting on the bellows tends to collapse it and shorten its length. Opposing this pressure is the heavy spring inside the bellows. However, changes in atmospheric pressure will change the length of the bellows. High altitudes, where atmospheric pressure is lower, will allow the bellows to lengthen. This effect counterbalances the loss of engine power due to the lower atmospheric pressure.

The flexible diaphragm is open to atmospheric pressure on one side and to intake-manifold vacuum on the other side. As this vacuum changes, the diaphragm moves, thereby moving the modulator valve.

In operation the bellows and the external spring (to the right in Fig. 4-11) act to move the modulator valve so as to increase modulator pressure. The flexible diaphragm acts on the modulator valve so as to reduce modulator pressure as engine intake-manifold pressure increases. For example, suppose the driver is cruising along at part throttle but gaining speed so that the hydraulic system signals for an upshift. The intake-manifold vacuum is high so that the flexible diaphragm acts to reduce modulator pressure. With lower modulator pressure, the shift takes place more slowly and is less noticeable. If the driver opens the throttle to demand full engine power, intake-manifold vacuum drops and the flexible diaphragm has less effect. This means that the modulator pressure increases and the shift takes place more quickly. With an open throttle and high engine speed, the shift must take place fast to prevent slippage and destructive heat buildup in the clutch. A quick shift under the circumstances is least noticeable and most desirable.

The combination has another effect. Engine power is lost at high altitudes due to the lower atmospheric pressure. With reduced engine power, fast shifts would be harsh and noticeable. So shifts should slow down. This slowing-down action is produced by the combined effect of the bellows and the diaphragm. The diaphragm, being larger in effective area than the bellows, acts to reduce modulator pressure. The bellows expands at higher altitudes, and this tends to increase modulator pressure. But the lowered atmospheric pressure, acting on the flexible diaphragm, tends to reduce modulator pressure. Because the diaphragm is larger, it overcomes the bellows action and does cause a decrease in modulator pressure.

NOTE: "Fast" and "slow" shifts are mentioned above. Actually, the critical part of the shift is over in about one second. This critical time includes only the moment when the band or clutch is actually taking hold.

✿ 4-12 Governor Governor pressure is critical to the operation of the transmission. The governor is driven by the output shaft of the transmission and is directly related to car speed (not engine speed). The governor contains two pairs of weights, as shown in Fig. 4-12. These weights are connected by springs. When car speed, and governor speed, increases, centrifugal force acts on the weights, forcing them to move outward against the spring force. This action pushes the governor valve upward in the assembly, tending to close off the exhaust port and partly open the drive port to line pressure. The effect is that as car speed increases, governor pressure increases. Therefore, governor pressure is a function of car speed alone. As governor pressure increases, it acts on other valves, getting them ready to produce shifts when the proper engine and car speed and throttle opening are reached. How this happens is explained later.

Governor pressure also has an effect on the vacuum-modulator valve. Look at Fig. 4-11. Note that governor pressure is introduced into a port that is located at an undercut in the modulator valve. The land on one side of the valve undercut is larger than the land on the

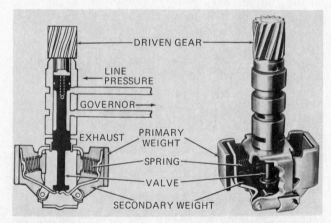

Fig. 4-12 Sectional and cutaway views of the governor assembly. (*Cadillac Motor Car Division of General Motors Corporation*)

other side. This means that the governor pressure tends to move the valve in the direction of the larger land (or to the right in Fig. 4-11). The result is that as governor pressure increases, signaling higher car speed, the modulator valve tends to move so as to decrease modulator pressure. The effect is to slightly modify the shifting speed and line pressure at high car speeds, thereby preventing harsh shifts.

Note that the governor has two sets of weights: The primary weights are effective at low speeds and allow the governor pressure to build up rapidly until about 20 mph [32 km/h]. At this speed, the primary weights have moved out as far as possible. Now, the secondary weights take over as car speed further increases. They produce a slow increase in governor pressure. Therefore, there is a rapid increase in governor pressure at low car speeds and a slower increase at higher speeds. This two-step curve of pressure increase allows pressure to build up rapidly at first but prevents it from going too high in the high car-speed ranges.

☼ 4-13 Pressure-regulator valve The pressure-regulator valve, shown in Fig. 4-13, regulates the pressure from the pump so as to avoid excessive pressures in the lines going to the other valves and to the clutches. The valve is shown in Fig. 4-9 and is numbered 2. The pump pressure is applied to one end (left in Fig. 4-13) of the valve. Opposing this is the spring force at the other end of the valve. The valve therefore positions itself so as to give the pressure designated by the spring force. Note that the pump pressure is reduced as necessary by the valve position, which allows some of the oil to flow to the oil sump, or oil pan. Dumping oil in this manner reduces the pressure.

The pressure is also affected by the modulator pressure during intermediate-, or second-, gear operation. The modulator pressure tends to raise the line pressure

so as to provide a more rapid shift from second to high. This prevents excessive slippage of the clutch during this shift. Also, pressure is affected when the transmission is shifted into reverse. During reverse operation, very firm engagement of the low-and-reverse clutch is necessary. This is produced by raising the line pressure. When the shift is made, oil pressure from the line is introduced into the reverse section of the pressure-regulator-valve train. This pressure acts on the larger land of the reverse boost valve (labled REV in Fig. 4-13 and numbered 4 in Fig. 4-9), adding to the spring force to obtain an increase in line pressure.

☼ 4-14 Manual valve The manual valve was discussed in ☼ 3-5. This valve directs oil pressure to different parts of the hydraulic circuit, according to where the valve has been set by the driver. The valve is shown in Fig. 4-14 in sectional view. The actual valve and link can be seen in Fig. 4-9, where it is numbered 1. Line pressure enters the valve body and flows to drive, intermediate, low, or reverse, according to the position of the valve. Figure 3-11 shows a manual valve in neutral, drive, and low.

☼ 4-15 Intermediate-clutch accumulator The purpose of the accumulator is to cushion the engagement of a clutch or application of a band (Fig. 2-16). The accumulator shown in Fig. 4-15 cushions the engagement of the intermediate clutch during shifts from first to second. Therefore, it is known as the *1-2 accumulator*.

When the hydraulic system calls for a shift from first to second, oil pressure is applied to the intermediate clutch. Pressure enters the accumulator through the line marked "1-2 CL" in Fig. 4-15. This stands for intermediate, or 1-2, clutch. The 1-2 clutch pressure is applied to one side of the piston. Line pressure is applied to the other side of the piston. Before the shift is called for, and without 1-2 clutch pressure, line pressure pushes the piston far to the right in the accumulator body. This compresses the spring. But when 1-2 clutch pressure comes on, it combines with the spring force to push the piston back against line pressure. This action gives the 1-2 clutch oil a place to go besides the intermediate clutch. The result is to allow the clutch to engage smoothly. The accumulator provides a cushion, in effect, so that all the 1-2 clutch oil does not suddenly ram into the clutch and cause it to engage suddenly. Instead, the oil pressure builds up slowly so that the clutch engagement is smooth.

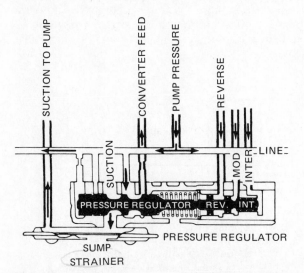

Fig. 4-13 Sectional view of the pressure-regulator valve train. There are three parts to the train, aside from the spring: the pressure-regulator valve, the reverse-and-modulator boost valve (labeled REV), and the intermediate boost valve (labeled INT). (*Oldsmobile Division of General Motors Corporation*)

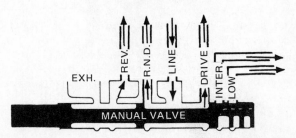

Fig. 4-14 Sectional view of the manual valve. (*Oldsmobile Division of General Motors Corporation*)

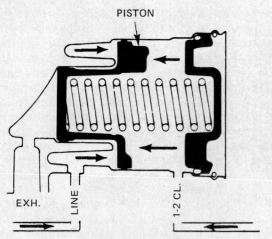

Fig. 4-15 Sectional view of the 1-2 accumulator. (*Oldsmobile Division of General Motors Corporation*)

☼ **4-16 Direct-clutch accumulator** The direct-clutch, or 2-3, accumulator does the same job for the direct clutch that the intermediate-clutch accumulator does for the intermediate clutch. It cushions the engagement of the direct clutch during the shift from 1-2 to 2-3. Figure 4-16 is a sectional view of the direct-clutch accumulator. Number 40 in Fig. 4-9 shows the piston.

The accumulator works this way: Before the shift to direct is called for, there is no 2-3 clutch pressure. The RND (reverse, neutral, drive) pressure pushes down on the piston. Also, 1-2 clutch pressure acts between the two parts of the piston. These pressures push the piston down and compress the lower spring. At the same time, 1-2 pressure tends to separate the two parts of the piston, acting almost like an auxiliary spring. Then, when the hydraulic system signals for a shift from second to high, or direct, the 2-3 clutch pressure is applied to the lower part of the piston. This pressure forces the piston up against the RND pressure. At the

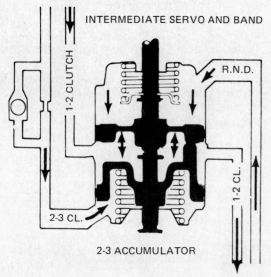

Fig. 4-16 Sectional view of the 2-3 accumulator. (*Oldsmobile Division of General Motors Corporation*)

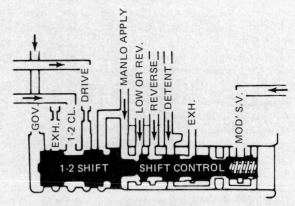

Fig. 4-17 Sectional view of the 1-2 shift valve train. There are two parts to the train, aside from the spring: the shift valve and the shift-control valve. (*Oldsmobile Division of General Motors Corporation*)

same time, 1-2 clutch pressure is released. Therefore, there is a combined upward pressure that allows the piston to move up. This action produces a relatively gradual buildup of 2-3 clutch pressure so that the direct clutch engages smoothly.

Note that the 2-3 accumulator also includes the servo which applies the intermediate overrun band under certain conditions, as explained later.

☼ **4-17 1-2 shift valve** The 1-2 shift valve, shown in Fig. 4-17, directs oil pressure to the transmission to cause it to shift from first to second or from second to first. It is numbered 14 in Fig. 4-9. The operation of this valve is controlled by governor pressure, detent pressure, modulator pressure, and spring force. Governor pressure, which is directly related to car speed, acts on the left end (Fig. 4-17) of the 1-2 shift valve and increases with car speed. Opposing this pressure is a spring, the modulator pressure, and the detent pressure. How these opposing forces operate to allow upshifting—or downshifting under certain conditions is covered in later sections on the complete hydraulic system.

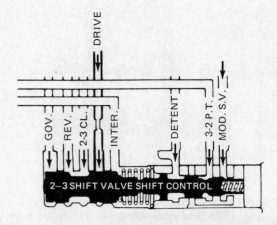

Fig. 4-18 Sectional view of the 2-3 shift-valve train. There are two parts to the train, aside from the spring: the shift valve and the shift-control valve. (*Oldsmobile Division of General Motors Corporation*)

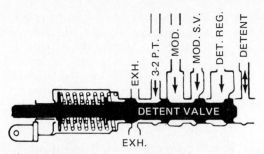

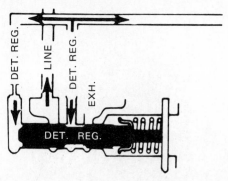

Fig. 4-19 Sectional view of the detent valve. (*Oldsmobile Division of General Motors Corporation*)

Fig. 4-20 Sectional view of the detent regulator valve. (*Oldsmobile Division of General Motors Corporation*)

✿ 4-18 2-3 shift valve The 2-3 shift valve, shown in Fig. 4-18, is numbered 8 in Fig. 4-9. It routes oil pressure in the transmission to cause a shift from second to direct or from direct to second. Its operation is controlled by several pressures—governor, modulator, and detent—and by spring force.

✿ 4-19 Detent valve The detent valve, shown in sectional view in Fig. 4-19 and numbered 28 in Fig. 4-9, is actuated by the downshift cable, which is linked to the accelerator. When the accelerator pedal is pushed all the way down, the linkage moves the detent valve so that it changes the routing of the oil through the transmission. If the transmission is in direct drive and the car is traveling at a high speed, the new oil routes cause the transmission to downshift from high to second. If the transmission is in either second or direct and the car is traveling at a low speed, the new oil routes cause the transmission to shift to first. This means that the shift is either 2-1 or 3-1 (✿ 4-33).

✿ 4-20 Detent regulator valve The detent regulator valve and spring are shown in sectional view in Fig. 4-20. They are numbered 23 and 25 in Fig. 4-9. This valve changes line pressure into detent regulator oil pressure, which is then used to control the car speed at which 1-2 and 2-3 upshifts will occur with full throttle.

✿ 4-21 Oil-cooler bypass valve The oil-cooler bypass valve (Fig. 4-1) is a spring-loaded valve which

permits oil to be fed from the torque converter directly into the hydraulic system when the oil is cold or if there is a restriction in the cooler or cooler lines. When the oil is cold, it is thicker and flows more slowly. Higher pressures result, which unseat the valve ball so that the oil can bypass the cooler. The same actions occur if the cooler circuit is restricted. This ensures adequate oil circulation to transmission parts.

✿ 4-22 Manual-low-control valve The manual-low-control valve is numbered 19 in Fig. 14-9. When the manual valve is positioned by the driver in low and the car is traveling below 45 mph [72 km/h], oil is directed to the 1-2 shift valve so that it moves to the downshifted position. At the same time, oil moves the 1-2 shift control valve to the upshifted position, which sends low apply oil to the low-and-reverse clutch to cause this clutch to engage. Now the transmission is shifted to low and will remain there. It cannot upshift regardless of engine or car speed.

At speeds above 45 mph [72 km/h], shifting to low is prevented by the manual-low-control valve. This is because the high governor pressure prevents the valve from moving.

✿ 4-23 Torque-converter clutch The torque-converter clutch assembly is shown disassembled in Fig. 4-21. Figure 4-22 shows the torque converter and clutch

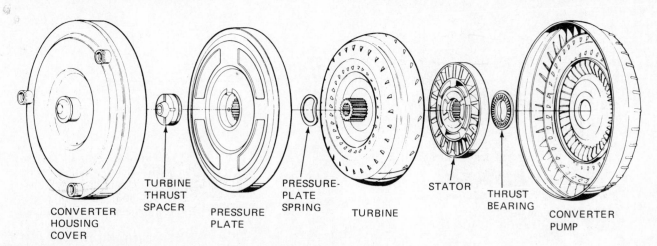

Fig. 4-21 Disassembled torque converter with clutch, used on the 350C and other automatic transmissions. (*Oldsmobile Division of General Motors Corporation*)

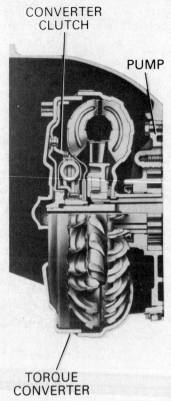

Fig. 4-22 Partial cutaway view of the torque converter with clutch. (*Oldsmobile Division of General Motors Corporation*)

assembly in sectional view. The converter-housing cover is attached to and turns with the pump assembly. The pressure plate is splined to the turbine shaft so they turn together. When the clutch is not engaged, the turbine and pressure plate are driven by the torque-converter action in the usual manner. Fluid passes from the pump to the turbine, causing the turbine to spin. However, when the clutch is engaged, the pressure plate is forced (to the left in Fig. 4-22) against the converter-housing cover. This locks the pressure plate and turbine to the converter housing and pump. The torque converter is locked up.

✿ 4-24 Torque-converter-clutch controls Figure 4-23 shows the part of the hydraulic system that controls the converter clutch. It includes a governor pressure switch (lower right), a solenoid, and converter-clutch valves. As long as the car speed is below the actuating speed, the clutch will not engage. When the car reaches 30 to 45 mph [48 to 72 km/h] the control system causes the clutch to engage. The actual speed depends on the engine-transmission combination.

When the engagement speed is reached, the governor pressure has increased enough to cause the governor pressure switch to close its points. This connects the solenoid to the battery so that the solenoid operates to close the exhaust line from the direct clutch. Now, the fluid that has been flowing to the direct clutch, and keeping it closed, is shut off.

Normally, the applying fluid that keeps the direct clutch engaged flows through the clutch and exhausts. When the exhaust line is closed by the solenoid, fluid pressure is applied to the converter-clutch valves. They, in turn, redirect the converter feed fluid so it flows into the apply side of the clutch. The clutch engages and locks up the torque converter.

When car speed falls below the engagement speed, the governor pressure falls enough to allow the pressure switch to open its points. This disconnects the solenoid from the battery, opening the exhaust line. The pres-

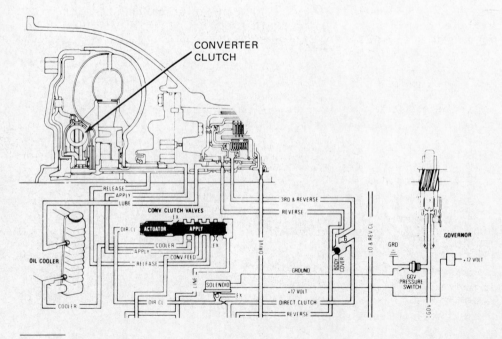

Fig. 4-23 Part of the hydraulic circuit showing the actuator-apply valve, solenoid, and governor pressure switch, which control the converter clutch. (*Oldsmobile Division of General Motors Corporation*)

sure on the converter-clutch valves is released so they move to open the line from the apply side of the clutch. The pressure on the clutch pressure plate is relieved and the clutch disengages.

The pressure plate has a damper assembly which is similar to the damper assembly in the clutch for a manual transmission. The damper assembly has springs between the splined hub and the pressure plate itself. These springs—the dampers—allow the pressure plate to rotate as much as 45° independently of the splined hub. This action damps out the feel of the clutch engagement. When the clutch engages, the springs compress to take up the shock. Therefore, the clutch application is smooth.

There are additional controls to aid in the engagement and disengagement of the clutch during different driving conditions. These controls are outside the transmission:

1. **Brake release switch** The brake release switch disengages the clutch whenever the brakes are applied. This avoids the possibility of stalling the engine. If the brakes are applied and the car is slowed or stopped, the engine will also slow or stop because the clutch is engaged.

2. **Thermal vacuum valve** The thermal vacuum valve, used with gasoline engines, prevents the clutch from engaging at low coolant and engine temperatures. The engine is unstable at this time and engagement of the clutch might cause the engine to stumble or stall.

3. **Low-vacuum switch** The low-vacuum switch disengages the clutch when the throttle is closed and the car is coasting. The vacuum switch also disengages the converter clutch when the engine vacuum drops way down during moderate acceleration, just prior to a part-throttle or detent downshift.

4. **Vacuum delay valve** The vacuum delay valve, used with gasoline engines, slows the vacuum-switch response to vacuum changes. The purpose is to prevent sudden clutching or declutching that could produce a harsh feel and might cause the engine to stumble or stall.

5. **High-vacuum switch** The high-vacuum switch is used with diesel engines. It disengages the clutch during a closed throttle shut-down of the engine.

6. **Vacuum relay valve** The vacuum relay valve prevents vacuum from reaching the vacuum switch at low engine and coolant temperatures.

☼ 4-25 Complete hydraulic system Transmission gearing with its related clutches and band, and the various valves in the hydraulic system, were discussed earlier. The following sections discuss the actions during the different phases of transmission operation.

In a simplified transmission (☼ 3-5), shifts are controlled by two factors: car speed, which provides a signal through governor pressure, and engine load, which provides a signal through intake-manifold vacuum. The sections below cover the following eight conditions in the transmission:

1. Neutral, engine running
2. Drive range, first gear
3. Drive range, second gear
4. Drive range, third, or direct, gear
5. Low range, first gear
6. Second range, second gear
7. Drive-range detent downshift
8. Reverse

NOTE: In the following discussions of the hydraulic systems, references are made to plate numbers. These plates are in the *Color Plates of Automatic-Transmission Hydraulics Circuits,* sixth edition. This 32-page booklet, which is wrapped with the *Automotive Automatic Transmissions* textbook, is to be used while you study the hydraulic circuits of the transmissions described in the textbook. The following sections describe the actions in the hydraulic circuits. The color plates show the positions of the valves and the clutches engaged or bands applied during the various operating conditions discussed.

☼ 4-26 Neutral, engine running The conditions in the transmission are as shown in Fig. 4-2 when the manual valve is in neutral and the engine is running. The conditions of the clutches and band are as follows:

Intermediate clutch—off

Direct clutch—off

Forward clutch—off

Low-and-reverse clutch—off

Intermediate overrun roller clutch—freewheeling

Low-and-reverse roller clutch—freewheeling

Intermediate overrun band—off

When the engine starts, with the manual valve in neutral, oil flows from the pump to the pressure-regulator valve, where the oil is controlled at line pressure. Oil flows from the pressure-regulator valve to the converter to fill it. Oil begins to circulate through the oil cooler and flows from the oil cooler to the transmission lubricating lines so that all bearing surfaces receive an immediate supply of oil. Line oil also flows from the pressure-regulator valve to the manual valve, which directs the oil, now called *reverse-neutral-drive* (RND) *oil,* to the servo part of the 2-3 accumulator.

At the same time, *line oil* flows to the 1-2 accumulator and strokes it in preparation for the 1-2 shift. "Stroke" means that the piston is moved by the oil pressure. *Line oil* also flows to the vacuum-modulator valve. It emerges from this valve as *modulator oil* and flows to the pressure-regulator-valve train and to the detent valve. It emerges from this valve as *modulator shift-valve oil* and flows to the 1-2 and 2-3 shift valves.

None of the valves reacts at this time. However, the valves are in readiness to react just as soon as the manual valve is shifted from neutral.

☼ 4-27 Drive range, first gear The conditions in the transmission are as shown in Plate 1 with the manual valve in drive range and the transmission in first gear. These conditions are shown in Fig. 4-3. The *line oil* entering the manual valve emerges as *drive oil*. The manual valve has already passed oil, called *reverse-neutral-drive oil,* to the 2-3 accumulator, stroking it in readiness for an upshift. The drive oil emerging from the manual valve flows to the 1-2 and 2-3 shift valves, where it has no immediate effect. Drive oil also flows to the governor and forward clutch, causing the forward clutch to engage. The conditions of the clutches and band are then as follows:

Intermediate clutch—off

Direct clutch—off

Forward clutch—on

Low-and-reverse clutch—off

Intermediate overrun roller clutch—locked

Low-and-reverse roller clutch—locked

Intermediate overrun band—off

Drive oil enters the governor and emerges as a variable-pressure oil, called *governor oil*. This variable pressure is applied to the ends of the 1-2 and 2-3 shift valves and to the modulator valve.

In summary, line oil is fed to the modulator valve, 1-2 accumulator, detent pressure-regulator valve, and manual valve. Line oil emerges from the manual valve as reverse-neutral-drive oil and as drive oil. This drive oil flows to the forward clutch, engaging it, so that the transmission is in first. The drive oil also flows to the governor and the 1-2 and 2-3 shift valves. RND oil emerges from the manual valve and flows to the servo part of the 2-3 accumulator. This oil strokes the 2-3 accumulator in preparation for it to cushion the 2-3 clutch (the direct clutch) when the shift is made from second to third. The 1-2 accumulator has already been stroked by line pressure to prepare it to cushion the 1-2 clutch (the intermediate clutch) when the shift is made from first to second.

☼ 4-28 Drive range, second gear The conditions in the transmission are as shown in Plate 2 with the manual valve in drive range and the transmission in second gear. These are the conditions shown in Fig. 4-4. As vehicle speed and governor pressure increases, the force of the governor pressure acting on the end of the 1-2 shift valve overcomes the valve-spring force. This moves the 1-2 shift valve so that drive oil is admitted to the 1-2, or intermediate, clutch, causing it to engage. As a result, the transmission upshifts to second gear. The conditions of the clutches and band are then as follows:

Intermediate clutch—on

Direct clutch—off

Forward clutch—on

Low-and-reverse clutch—off

Intermediate overrun roller clutch—locked

Low-and-reverse clutch—freewheeling

Intermediate overrun band—off

Note that modulator shift-valve pressure also helps the 1-2-shift-valve spring. Therefore, governor pressure must overcome both the spring force and modulator shift-valve pressure before it can move the shift valve. Modulator shift-valve pressure starts from the modulator valve as simply modulator pressure. Modulator pressure (☼ 4-11) is related to engine intake-manifold vacuum and atmospheric pressure. The major influence is from intake-manifold vacuum. With part-throttle, high-vacuum conditions, modulator pressure is low. This low pressure passes through the detent valve and emerges as modulator shift-valve pressure, which in this case is the same pressure. The low modulator (or modulator shift-valve) pressure adds little force to the 1-2 shift-valve spring. Therefore, a relatively low car speed, meaning a relatively low governor pressure, can produce the upshift from first to second.

If the throttle is wide open so that the intake-manifold vacuum is low, then the modulator pressure will be high. This high pressure, added to the 1-2 shift-valve spring, makes it necessary for the car to reach a higher speed before the shift can take place. This higher car speed is required so that the higher governor pressure needed to make the shift is reached. The governor pressure must go higher in order to overcome both the spring force and higher modulator pressure.

Notice that the 1-2, or intermediate, clutch oil also flows to the 2-3 accumulator and from there to the 1-2 accumulator. This oil cushions the intermediate-clutch engagement. A typical application has these shift points: When in drive range with the throttle wide open, the upshift will occur at about 50 mph [80 km/h]. However, with a minimum throttle, the upshift will occur at about 12 mph [19 km/h]. Modulator and governor pressures play a very important role in controlling the shift point.

☼ 4-29 Drive range, third, or direct, gear The conditions in the transmission are as shown in Plate 3 with the manual valve in drive range and the transmission in third, or direct, gear. These conditions are shown in Fig. 4-5. As car speed and governor pressure increase, the increasing pressure, acting on the end of the 2-3 valve, overcomes the force of the 2-3 shift-valve spring and modulator pressure to move the shift valve. This movement allows drive oil to emerge from the 2-3 shift valve—as 2-3 clutch oil. This action causes the 2-3, or direct, clutch to engage so that the transmission goes into direct drive. This produces the following conditions of the clutches and band:

Intermediate clutch—on

Direct clutch—on

Forward clutch—on

Low-and-reverse clutch—off

Intermediate overrun roller clutch—freewheeling

Low-and-reverse roller clutch—freewheeling

Intermediate overrun band—off

Note that modulator pressure (modulator shift-valve pressure) is also helping the 2-3 shift-valve spring.

Therefore, governor pressure must overcome both the spring force and modulator pressure before it can move the 2-3 shift valve. With part-throttle, high-intake-manifold-vacuum conditions, the modulator pressure is low. With a wide-open throttle and low intake-manifold vacuum, the modulator pressure is high. This variation in pressure makes a great difference in the shift point in shifting from second to direct. For example, on one car, the upshift to direct will occur at about 85 mph [137 km/h] with a wide-open throttle. It will occur at about 20 mph [32 km/h] with a minimum throttle opening.

It is the increasing governor pressure that produces the upshift from first to second and from second to third. There must be a considerable difference in the two shifting pressures. Pressure must be relatively low to produce the 1-2 shift and relatively high to produce the 2-3 shift. Typically, a governor pressure of 46 psi [317 kPa] is required to produce the 1-2 shift with wide-open throttle. A governor pressure of 83 psi [572 kPa] is required to produce the 2-3 shift with a wide-open throttle.

Now look at what happens at the 2-3 accumulator. When the 2-3 clutch oil flows to the direct clutch, it also flows to the lower part of the 2-3 clutch, forcing the piston upward against the RND pressure. This movement cushions the engagement of the direct clutch.

☼ 4-30 Drive range, direct, with lockup
When the selector lever is in D, or drive range, and lockup car speed is reached, the torque-converter clutch will engage as explained in ☼ 4-23 and 4-24. This locks the torque converter. The hydraulic-system conditions, including the clutches and band, are as shown in Plate 4 and described in ☼ 4-29. The only difference in the system is that the governor pressure switch has closed its points. This operates the solenoid and converter-clutch valves (Fig. 4-23).

☼ 4-31 Low range, first gear
When the driver moves the selector lever to L, or low, the conditions in the transmission are as shown in Plate 5. These are the conditions shown in Fig. 4-6. The transmission is in low, or first, gear, with the following conditions of the clutches and band:

Intermediate clutch—off

Direct clutch—off

Forward clutch—on

Low-and-reverse clutch—on

Intermediate overrun roller clutch—locked

Low-and-reverse roller clutch—locked

Intermediate overrun band—off

When the driver shifts to L, or low, line oil entering the manual valve emerges as low oil, which passes through the manual-low-control valve. The oil then passes through the 1-2 shift valve, which directs it, as low or reverse oil, to the low-and-reverse clutch so that it engages. Note that the manual valve also directs line

oil, in this case called *intermediate oil*, to the 2-3 shift valve. In both the 1-2 and 2-3 shift valves, the valves are moved against governor pressure to shut off the oil pressure to the intermediate and direct clutches so that they can disengage.

Governor pressure and throttle opening also affect the timing of the downshift. For example, if the car speed is above about 45 mph [72 km/h] or if engine speed is above about 3600 rpm, then the governor or modulator pressure will not permit the downshift. This is because governor pressure acts against the ends of the 1-2 and 2-3 shift valves. Modulator pressure acts against the ends of the shift-control valves. Both of these pressures must be overcome by the low, or the manual-low-apply, pressure before downshift will occur. This condition protects the engine. If it were not for this protection, and if a downshift did occur at high car speed, then the transmission could spin the engine so fast that it might be severely damaged.

☼ 4-32 Second range, second gear
Second or super range, second gear, is also known as low 2 (L2) or drive 2 (D2). When the driver moves the selector lever to this position, the conditions in the transmission are as shown in Plate 6. These are the conditions shown in Fig. 4-7. The transmission is in second gear, with the following clutch and band conditions:

Intermediate clutch—on

Direct clutch—off

Forward clutch—on

Low-and-reverse clutch—off

Intermediate overrun roller clutch—locked

Low-and-reverse roller clutch—freewheeling

Intermediate overrun band—on

When the driver moves the selector lever to this position, line oil entering the manual valve now exits as intermediate oil. This oil is then directed to the 2-3 shift valve, and to the end of the intermediate boost valve. This valve is marked INT in Plate 6 and is to the right of the pressure-regulator valve in the illustration. This action increases minimum line pressure to 95 psi [655 kPa]. The intermediate oil moves the 2-3 shift valve to the downshifted position, and this disengages the direct clutch. Note that when the manual valve is moved to the intermediate position, the RND oil is exhausted from the top of the servo, which is the upper part of the 2-3 accumulator. Then, 1-2 oil, acting through the 1-2 accumulator, flows to the lower part of the servo and forces the servo piston up. This movement applies the intermediate overrun band so that the shell and sun gear of the front gearset are locked (stationary). Now, as shown in Fig. 4-7, power flows through the forward clutch to the ring gear of the front gearset. The planet gears walk around the sun gear to produce the gear reduction.

Note that the situation here is different from the conditions when the manual valve is in D, or drive, and the transmission has upshifted from 1 to 2. Here, the sun gear is being held by the intermediate overrun band,

and it cannot rotate. When the manual valve is in D with the transmission in intermediate or second gear, an upshift will occur, when the 2-3, or direct, clutch engages. But when the manual valve is in intermediate, an upshift cannot occur because the 2-3 oil is exhausted through the 2-3 shift valve. The 2-3 oil drains back through the reverse line.

The transmission will remain in second as long as the manual valve remains in intermediate unless the car slows down to about 9 mph [15 km/h]. If this should happen, then the governor pressure will drop so low that the 1-2 shift valve will move. This movement allows the 1-2 clutch oil to exhaust so that the clutch disengages, the band releases, and the low-and-reverse clutch engages. Now the transmission drops back into first.

☼ **4-33 Drive-range detent downshift** When operating below 75 mph [121 km/h], a forced downshift can be made from third to second if the accelerator is fully depressed. When this happens, the detent valve is moved by the cable that connects it to the throttle linkage. The detent valve is moved to its extreme inner position (to the right in Plate 7). Therefore the hydraulic system is set up in the situation shown in Plate 7. Modulator oil now goes to the 3-2 part-throttle line, and detent regulator oil is routed to the modulator shift valve and detent passages. Detent regulator oil therefore acts on both the 1-2 and 2-3 shift control valves. Also, modulator pressure acts on the 2-3 shift control valve through the 3-2 part-throttle passage. Detent regulator oil is also routed to the modulator valve through the detent passage.

Modulator oil, detent regulator oil, and the force of the 2-3 shift-control-valve spring move the 2-3 shift valve to the downshifted position below approximately 75 mph [121 km/h]. Now the transmission downshifts into second gear.

The transmission can also be downshifted from second to first (2-1) or from third to first (3-1) below about 35 mph [56 km/h]. This is because detent regulator oil is directed to the 1-2 shift control valve. This action allows detent regulator oil plus the 1-2 shift control spring to move the 1-2 shift valve to the downshifted position to place the transmission into first gear.

In the detent downshifted position (to second), these are the conditions of the clutches and band:

Intermediate clutch—on

Direct clutch—off

Forward clutch—on

Low-and-reverse clutch—off

Intermediate overrun roller clutch—locked

Low-and-reverse roller clutch—freewheeling

Intermediate overrun band—off

☼ **4-34 Reverse** When the driver moves the selector lever to R, or reverse, the conditions in the transmission are as pictured in Plate 8. These are the conditions shown in Fig. 4-8. The conditions of the clutches and band are as follows:

Intermediate clutch—off

Direct clutch—on

Forward clutch—off

Low-and-reverse clutch—on

Intermediate overrun roller clutch—freewheeling

Low-and-reverse roller clutch—freewheeling

Intermediate overrun band—off

With the manual valve in the reverse position, line pressure enters the reverse circuit. Reverse oil then flows to the following:

Direct clutch

Low-and-reverse clutch

1-2 shift valve

2-3 shift valve

Reverse boost valve

The reverse boost valve, numbered 4 in Fig. 4-9, acts to increase the line pressure to about 250 psi [1724 kPa]. This higher pressure is needed to hold the clutches during reverse operation. Strong holding power is required at this time.

Know – CHP-4 AND CHP3 WEll

Chapter 4 review questions   *350*

Select the *one* correct, best, or most probable answer to each question. Then check your answers against the correct answers given at the end of the book.

1. The 350C transmission includes a torque converter, two planetary gearsets, two roller clutches, a band, and:
   a. two multiple-disk clutches
   b. three multiple-disk clutches
   c. four multiple-disk clutches
   d. five multiple-disk clutches

2. The 350C transmission provides reverse and:
   a. two forward speeds
   b. three forward speeds
   c. four forward speeds
   d. two forward speeds and overdrive

3. When the transmission is in low gear in drive range, both roller clutches are locked and the following is engaged:
   a. forward clutch
   b. low-and-reverse clutch
   c. direct clutch
   d. intermediate clutch

4. When the transmission is in second gear in drive range, the intermediate overrun roller clutch is locked and the following are on:
   a. low-and-reverse clutch and forward clutch
   b. intermediate clutch and intermediate band
   c. direct clutch and forward clutch
   d. intermediate clutch and forward clutch

5. With the transmission in direct drive, neither roller clutch is locked and the following are on:
   a. intermediate band and direct clutch
   b. forward and low clutches
   c. forward, intermediate, and direct clutches
   d. intermediate band and direct and forward clutches

6. With the transmission in low gear in L range, both the roller clutches are locked and the following are on:
   a. forward clutch and low-and-reverse clutch
   b. forward and direct clutches
   c. low, forward, and direct clutches
   d. intermediate and low bands

7. With the transmission in D and second gear, the intermediate overrun roller clutch is locked and the following are on:
   a. intermediate and forward clutches and intermediate band
   b. intermediate and low clutches
   c. direct, forward, and intermediate clutches
   d. both bands and direct clutch

8. With the transmission in reverse, neither roller clutch is locked, the low-and-reverse clutch is on, and also the following is effective:
   a. forward clutch
   b. direct clutch
   c. intermediate band
   d. reverse band

9. When the intermediate overrun band is applied, it locks the:
   a. ring gear
   b. sun gear
   c. planet-pinion carrier
   d. output shaft

10. When the forward clutch is engaged, the ring gear of the front gearset is locked to the:
    a. sun gear
    b. output shaft
    c. input shaft
    d. planet-pinion carrier

11. Counting all the clutches in the 350C transmission, including the roller clutches and the multiple-disk clutches, there are a total of:
    a. four
    b. five
    c. six
    d. seven

12. In the 350C automatic transmission, the sun gear is made up of:
    a. a single piece
    b. two sun gears
    c. three sun gears
    d. a sun gear with integral planet pinions

13. Mechanic A says the purpose of the vacuum modulator is to produce a modulated fluid pressure based on engine speed. Mechanic B says it is based on intake-manifold vacuum. Who is right?
    a. A only
    b. B only
    c. both A and B
    d. neither A nor B

14. The purpose of the governor is to produce an oil pressure correlated with:
    a. engine speed
    b. car speed
    c. intake-manifold vacuum
    d. throttle position

15. The purpose of the pressure-regulator valve is to:
    a. reduce pump pressure to line pressure
    b. maintain pump pressure
    c. vary pressure with car speed
    d. increase line pressure with throttle opening

16. The purpose of the reverse-boost valve is to:
    a. aid shifting into reverse
    b. increase line pressure in reverse
    c. boost the low-reverse valve
    d. engage the reverse clutch

17. The purpose of the accumulator is to cushion the:
    a. selector lever
    b. manual valve
    c. 2-3 shift valve
    d. clutch

18. The purpose of the detent valve is to:
    a. reroute fluid pressure to produce a downshift
    b. reroute fluid pressure to prevent a downshift
    c. hold the manual valve in detent
    d. force engagement of the direct clutch

19. The purpose of the 1-2 shift valve is to route fluid pressure so as to:
    a. hold gears in 1-2 position
    b. produce an upshift
    c. engage the direct and intermediate clutches
    d. lock both one-way clutches in drive

20. The purpose of the 2-3 shift valve is to route fluid pressures so as to:
    a. engage the roller clutches
    b. hold gears in 2-3 position
    c. produce an upshift or downshift
    d. none of the above

CHRYSLER TORQUEFLITE AUTOMATIC TRANSMISSIONS

After studying this chapter, you should be able to:

1. Describe the construction and operation of the planetary gearsets in the TorqueFlite transmissions.
2. Explain the power flow through the transmission in D breaking away, D upshifted to second, D direct, D2, D1, and reverse.
3. Identify and name the transmission parts when they are laid out on the bench before you.
4. Explain how the shift controls work to operate the clutches and bands.
5. Explain how the automatic lockup torque converter works.

5-1 Introduction to Chrysler automatic transmissions This chapter describes the construction and operation of the various Chrysler TorqueFlite automatic transmissions used in rear-wheel-drive cars. These transmissions are all basically similar in construction and operation. Some models are smaller and lighter and are for the smaller Chrysler Corporation automobiles. Others are for standard cars. Cars with high-performance engines use a TorqueFlite with additional parts to handle the higher engine power and torque. However, the planetary gearsets and hydraulic controls are similar for all TorqueFlite models.

5-2 TorqueFlite automatic transmissions Figures 5-1 to 5-3 show, in sectional view, three variations of the basic TorqueFlite. Figure 5-1 shows the model A-904, which is used in the intermediate cars. Figure 5-2 shows the model A-727, which is more heavily constructed for the larger cars. Figure 5-3 shows the model MA-904A, which is used on smaller cars with four-cylinder engines. The model MA-904A, for example, has a 9.5-inch [241.3-mm] diameter converter. The model A-904 has a 10.75-inch [273.5-mm] diameter converter. The model A-727 used on the higher-performance engines has an 11.75-inch [298.45-mm] diameter converter.

The larger-diameter converters are used on the larger cars with the larger engines. However, the basic design for all models is very similar. But only the A-904 and the A-727 have the automatic lockup torque converter. This feature is described in ✿ 5-8.

The TorqueFlites have two multiple-disk clutches, an overrunning clutch, two bands operated by servos, and a compound planetary gearset. The transmission has a hydraulic system which controls the application of the bands and engagement of the clutches.

The compound planetary gearset, which is the same design for all models, has one long sun gear, two sets of planetary pinions with separate carriers, and two internal, or ring, gears. The long sun gear meshes with both sets of planetary pinions. The sun gear is connected to the front clutch by a driving shell which is splined to the sun gear. The driving shell extends around and in front of the rear clutch. Figures 5-1 to 5-4 show these relationships.

5-3 TorqueFlite gearshifts Figure 5-5 shows, in chart form, the various patterns of clutch engagement and band application to obtain the various gear ratios through the gearsets. Figures 5-4 to 5-10 show these arrangements. Figure 5-4, for example, shows the power flow through the transmission with the selector lever in D (drive) when first breaking away from a standing start. The power flows from the input shaft through splines to the hub on which the inner drum of the front clutch and the outer drum of the rear clutch are mounted. These parts all turn as a unit at all times. When the front clutch is disengaged and the rear clutch is engaged, as shown in Fig. 5-4, then the ring gear of the front gearset must turn also. This movement drives the planet pinions of the front gearset, and they, in turn, drive the sun gear. Note that there is gear reduction through the front gearset.

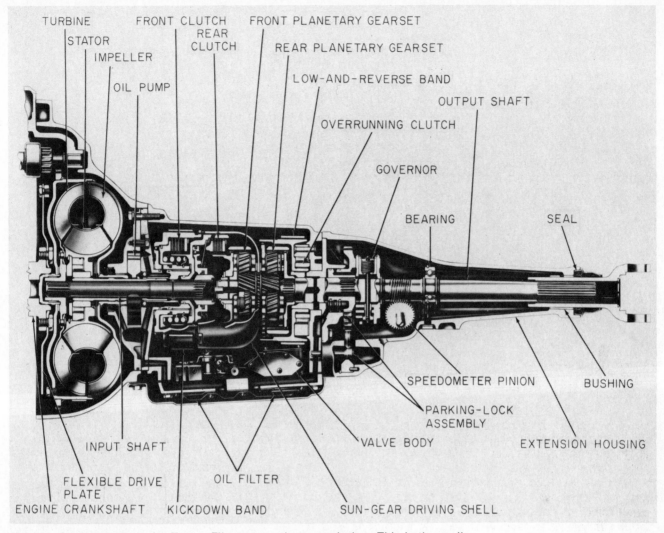

TURBINE
STATOR
IMPELLER
OIL PUMP
FRONT CLUTCH
REAR CLUTCH
FRONT PLANETARY GEARSET
REAR PLANETARY GEARSET
LOW-AND-REVERSE BAND
OVERRUNNING CLUTCH
GOVERNOR
OUTPUT SHAFT
BEARING
SEAL
SPEEDOMETER PINION
BUSHING
PARKING-LOCK ASSEMBLY
VALVE BODY
EXTENSION HOUSING
INPUT SHAFT
FLEXIBLE DRIVE PLATE
ENGINE CRANKSHAFT
OIL FILTER
KICKDOWN BAND
SUN-GEAR DRIVING SHELL

Fig. 5-1 Sectional view of a TorqueFlite automatic transmission. This is the earlier version without the torque-converter lockup clutch. (*Chrysler Corporation*)

As the sun gear turns, it drives the planet pinions of the rear gearset, and they, in turn, drive the ring gear. The planet-pinion carrier is mounted on an overrunning clutch, and this clutch holds the carrier stationary. The arrangement produces an additional gear reduction so that the total gear reduction through the two gearsets is 2.45:1. This means that the input shaft must turn 2.45 times to cause the output shaft to turn once.

NOTE The gear ratios may vary in different TorqueFlite models.

Now note the situation when the transmission upshifts to second, as shown in Fig. 5-6. The front, or kickdown, band is applied and the rear clutch is engaged. Applying the kickdown band locks the sun gear in a stationary position. Now, the power flow is through the rear clutch, ring gear, planet pinions, and carrier to the output shaft. Gear reduction is achieved in the front gearset only. The total gear reduction is 1.45:1.

In direct drive, shown in Fig. 5-7, both front and rear clutches are engaged, locking the front gearset together (sun gear to ring gear). Therefore, the assembly turns as a unit to give direct drive.

Figure 5-8 shows the situation with the selector lever in second. The transmission is held in the downshifted position. Figure 5-9 shows the conditions when the selector lever is in L (low). They are similar to the conditions shown in Fig. 5-4, which is low with the selector lever in D when first starting out. However, with the selector lever in L, the low-and-reverse band is applied to lock the rear planet carrier (along with the overrunning clutch holding it). The reason for this change is that the low-and-reverse band will remain applied as long as the selector lever is in L and will prevent freewheeling. The 2.45:1 gear reduction will therefore remain in effect. But if the selector lever were in D, then the overrunning clutch would hold the rear planet carrier only until upshifting occurred. The overrunning clutch would then permit the carrier to overrun, as shown in Figs. 5-6 and 5-7.

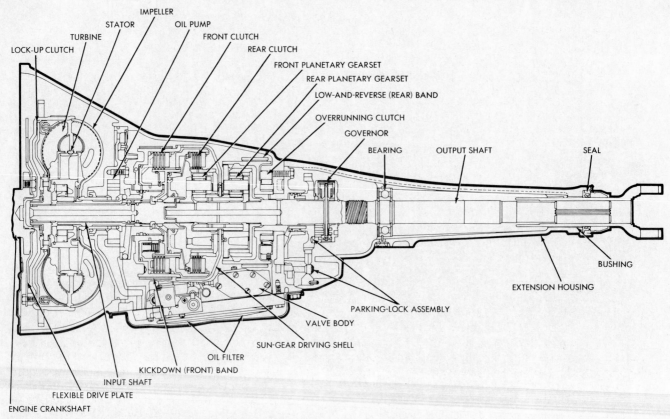

Fig. 5-2 Sectional view of the type A-727 TorqueFlite transmission. It is the later version with the torque-converter lockup clutch. (*Chrysler Corporation*)

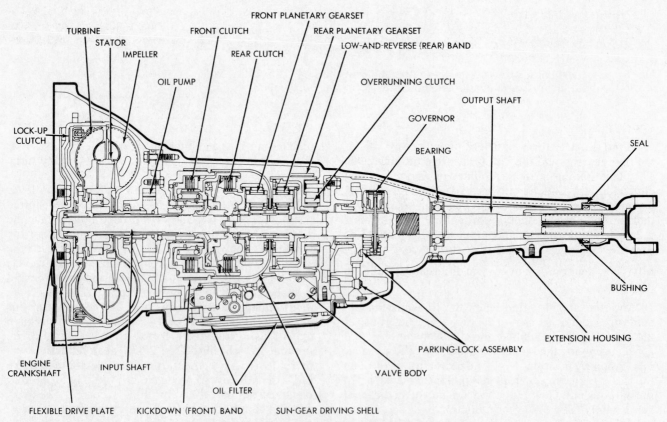

Fig. 5-3 Sectional view of the type A-904 TorqueFlite automatic transmission. This later version has the torque-converter lockup clutch. (*Chrysler Corporation*)

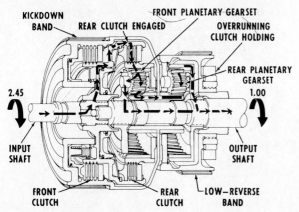

Fig. 5-4 Power flow in D (drive range) when first breaking away from a standing start with gears in first, or low. (*Chrysler Corporation*)

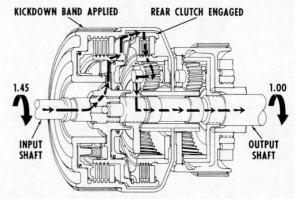

Fig. 5-6 Power flow in D after an upshift to second gear or after a kickdown from direct. (*Chrysler Corporation*)

In reverse, shown in Fig. 5-10, the front clutch is engaged and the low-and-reverse band applied. Now the power flow is through the front clutch, driving shell, sun gear, and rear planetary gearset to the output shaft. Note that the low-and-reverse band, when applied, holds the rear-planetary-gearset carrier stationary. The planetary gears therefore act as idlers and cause the ring gear and output shaft to turn in the reverse direction so that the car is backed. The gear ratio is 2.20:1.

✿ **5-4 TorqueFlite hydraulic circuits** The hydraulic circuits, with the various valves, that control the shifting of the automatic transmission are shown in Plates 9 to 13 in the *Color Plates of Automatic-Transmission Hydraulic Circuits*, sixth edition.

Plate 9 shows the hydraulic circuit and the four active gearshift controls. These are the two clutches and the two servos. Locate the various valves and the clutches, servos, and accumulator in Plate 9. The shuttle valve cushions certain shifts, as explained later. The T/C (torque-converter) relief valve (lower left) prevents excessive pressures in the torque converter. The lockup valve and fail-safe valve, shown in a box to the right, control the lockup piston, which locks the torque converter in direct drive (Plate 11). All valves and their operation are described in following sections.

✿ **5-5 Low gear, or breakaway, in drive** Chrysler calls low gear the "breakaway" gear. The conditions are shown in Fig. 5-4 and Plate 9. The manual valve has been moved to D. This feeds line pressure to the rear clutch and the accumulator. The rear clutch en-

LEVER POSITION DRIVE RATIO	FRONT CLUTCH	REAR CLUTCH	FRONT (KICKDOWN) BAND	REAR LOW-REVERSE BAND	OVER-RUNNING CLUTCH
N—neutral	Disengaged	Disengaged	Released	Released	No movement
D—drive					
(Breakaway) 2.45:1	Disengaged	Engaged	Released	Released	Holds
(Second) 1.45:1	Disengaged	Engaged	Applied	Released	Overruns
(Direct) 1.00:1	Engaged	Engaged	Released	Released	Overruns
Kickdown					
(To second) 1.45:1	Disengaged	Engaged	Applied	Released	Overruns
(To low) 2.45:1	Disengaged	Engaged	Released	Released	Holds
2—second 1.45:1	Disengaged	Engaged	Applied	Released	Overruns
1—low 2.45:1	Disengaged	Engaged	Released	Applied	Partial hold
R—reverse 2.20:1	Engaged	Disengaged	Released	Applied	No movement

Fig. 5-5 Chart of clutch engagements and band applications for all transmission operating conditions. (*Chrysler Corporation*)

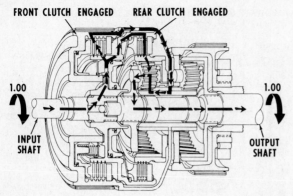

Fig. 5-7 Power flow in D after an upshift to direct drive. (*Chrysler Corporation*)

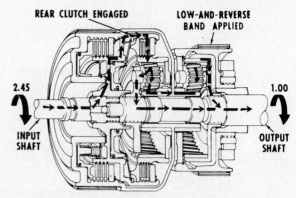

Fig. 5-9 Power flow in D1, or with the selector lever moved to 1. (*Chrysler Corporation*)

gages to produce the power flow as shown in Fig. 5-4. The regulator valve, to the lower left, in Plate 9, controls line pressure relative to the kickdown valve and the throttle valve. The throttle valve is a balanced valve (⚙ 2-7). It cuts down the line pressure coming from the manual valve. The amount the throttle valve reduces line pressure depends on the position of the kickdown valve, which is mechanically linked to the throttle.

As the driver pushes down on the accelerator pedal, the throttle opens and the kickdown valve is moved inward (to the left in Plate 9). This movement increases the spring force on the throttle valve, forcing it to move. The throttle valve then passes more line pressure so that the pressure exiting from the throttle valve, called *throttle-valve pressure*, increases. This increasing pressure acts on the plug to the right of the regulator valve. The plug, in turn, reacts on the regulator valve so that it tends to cut off the dump, or "suction," line to the pump. Therefore the regulated pressure increases. This provides greater holding power for the clutches and bands at higher speeds and with higher torques passing through the transmission.

In breakaway—low gear in drive range—the manual valve has been moved so that line pressure is directed to the rear clutch, engaging it. The power flow is shown in Fig. 5-4. Note that governor pressure is applied to the governor plugs behind the 1-2 and 2-3 shift valves. Opposing this governor pressure, at the opposite ends

of the 1-2 and 2-3 shift valves, is spring force plus throttle-valve pressure. Throttle-valve pressure increases with increased throttle opening. This increase in pressure is due to the movement of the kickdown valve, which is linked to the throttle linkage.

As car speed and governor pressure increase, the increasing governor pressure will overcome the 1-2 shift-valve spring and throttle-valve pressure. Then, the transmission will upshift to second.

⚙ 5-6 Drive range, second gear When the upshift occurs, the governor plug, with governor pressure behind it, has pushed the 1-2 shift valve to the right. As the 1-2 shift valve moves, it allows line pressure to flow to the front servo so that the front, or kickdown, band is applied. Now, the power flow is as shown in Fig. 5-6.

Note that the 1-2 shift-valve movement also allows line pressure to flow to the accumulator. The accumulator cushions the application of the kickdown band. Accumulator action is discussed in ⚙ 2-11.

The point at which the upshift occurs depends on two factors: car speed and throttle opening. With a small throttle opening, there will be a relatively small throttle-valve pressure. Therefore a relatively low car speed and governor pressure can produce the upshift. The actual speed at which upshifts occur with closed and wide-open throttle varies greatly with different cars and engines. This variation in different transmission

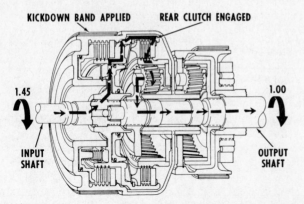

Fig. 5-8 Power flow in D2, or with the selector lever moved to 2. (*Chrysler Corporation*)

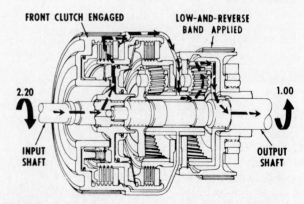

Fig. 5-10 Power flow in R, or reverse. (*Chrysler Corporation*)

models is provided by using different valve springs, clutch plates and other parts. A typical pattern would be for a closed-throttle 1-2 upshift to take place at about 12 mph [19 km/h]. With wide-open throttle, the 1-2 upshift would then take place at about 40 mph [64 km/h]. For this particular transmission, the upshift would vary between those two extremes, depending on the amount of throttle opening.

❀ 5-7 Drive range, direct, or third, gear

With further increase in car speed and therefore in governor pressure, the governor pressure reaches a value sufficient to overcome the spring force and throttle-valve pressure acting on the 2-3 shift valve. The 2-3 shift valve is forced to move to the right, as shown in Plate 10. Now, line pressure is directed through the 2-3 shift valve to the front clutch, causing it to engage. At the same time, line pressure is admitted to the upper part of the kickdown servo, forcing the servo piston to move down and release the kickdown band. The situation is then as shown in Fig. 5-7.

The shuttle valve, to the upper right in Plate 10, acts to smooth out the engagement of the front clutch and the release of the kickdown band. It does this by modulating the line pressure going to the release side of the kickdown piston and to the front clutch. The modulation is related to throttle-valve pressure, which acts on the throttle plug to the left of the shuttle valve. This action positions the shuttle valve according to throttle-valve pressure. The shuttle valve then modulates the pressure going to the release side of the kickdown piston and to the front clutch.

For example, consider a "lift-foot" upshift, which is made by accelerating the car in second and then lifting the foot to move the throttle toward a closed position. When this happens, throttle-valve pressure on the 2-3 shift valve is reduced so that governor pressure can move the 2-3 valve to produce the upshift. The reduced throttle-valve pressure on the throttle plug eases the pressure on the shuttle valve, allowing it to reduce line pressure to the kickdown piston and front clutch. As a result, the band releases more slowly and the front clutch engages more softly. The upshift is mild and without any noticeable jerk. If the throttle opening is maintained, the kickdown band must release rapidly and the front clutch must engage quickly to produce a smooth 2-3 upshift.

The point at which a 2-3 upshift takes place depends upon car speed and throttle opening. The greater the throttle opening, the higher the car speed at which the upshift will take place. For example, with a closed throttle, the upshift could take place at 18 mph [29 km/h]. But on the same car, with a wide-open throttle, the upshift would not take place until 70 mph [113 km/h].

❀ 5-8 Automatic-lockup torque converter

Most late models of the A-747 and the A-904 TorqueFlite have an automatic-lockup torque converter. At road speed, the impeller in a torque converter without lockup is turning slightly faster than the turbine. The impeller must turn faster so that it can continue to

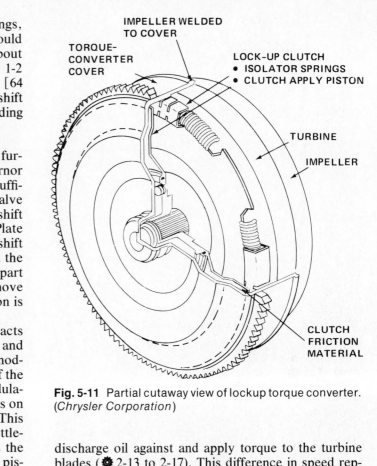

Fig. 5-11 Partial cutaway view of lockup torque converter. (*Chrysler Corporation*)

discharge oil against and apply torque to the turbine blades (❀ 2-13 to 2-17). This difference in speed represents a power loss. For this reason, many Torque-Flite transmissions were redesigned to include a means of automatically locking the torque converter when the car reaches cruising speed. This results in better fuel economy. Also, the transmission oil does not get as hot, because it does no work in the lockup mode.

Figure 5-11 is a cutaway view of a lockup torque converter. It has a clutch and a clutch-apply piston, which is an annular ring. The isolator springs on the clutch help dampen the clutching action as the torque converter goes into the lockup mode. These isolator springs also dampen out the power pulses from the engine when the transmission is in direct drive and the converter is locked. They do the same job as the torsional springs in the standard-clutch friction disk.

Figure 5-12 shows how the lockup works. In the center illustration, the torque converter is not locked. The piston is released. The right side of Fig. 5-12 shows the locked position. The torque-converter cover has a ring of friction material bonded to it. When the converter goes into lockup, oil pressure is applied back of the piston. This forces the piston to the left (in Fig. 5-12). Now, the piston and the output shaft must turn together. Since the power input is to the converter cover, the torque converter turns as a unit. The arrows in Fig. 5-12 show the power flow through the converter in the two modes, unlocked and locked.

Plate 11 shows the hydraulic circuits in action when the converter is in the lockup mode. Note how the lockup valve has been forced to the left by the increasing governor pressure. This admits line pressure to the

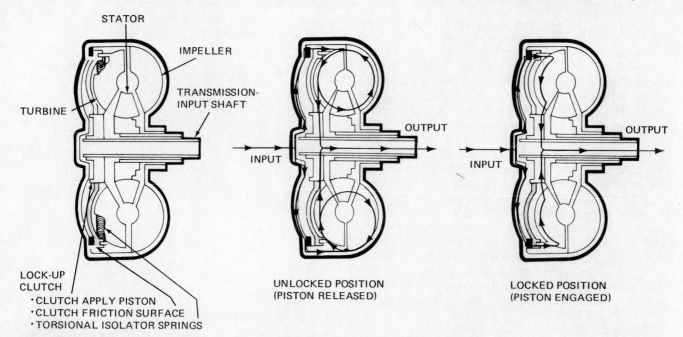

Fig. 5-12 Operation of the lockup torque converter. To left, parts named. Center, torque-converter clutch in the unlocked position. Right, piston locking up the torque converter. (*Chrysler Corporation*)

fail-safe valve. The pressure passes through the fail-safe valve to the torque converter, causing the lockup clutch to engage. Feed to the lockup clutch is restricted by the fail-safe valve if front-clutch pressure drops. The fail-safe valve permits lockup only in direct drive and provides a fast lockup release during a kickdown.

✿ 5-9 Kickdown in drive range If quick acceleration is required to pass another car, the driver pushes the accelerator down all the way to wide-open throttle. The linkage from the throttle to the kickdown valve, shown to the lower right in Plate 12, pushes the kickdown valve to the left. This movement allows throttle-valve pressure to pass through the kickdown valve. Now called *kickdown pressure,* this pressure passes to the 2-3 and 1-2 shift valves. If the kickdown pressure is great enough to overcome governor pressure, a downshift will occur. Throttle-valve pressure must overcome governor pressure, or car speed. When the downshift occurs, the oil is dumped from the front clutch and the upper, or release, side of the kickdown-servo piston. The front clutch disengages and the kickdown servo applies the kickdown band. The situation is as shown in Fig. 5-8. The transmission has downshifted into second gear.

The shuttle valve helps smooth the downshift. It permits the kickdown to take place rapidly at low car speed but slows the kickdown at higher car speed. At low car speed, very little time is required to complete the shift. This is because there is a comparatively small change in engine speed between direct drive and kickdown, or second gear. However, at high car speed, the engine speed must change considerably to pick up the difference between direct gear and second. Therefore, the shuttle valve slows the downshift at higher car

speeds. If it were not for shuttle-valve action, there would be a very noticeable jerk as the downshift occurred.

If the car is in second gear and moving slowly, the kickdown will shift the gears from second to first. The throttle-valve pressure will be sufficient to overcome governor pressure at the 1-2 shift valve, causing it to move and produce the downshift to first. As the 1-2 shift valve moves, it cuts off the pressure to the kickdown servo so that the band releases. This produces the condition shown in Fig. 5-4.

✿ 5-10 Reverse When the driver shifts the selector lever to R (reverse), the hydraulic circuit is as shown in Plate 13. Power flow is as shown in Fig. 5-10. Note that everything in the hydraulic circuit becomes ineffective except the pump, regulator valve, torque-converter control valve, and manual valve. The manual valve feeds line pressure to the pressure-regulator valve. Therefore the pressure regulator regulates to a higher pressure—as much as 230 to 260 psi [1586 to 1793 kPa]. This high pressure ensures good, firm application of the low-and-reverse band and engagement of the front clutch.

✿ 5-11 Drive 2 range This is the condition that results if the driver places the selector lever in drive 2 position. The manual valve is moved so that line pressure is applied to the rear clutch and the apply side of the kickdown-servo piston (Plate 7). With the rear clutch engaged and kickdown band applied, the situation shown in Fig. 5-8 results.

At high car speed, with the transmission in direct, the downshift will not occur until governor pressure (car speed) drops enough to permit line pressure to overcome governor pressure and move the governor plug (to the left in Plate 12). This action allows the spring to move the 2-3 shift valve so that the line pressure is cut off from the front clutch and the release side of the kickdown-servo piston.

At low car speed, the transmission can downshift from second to first. It can then upshift to second with increasing car speed. But it will not upshift to third as long as the selector lever and the manual valve are in the drive 2 range.

✿ 5-12 Drive 1 range This is the condition that results if the driver places the selector lever in drive 1 position. The transmission stays in low gear. The power flow is as shown in Fig. 5-9. The manual valve moves so that line pressure is admitted back of the 1-2 gov-

ernor plug and the 2-3 governor plug. This action prevents any upshift.

If car speed is above a certain minimum and the transmission is in second or direct gear, the downshift to first will not take place until the car speed, and governor pressure, drops to the required minimum. When this happens, governor pressure, working against the governor plugs, is overcome by line pressure directed to the shift valves. The downshift take place.

Drive 1 and 2 ranges are for pulling heavy loads up long hills where the accelerator is almost or fully open for ½ mile [0.8 km] or longer. These lower gears prevent transmission overheating on the hard pull. They also permit engine braking of the car when going down steep hills.

Chapter 5 review questions

Select the *one* correct, best, or most probable answer to each question. Then check your answers against the correct answers given at the end of the book.

1. Counting all the active control members in the TorqueFlite transmission, including the clutches and bands, there are:
   a. four
   b. five
   c. six
   d. seven
2. The compound planetary gearset has two sets of planet pinions, two internal (or ring) gears, and:
   a. three sun gears
   b. two sun gears
   c. one planet-pinion carrier
   d. one sun gear
3. The front clutch is connected by a driving shell to the:
   a. output shaft
   b. rear ring gear
   c. sun gear
   d. front ring gear
4. In D range on breakaway in first gear, the overrun clutch holds, and the:
   a. front clutch is engaged
   b. rear clutch is engaged
   c. kickdown band is applied
   d. reverse band is applied
5. In D range in second gear, the overrun clutch overruns, the rear clutch is engaged, and the:
   a. front clutch is engaged
   b. rear band is applied
   c. front band is applied
   d. torque converter is locked up

6. In direct drive, the overrun clutch overruns, both clutches are engaged, and:
   a. the front band is applied
   b. the rear band is applied
   c. both bands are released
   d. both bands are applied
7. In second range, second gear, the overrun clutch overruns, the kickdown or front band is applied, and:
   a. the rear clutch is engaged
   b. the rear band is applied
   c. the front clutch is engaged
   d. both clutches are engaged
8. In low range, low gear, the rear band is applied and:
   a. the front clutch is engaged
   b. the front band is applied
   c. the rear clutch is engaged
   d. both bands are applied
9. In reverse, the rear band is applied and:
   a. the rear clutch is engaged
   b. the front clutch is engaged
   c. the front band is applied
   d. both bands are applied
10. The sun gear is locked in a stationary position by:
   a. front-clutch engagement
   b. front-band application
   c. rear-band application
   d. both bands applying

FORD C4 AND C6 AUTOMATIC TRANSMISSIONS

After studying this chapter you should be able to:

1. Explain the construction of the planetary gearset and how it produces three forward speeds and one reverse speed.
2. Name the holding members in action in all gear positions and describe their effect on the gearset.
3. List six major valves in the hydraulic system and describe their purposes.
4. Explain the purpose of the accumulator.
5. Describe the actions of the basic valves in the various gear positions and how they act to produce shifts.

6-1 Introduction to Ford automatic transmissions This chapter describes the construction and operation of two widely used Ford transmissions, the C4 and the C6. These two transmissions are similar in construction and operation. The major difference is that the C4 is smaller and uses a low-and-reverse band while the C6 is larger and uses a low-and-reverse clutch to perform the same function. Both transmissions are three-speed units that, aside from the low-and-reverse arrangement, use similar clutches, bands, and planetary gearsets. The hydraulic systems and valve arrangements are also very similar.

6-2 C4 and C6 transmissions Figure 6-1 is a cutaway view of the C6 transmission. It has three multiple-plate clutches, one band, and one one-way, or overrunning, clutch. The operation of servos that operate the bands, overrunning clutches, multiple-plate clutches, and planetary-gear systems is covered in other chapters. This chapter explains how the clutches and bands operate to produce the various gear ratios, and points out the special features of the hydraulic system used in the C4 and C6 transmissions.

6-3 Transmission construction Figure 6-2 is a cutaway view of the transmission gear train for the C4 transmission. Figure 6-3 is a sectional view of the gear train for the C6 transmission. The main difference between the two is that the C4 has a low-and-reverse band while the C6 has a low-and-reverse multiple-plate clutch. The effect is the same with both. When the band or clutch is on, the reverse planet carrier is held stationary.

Figures 6-4 to 6-14 build the C6 transmission, part by part. Then the conditions in the transmission in the different gear ratios are shown.

First, the input and output shafts to the converter are placed in the case, as shown in Fig. 6-4. Next, the forward-clutch cylinder is installed, as shown in Fig. 6-5. It is splined to the input shaft and therefore must turn with it.

Next, the sun gear is added, as shown in Fig. 6-6. The sun gear is installed on the output shaft and has two bushings so that it can turn freely on the shaft.

NOTE: This is not the way the transmission is actually assembled. The transmission is being built up in Figs. 6-4 to 6-14 to show how the parts are related and how they work together.

Now the forward planetary gear unit is installed, as shown in Fig. 6-7. It consists of the ring-gear pilot, ring gear (also called the *internal gear*), and planet carrier with planet pinions. The splines cut on the outside of the ring gear engage with the clutch plates. Therefore, the ring gear is also the forward-clutch hub. The planet-pinion carrier is splined to the output shaft.

Next, the forward-clutch plates are installed, as shown in Fig. 6-8. The plates are alternately splined to the forward-clutch cylinder and the clutch hub (the outside of the ring gear installed in Fig. 6-7). When the forward clutch is engaged, the forward ring gear is driven at shaft-input speed.

Now, the reverse-and-high clutch is added, as shown in Fig. 6-9. Notice that the reverse-and-high clutch is located ahead of the forward clutch in the gear train. The reverse-and-high clutch includes the reverse-and-

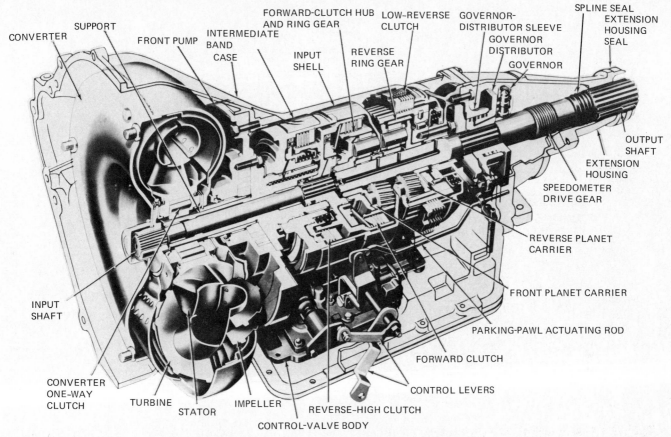

Fig. 6-1 Cutaway view of the Ford C6 automatic transmission. (*Ford Motor Company*)

high drum, piston, clutch plates, pressure plate, and input shell. The input shell is splined to the center of the sun gear. The reverse-and-high drum is free-running on the stationary front-pump hub. The clutch plates are alternately splined to the reverse-and-high-clutch drum and the hub of the forward-clutch cylinder.

Next, the intermediate band and servo are added, as shown in Fig. 6-10. The band is positioned around the outside of the reverse-and-high drum. When hydraulic apply pressure is directed to the servo, it tightens the band on the drum, bringing the drum to a halt. The

Figure 6-2 Cutaway view of the C4 transmission gear train. (*Ford Motor Company*)

drum is connected to the sun gear through the lugs on the drum that fit the notches in the input shell. The input shell is splined to the center of the sun gear. Therefore, the reverse-and-high drum, input shell, and sun gear always turn or are stationary together. They are, in effect, one part.

The transmission is almost completely assembled. Now, how various gear ratios are obtained can be discussed. Second gear is achieved by applying the band and engaging the forward clutch. This produces the situation shown in Fig. 6-11. The sun gear is held stationary by the band, and the forward ring gear is driven through the forward clutch by the input shaft. The planet pinions are forced to walk around the stationary sun gear, carrying the pinion carrier around with them. The pinion carrier is splined to the output shaft. Therefore, the output shaft is turned. But there is gear reduction through the planetary-gear system.

Direct drive can be obtained through the transmission, as shown in Fig. 6-12. This will occur when both the reverse-and-high and forward clutches are engaged. This action locks the input shaft to the sun gear and the ring gear. Therefore the planetary-gear system is locked up so it turns as a unit. The output shaft turns at the same speed as the input shaft.

To get low gear and reverse, more parts must be added to the transmission. First, the low-and-reverse ring gear is added, as shown in Fig. 6-13. This ring gear is splined through a hub to the output shaft. Next, the low-and-reverse pinion carrier and clutch are added, as

73

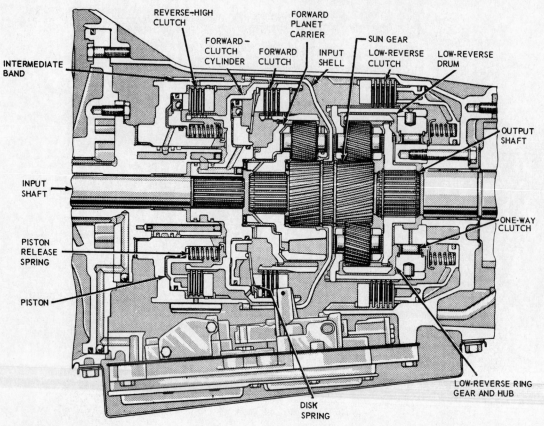

Fig. 6-3 Sectional view of the planetary-gear train and clutches of the C6 transmission. (*Ford Motor Company*)

shown in Fig. 6-14. The clutch plates are alternately splined to the transmission case and to the low-and-reverse-clutch hub. The hub is attached to the carrier by alternating tabs or lugs. When the low-and-reverse clutch is engaged, the pinion carrier is locked to the case.

To get reverse, the low-and-reverse clutch is engaged to hold the planet-pinion carrier stationary. At the same time, the reverse-and-high clutch is engaged so that the input shaft drives the sun gear. This is shown in Fig. 6-15.

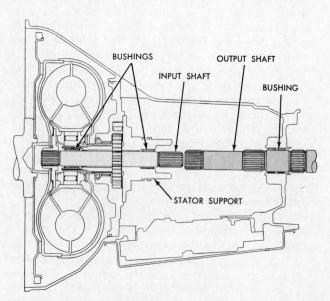

Fig. 6-4 Input and output shafts in place in the C6 transmission. (*Ford Motor Company*)

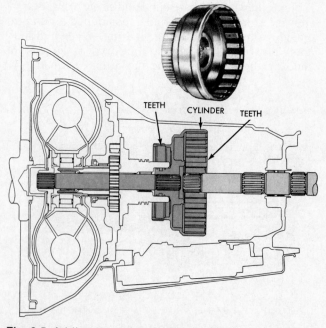

Fig. 6-5 Adding the forward-clutch cylinder. (*Ford Motor Company*)

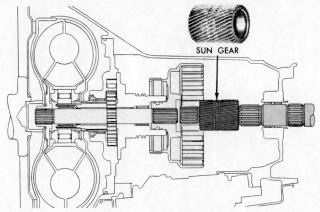

Fig. 6-6 Adding the sun gear. (*Ford Motor Company*)

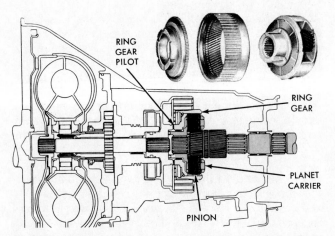

Fig. 6-7 Adding the forward planetary-gear unit. (*Ford Motor Company*)

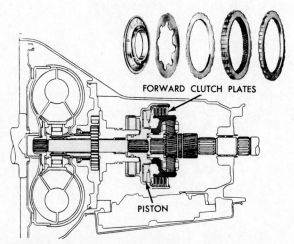

Fig. 6-8 Adding the forward-clutch plates. (*Ford Motor Company*)

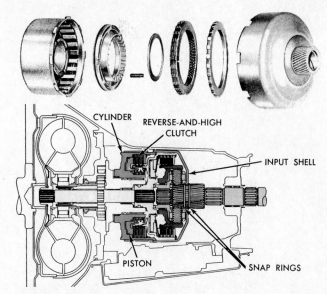

Fig. 6-9 Adding the reverse-and-high clutch. (*Ford Motor Company*)

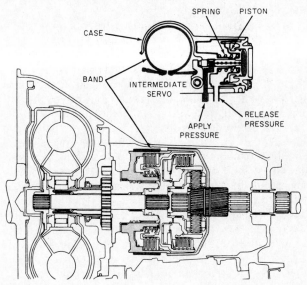

Fig. 6-10 Adding the intermediate band and servo. (*Ford Motor Company*)

THE INTERMEDIATE BAND IS APPLIED. THE REVERSE-AND-HIGH CLUTCH DRUM, THE INPUT SHELL AND THE SUN GEAR ARE HELD STATIONARY.

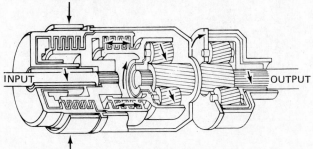

THE FORWARD CLUTCH IS ENGAGED. THE FRONT RING GEAR IS LOCKED TO THE INPUT SHAFT.

Fig. 6-11 Transmission in second gear, drive range. (*Ford Motor Company*)

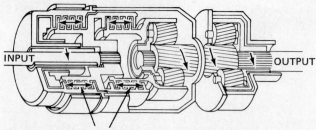

INPUT → ← OUTPUT

BOTH THE FORWARD AND THE REVERSE-AND-HIGH CLUTCH
ARE ENGAGED. ALL PLANETARY-GEAR MEMBERS ARE
LOCKED TO EACH OTHER AND TO THE OUTPUT SHAFT.

Fig. 6-12 Transmission in high gear, or direct drive. (*Ford Motor Company*)

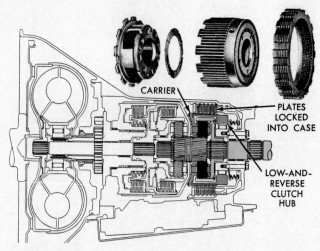

CARRIER — PLATES LOCKED INTO CASE — LOW-AND-REVERSE CLUTCH HUB

Fig. 6-14 Adding the low-and-reverse pinion carrier and clutch. (*Ford Motor Company*)

When the carrier held and sun gear driven clockwise, the pinions act as idlers and reverse the direction of rotation. Therefore, the ring gear is driven counterclockwise at a reduced speed. The ring gear is attached to the output shaft so that it turns in the reverse direction.

In the following sections, each gear condition is discussed in detail.

✿ 6-4 Low gear in D   Low, or first, gear is shown in Fig. 6-16. The forward clutch is engaged, locking the forward ring gear to the input shaft. This drives the sun gear through the forward planet pinions. The sun gear then drives the low-and-reverse pinions, which drive the low-and-reverse ring gear. The one-way, or overrunning, clutch holds the reverse planet carrier stationary. Since the ring gear is splined to the output shaft, the output shaft turns. There is gear reduction through both planetary gearsets.

Note that the situation might be called a *double reverse*. The forward planetary gearset reverses input rotation and produces gear reduction. The low-and-reverse planetary gearset then re-reverses the rotation to give forward motion at a further gear reduction.

✿ 6-5 Low gear in L1   In low gear with the selector lever in L1, the forward and the low-and-reverse clutches are engaged. The low-and-reverse pinion carrier is held from turning by the low-and-reverse clutch. Power flow is the same as in low gear in D. The difference is that if the car starts down a hill in L1, the car can drive the engine to produce engine braking of the car. In low gear in D range, if the car tries to drive the engine, the one-way clutch unlocks and the car freewheels.

✿ 6-6 Second gear in D   The conditions in the transmission with the gears in second are shown in Fig. 6-11 and described in ✿ 6-3. The sun gear is held stationary by the band, and the forward ring gear is driven through the forward clutch by the input shaft. The planet pinions walk around the stationary sun gear, carrying the pinion carrier around with them. The pinion carrier is splined to the output shaft so that the output shaft turns, with a gear reduction.

✿ 6-7 High gear   The condition in high gear is shown in Fig. 6-12. Both the reverse-and-high and forward

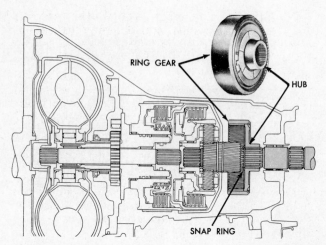

RING GEAR — HUB

SNAP RING

Fig. 6-13 Adding the low-and-reverse ring gear. (*Ford Motor Company*)

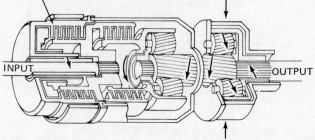

THE REVERSE-AND-HIGH CLUTCH IS ENGAGED. THE INPUT SHAFT IS LOCKED TO THE REVERSE-AND-HIGH CLUTCH DRUM, THE INPUT SHELL, AND THE SUN GEAR.

INPUT → ← OUTPUT

THE LOW-AND-REVERSE CLUTCH IS ENGAGED. THE REVERSE-UNIT PLANET CARRIER IS HELD STATIONARY.

Fig. 6-15 Transmission in reverse. (*Ford Motor Company*)

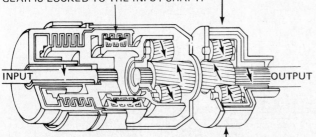

THE FORWARD CLUTCH IS ENGAGED. THE FRONT RING GEAR IS LOCKED TO THE INPUT SHAFT.

INPUT

OUTPUT

THE LOW-AND-REVERSE CLUTCH (LOW RANGE) OR THE ONE-WAY CLUTCH (D1 RANGE) IS HOLDING THE REVERSE-UNIT PLANET CARRIER STATIONARY

Fig. 6-16 Transmission in low—or first—gear, drive range. (*Ford Motor Company*)

clutches are engaged. This action locks up the planetary gear system so that it turns as a unit and the gear ratio is 1:1.

✹ 6-8 Reverse Figure 6-15 shows the conditions with the transmission in reverse. The low-and-reverse clutch is engaged. The reverse-and-high clutch is also engaged. This action holds the carrier and causes the sun gear to be driven clockwise. The pinions act as idlers and drive the ring gear counterclockwise so that the output shaft is driven counterclockwise for reverse.

✹ 6-9 Hydraulic control system The complete hydraulic system for the C6 transmission is shown in Fig. 6-17. The multiple-plate clutches, overrunning clutch, intermediate servo, and band are very similar to those used on transmissions described in other chapters. These parts are not discussed here. This also applies to the pump, main oil-pressure regulator, main oil-pressure booster, and governor.

The 1-2 and 2-3 shift valves are similar to the upshift valves used in the TorqueFlite transmissions. They have governor pressure working at one end, and throttle pressure plus a spring force at the other. In the following sections, the operation of the other valves in the C6 hydraulic system is discussed.

✹ 6-10 Manual valve The manual valve, similar to those in other transmissions, has the usual six positions, as shown in Fig. 6-17: P, R, N, D, 2, and 1. Positioning of the manual valve is controlled by movement of the selector lever on the steering column or on the floor.

✹ 6-11 Throttle control valve The throttle control valve produces a varied pressure dependent on intake-manifold vacuum. As the accelerator pedal is pushed down and the throttle is opened, the throttle control valve passes more and more line pressure. This becomes throttle pressure.

The throttle control valve contains an airtight diaphragm against which intake-manifold vacuum is applied. This vacuum causes atmospheric pressure to move the diaphragm and the spool valve connected to

it. Opposing this pressure is the force of a spring. When the valve is moved, it passes more line pressure, and this becomes throttle pressure.

✹ 6-12 Main oil-pressure booster This valve boosts line pressure as necessary to provide sufficient holding power at the clutches and band servo. With increased throttle opening and higher engine torques, the throttle pressure increases. This increasing pressure, as it acts on the oil-pressure booster valve, puts additional spring force on the main oil-pressure regulator valve. Therefore regulated pressure increases.

During reverse operation, the manual valve directs line pressure to the oil-pressure booster valve so that the booster valve forces the regulated pressure to increase. High holding power is necessary in reverse to prevent band or clutch slippage.

✹ 6-13 Throttle-pressure booster valve Throttle openings above about 50 mph [80 km/h] provide very little additional change in intake-manifold vacuum. The throttle-pressure booster valve boosts throttle pressure to provide the necessary shift delay for higher engine speeds. When throttle pressure increases above about 65 psi [448 kPa] (which varies with different transmissions), the higher pressure, acting on the throttle-pressure booster valve, causes it to move against its spring. This movement permits line pressure to enter and take over control of the booster valve. Now, with the higher line pressure at work, the booster valve increases the throttle pressure about 3 psi [21 kPa] for every 1 psi [6.9 kPa] that the pressure from the throttle control valve increases.

✹ 6-14 Governor The governor works in the same manner as the governors used with other transmissions. As car speed increases, so does governor pressure. This pressure, working against the 1-2 and 2-3 shift valves, causes upshifts when the right combination of car speed and throttle opening is reached. The governor is a dual unit, with the primary governor regulating at low speeds and the secondary governor regulating at high speeds.

✹ 6-15 Throttle downshift valve The throttle downshift valve is mechanically linked to the throttle so that when the driver opens the throttle wide, the downshift valve is moved. Then, if car speed is below a certain value, the line pressure will actuate the 2-3 shift valve to cause a downshift. Above this speed, the downshift will not occur, because the governor pressure will hold the 2-3 shift valve in the third-, or high-, gear position.

✹ 6-16 Hydraulic system The complete hydraulic system is shown in Plate 14 in the *Color Plates of Automatic-Transmission Hydraulic Circuits*, sixth edition. The conditions shown exist with the selector lever in the D range and the transmission in first gear. Note the manual valve at the bottom of Plate 14. An earlier version of the C6 transmission used P, R, N, D2, D1, and L for manual valve positions. The earlier

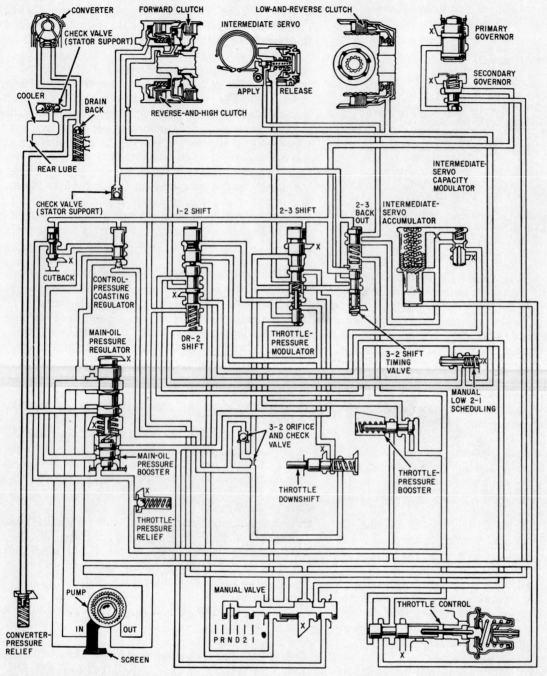

Fig. 6-17 Hydraulic control system for the C6 transmission. (*Ford Motor Company*)

version used D1 an the normal driving range. The transmission shifted automatically from low to second to direct or high. The later version conforms to the P R N D S L or P R N D 2 1 arrangement found on other transmissions (Fig. 6-17). This is about the only significant difference between the earlier and later versions of the C6.

Notice the names of the intermediate-servo accumulator and intermediate-servo capacity modulator in Fig. 6-17. In the earlier version these were called the *1-2 accumulator valve* and the *1-2 scheduling valve*. However, the hydraulic system in the later version (Plate 14) works the same way as in the earlier version.

✿ 6-17 Hydraulic system in drive range, low gear The conditions in the hydraulic system during operation in low gear in drive range are shown in Plate 14. With the selector lever and manual valve in drive range, the forward clutch is engaged and the power flow is as shown in Fig. 6-16. Governor pressure increases as car speed increases. Governor pressure is applied to the ends of the 1-2 and 2-3 shift valves. When this pressure has increased sufficiently to overcome the spring force and the throttle pressure working on the 1-2 shift valve, the shift valve will move, admitting pressure to the intermediate servo. The intermediate servo applies the intermediate band, locking the reverse-and-high clutch drum, input shell, and sun gear in a stationary position. Now, the transmission is in second gear. The power flow is as shown in Fig. 6-11. The 1-2 accumulator valve—or intermediate-servo accumulator, as it is called in Fig. 6-17—cushions the application of the band so that a smooth shift occurs.

❂ 6-18 Hydraulic system in drive range, direct or high gear The conditions in the hydraulic system during operation in direct drive are shown in Plate 15. To reach this condition, governor pressure—car speed—must increase enough to overcome the spring force and throttle pressure acting on the 2-3 shift valve. When this happens, the 2-3 shift valve moves so that pressure is admitted to the reverse-and-high clutch. At the same time, pressure is admitted to the release side of the intermediate servo so that the band is released. The transmission is in direct drive, and the power flow through the transmission is as shown in Fig. 6-12.

❂ 6-19 Hydraulic system in reverse The conditions in the hydraulic system during operation in reverse are as shown in Plate 16. Movement of the manual valve to R permits pressure to flow to the reverse-and-high clutch and the low-and-reverse clutch. Now, the transmission is in reverse, and the power flow is as shown in Fig. 6-15. Note that throttle pressure is applied to the pressure booster valve so that line pressure increases. This ensures the holding power that the clutches need in reverse.

2 BAnds
2 clutches Quzz
In lab 5-6 _two-clutch Packs_

Chapter 6 review questions

Select the *one* correct, best, or most probable answer to each question. Then check your answers against the correct answers given at the end of the book.

1. Counting all the active control members in the C6 transmission, including all clutches and the band, there are:
   a. four
   b. five
   c. six
   d. seven
2. The compound planetary-gear system has two sets of planetary pinions, two ring gears, and:
   a. one sun gear
   b. two sun gears
   c. three sun gears
   d. no sun gear
3. The forward-clutch cylinder is splined to the:
   a. output shaft
   b. input shaft
   c. input shell
   d. output shell
4. The reverse-and-high clutch and input shell are splined to the:
   a. input shaft
   b. output shaft
   c. planet carrier
   d. sun gear
5. The intermediate band is positioned around the:
   a. forward-clutch cylinder
   b. reverse-and-high clutch cylinder
   c. output ring gear
   d. input ring gear

6. When the low-and-reverse clutch is engaged, it locks the reverse-unit pinion carrier to the:
   a. output shaft
   b. input shaft
   c. transmission case
   d. sun gear
7. In low gear, drive range, the one-way clutch is holding the reverse-unit pinion carrier stationary and the:
   a. low-and-reverse clutch is engaged
   b. forward clutch is engaged
   c. band is applied
   d. rear clutch is engaged
8. In second gear, drive range, the forward clutch is engaged and the:
   a. intermediate band is applied
   b. low-and-reverse clutch is engaged
   c. one-way clutch is holding
   d. sun gear is stationary
9. In direct drive, the forward clutch is engaged and the:
   a. reverse-and-high clutch is engaged
   b. intermediate band is applied
   c. one-way clutch is holding
   d. rear band is applied
10. In reverse, the reverse-and-high clutch is engaged and the:
    a. intermediate band is applied
    b. forward clutch is engaged
    c. low-and-reverse clutch is engaged
    d. sun gear is locked to the output shaft

FORD FOUR-SPEED OVERDRIVE AUTOMATIC TRANSMISSION

After studying this chapter, you should be able to:

1. Explain how the upshifts and downshifts are made.
2. Describe how overdrive is achieved.
3. Explain the purpose of the damper assembly and how it works.
4. Discuss how the input torque is split between two shafts in third gear.
5. Describe the power flow through the transmission in the four speeds and reverse.

7-1 Introduction to Ford automatic overdrive transmissions This chapter describes the construction and operation of the Ford four-speed overdrive automatic transmission introduced in 1980 (Fig. 7-1). This transmission has a different planetary gearset from the C4 and C6 discussed in Chap. 6. The gearset is similar to the Ford FMX automatic transmission. It has two sun gears, three short pinions, three long pinions, and one ring gear. These components are controlled by six clutches and two bands. Two of the clutches are one-way clutches. They provide the additional control of the planetary gearset needed for overdrive.

7-2 Shift-selector positions Figure 7-2 shows the six positions for the shift selector. In P, or park, the transmission is in neutral and the output shaft is locked to the transmission case by a parking pawl. The engine can be started.

In R (reverse), the output shaft turns in the reverse direction to the input shaft so the car is backed. In N (neutral), the output shaft is not locked. The engine can be started. The brakes should be applied because the wheels are free to turn.

Overdrive ⒟ is the normal driving range. The transmission will upshift and downshift automatically through all the forward gears—first, second, third (direct), and overdrive.

Three is overdrive lockout. In this range, the transmission can upshift and downshift automatically through first, second, and third (direct). It will not shift into overdrive because overdrive is locked out. This corresponds to the normal drive range in other automatic transmissions.

One is manual low. The transmission is locked in low gear. This range is used for downhill braking or for steep uphill pulls.

7-3 Downshifts The transmission will downshift as the car coasts down or is braked. Demand downshifts will also occur if the driver opens the throttle to accelerate. This is the same as the kickdown downshift we described for the Chrysler TorqueFlite (✿ 5-9) or the throttle downshift for the Ford C6 (✿ 6-15). The transmission drops into a lower gear to provide more torque to the wheels. The transmission also downshifts when road speed decreases so that the governor pressure drops.

7-4 Torque converter The torque converter (Fig. 7-3) is similar to the torque converters for other automatic transmissions except for two things. It has automatic bypass of the torque converter in overdrive. Also, in third gear, the engine torque is split between two shafts. One is a solid shaft called the *direct-drive shaft* and the other is called the *turbine shaft* and is a hollow tube. This provides two power inputs to the gearset in third. These actions are explained later.

7-5 Planetary gearset The planetary gearset (Fig. 7-4) has two sun gears, three short pinions, three long pinions, one planet carrier, and one ring gear. The short pinions and long pinions are both carried on a single planet-pinion carrier. To understand how holding and driving different members of the planetary gearset can produce speed reduction, speed increase, and lockup, refer to the same numbered component in Fig. 7-4 as shown below.

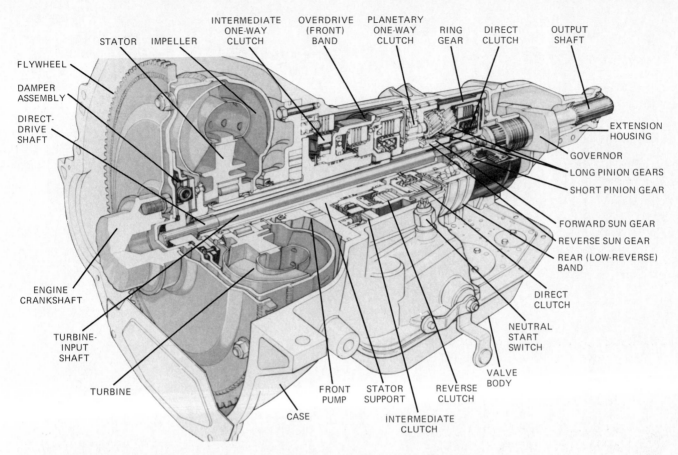

Fig. 7-1 Cutaway view of the automatic overdrive transmission. *(Ford Motor Company)*

1. The reverse sun gear is integral with the input shell. It is driven by the turbine (tube) input shaft under some conditions. Under other conditions, the input shell is held stationary.
2. The planet carrier carries both the long pinions (3) and the short pinions (4). The carrier can be held stationary under certain conditions.
3. The long pinions in the planet carrier are always in mesh with the ring gear (6), the reverse sun gear (1), and the short pinions (4).
4. The short pinions are in constant mesh with the forward sun gear (5) and with the long pinions (3).
5. The forward sun gear (5) is driven whenever the forward clutch is engaged. This clutch is always engaged in first, second, and third. It is actually located behind the reverse sun gear but is called the forward sun gear because it is in action in the forward speeds (except for overdrive). The forward sun gear meshes with the short planet pinions (4).
6. The ring gear meshes with the long planet pinions (3). It is splined to a flange on the output shaft.
7 to 14. These are the holding members that hold different components of the planetary gearset to produce the four forward speeds and reverse. Note that there are six clutches (two of which are one-way clutches) and two bands. Their operation is described later when transmission actions during different operating conditions are discussed.

☀ 7-6 Input and output shafts Figure 7-5 shows the shafts and driving members of the transmission. There are two inputs to the gearset, the direct-drive input shaft (1) and the turbine (tube) input shaft (3).

1. The direct-drive input shaft (1) is splined to the damper assembly in the converter cover (Fig. 7-3) at one end. At the other end, it is splined into the direct-clutch cylinder (2). The shaft and direct-clutch cylinder turn at engine speed.
2. The direct-clutch cylinder (2) is supported by a bushing (B13) on the output shaft (5). It can turn independently of the output shaft.
3. The turbine (tube) input shaft (3) is splined into the

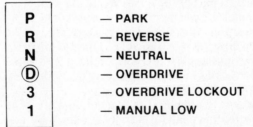

Fig. 7-2 Automatic overdrive shift-selector positions. *(Ford Motor Company)*

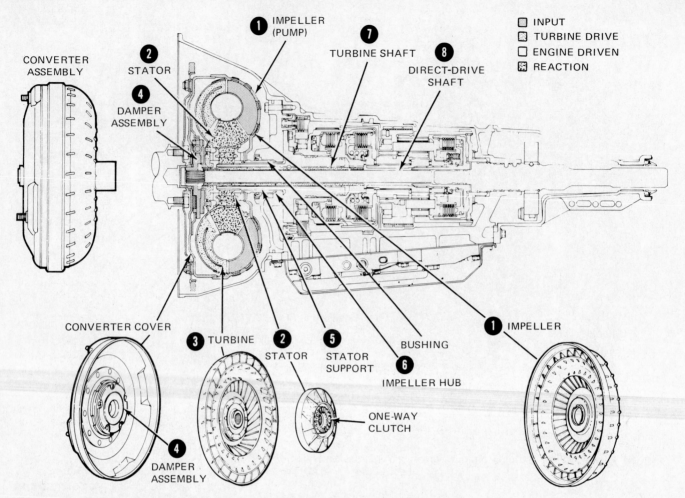

CONVERTER ASSEMBLY

❶ IMPELLER (PUMP)
❷ STATOR
❹ DAMPER ASSEMBLY
❼ TURBINE SHAFT
❽ DIRECT-DRIVE SHAFT

◻ INPUT
◻ TURBINE DRIVE
◻ ENGINE DRIVEN
▨ REACTION

CONVERTER COVER
❸ TURBINE
❷ STATOR
❺ STATOR SUPPORT
BUSHING
❻
IMPELLER HUB
ONE-WAY CLUTCH
❹ DAMPER ASSEMBLY
❶ IMPELLER

Fig. 7-3 Sectional view of the automatic overdrive transmission with the torque converter and its separate subassemblies shown separately. *(Ford Motor Company)*

turbine hub at one end and the forward-clutch hub at the other end. This shaft drives the forward-clutch cylinder (4) and the reverse-clutch hub at turbine (converter output) speed. It is supported by bushing B5 in the stator support and by bushing B1 on the direct-drive shaft (1).

4. The forward-clutch cylinder (4) turns on the rear end of the stator support and is supported by the turbine (tube) input shaft (3). It is splined to take two sets of clutch plates. The forward-clutch plates go inside the forward-clutch cylinder. The reverse-clutch plates fit around the outside of the reverse-clutch hub, which is part of the forward-clutch cylinder.

5. The output shaft is supported by bushing B14 in the case and by the slip yoke and bushing B15 in the extension housing. The slip yoke is part of the drive shaft front universal joint. The output shaft has a series of seal rings near its front to prevent oil leakage into the extension housing.

6. The ring gear is supported by the output shaft. The hub on the output shaft (5) is splined to the shaft and also has a flange which is splined to the ring gear. The teeth on the outside of the ring gear are for engagement with the parking pawl in P, or park.

❂ **7-7 Forward clutch** The forward clutch (11 in Fig. 7-4) is engaged when the transmission shifts into first gear in either 3 or Ⓓ (overdrive). Figure 7-6 shows the location of the forward clutch (1) and the clutch disassembled. The clutch has a piston and a series of steel and friction plates. The internally splined friction plates are splined to a driving hub, which is splined onto the front of the forward sun gear (2). The externally splined steel plates are splined to the forward-clutch cylinder.

When hydraulic pressure is directed back of the piston, it forces the plates together and the clutch engages. The turbine (tube) input shaft then drives the sun gear at turbine speed.

The center support (3) is splined to the case. It has a hub protruding to the rear that is in the inner race of the planetary one-way clutch. It also supports the planet-pinion carrier.

The one-way clutch (4) prevents the planet carrier from turning backward, or counterclockwise. If the carrier turns clockwise, the clutch overruns, or freewheels.

The planetary gearset (5) has the planet carrier supported by bushing B7 at the center support, and by B12 at the rear.

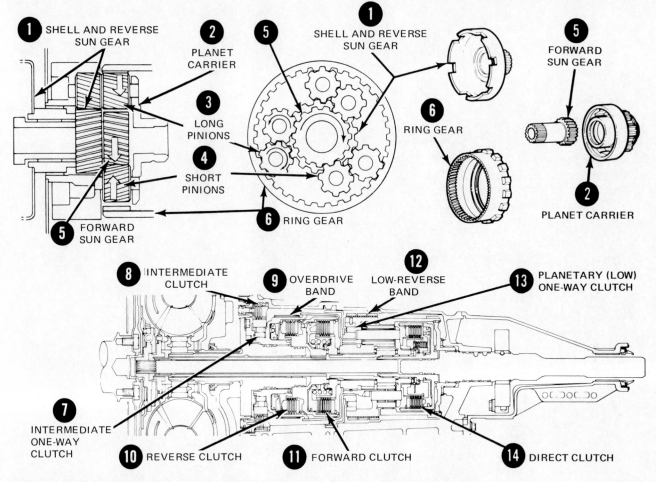

Fig. 7-4 Top, planetary gearset and component parts. Bottom, active members in the automatic overdrive transmission. *(Ford Motor Company)*

✿ 7-8 Power flow in first in Ⓓ or 3 range The power flow is shown in Figs. 7-7 and 7-8 in first in either overdrive or third range. The holding members are:

1. The forward clutch (1), which locks the turbine shaft to the forward sun gear (6)
2. The planetary one-way clutch (2), which holds the planetary carrier from turning backward

This is the power flow: The engine drives the impeller and turbine (4) clockwise, hydraulically. The turbine shaft (tube) drives the forward clutch (5) clockwise. The clutch (1) drives the forward sun gear (6) clockwise.

The sun gear drives the short pinions (7) counterclockwise. The short pinions drive the long pinions (8) clockwise. The long pinions drive the ring gear (9) clockwise at a reduced speed. The gear ratio in first is 2.4:1. In full reduction, the torque multiplication ratio is about 1:5. Torque is multiplied about 5 times through the torque converter and the planetary gearset.

✿ 7-9 Manual low gear and low-reverse band In first gear with the selector lever in Ⓓ or 3, there is no engine-braking effect on a coast-down. In a coast-down, the one-way clutch overruns, so the car wheels are, in effect, disconnected from the engine. To provide engine braking, range 1 can be used. This applies the low-reverse band, which is actuated by the low-reverse servo (Fig. 7-9).

The band (1) is installed around the outer drum surface of the planet carrier. It is operated by the servo (2, 3, 5, and 7). When the selector lever is moved to 1 or to R fluid pressure is sent to the piston. It moves up, tightening the band around the planet carrier. This locks the planet carrier stationary.

The power flow is the same as shown in Fig. 7-9 except that the band prevents the carrier from turning (instead of the one-way clutch). Therefore engine braking can take place.

✿ 7-10 Intermediate clutches and reverse sun gear The parts that are necessary for second gear are added in Fig. 7-10. These include:

1. The intermediate clutch (1), which is installed with its piston in the rear of the oil-pump housing and the plates in the case. The steel plates and pressure plate

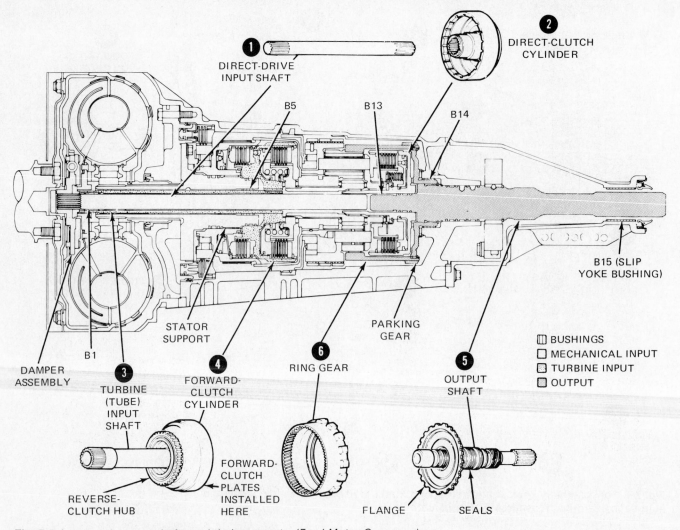

Fig. 7-5 Input and output shafts and their supports. *(Ford Motor Company)*

are splined to the case. The friction plates are splined to the outer race of the one-way clutch (2). When the intermediate clutch is engaged, it holds the one-way clutch outer race stationary by locking it to the case.

2. The one-way clutch (2) is the caged-roller type. Its inner race is locked onto the reverse-clutch drum (3). If the reverse-clutch drum tries to turn counterclockwise, the one-way clutch locks up and holds it stationary. This happens provided the intermediate clutch is engaged.

   If the reverse-clutch drum turns clockwise, the one-way clutch overruns. This allows for freewheeling coast-down in second gear (in Ⓓ or 3 range). It also allows the intermediate clutch to remain engaged in third and fourth gears.

3. The shell and reverse (front) sun gear (4) are, in effect, one part. Lugs on the reverse-clutch drum mate with lugs on the shell. Therefore, if the reverse-clutch drum is held stationary, the reverse sun gear is also stationary. If the reverse-clutch drum is driven, it drives the sun gear.

The reverse-clutch drum is supported on the stator support by bushings B3 and B4. The reverse sun gear is supported on the carrier by bushing B8.

⚙ 7-11 Second gear in Ⓓ or 3 range Figure 7-11 shows the members that are holding in second gear. These include:

1. The intermediate clutch (1), which locks the one-way clutch outer race stationary
2. The one-way clutch, which prevents the reverse-clutch drum and reverse sun gear from turning backward
3. The forward clutch, which locks the turbine shaft to the forward sun gear

The power flow and rotation of parts are shown in Figs. 7-11 and 7-12. The power flow is as follows:

1. The engine drives the converter impeller (4) clockwise.
2. The turbine (5) is driven clockwise hydraulically.
3. The turbine shaft (tube) drives the forward clutch (6) clockwise.

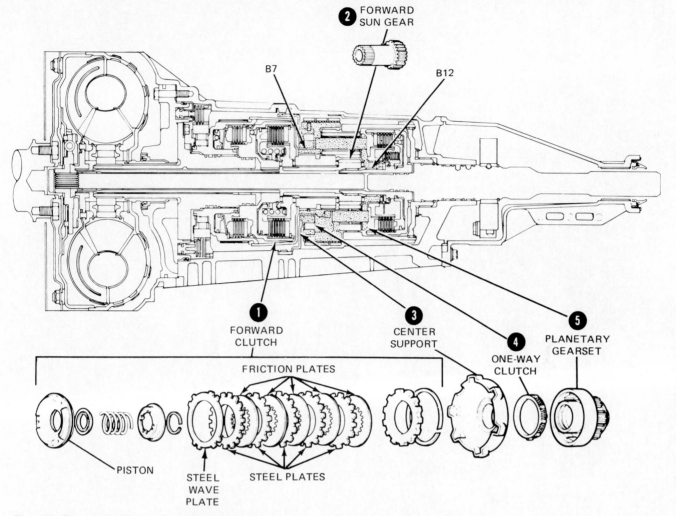

Fig. 7-6 Forward clutch, sun gear, planetary gearset, and one-way clutch. *(Ford Motor Company)*

4. The forward sun gear (7) drives the short pinions in the planetary gearset (8) counterclockwise.
5. The short pinions drive the long pinions (9) clockwise and they walk clockwise around the stationary reverse sun gear.
6. The long pinions drive the ring gear (10) and output shaft clockwise at a reduced speed.

Second gear has a 1.47:1 gear reduction and a torque multiplication of 1:1.47 (not including torque converter action).

❀ 7-12 Reverse-clutch assembly Figure 7-13 shows the members involved in reverse. The low-reverse servo, which applies the low-reverse band (Fig. 7-9), was covered earlier. The drum (1) drives the reverse sun gear. The hub for the reverse clutch is integral with the forward-clutch cylinder, and it is permanently joined to the turbine input shaft (2).

The reverse-clutch piston (3) is applied by hydraulic pressure through a passage in the stator support. It moves rearward against a Belleville (wavy) spring (4).

This forces the clutch plates (5) together and against the pressure plate (6). This is the only Belleville spring in the transmission clutches. It operates as a lever through a pressure plate (7). The other clutches are directly engaged by their pistons and disengaged by coil springs.

When the reverse clutch is engaged, the turbine input shaft (2) is coupled with the clutch drum (1). The drum, being locked to the shell-and-sun-gear assembly (8), causes the reverse sun gear to be driven at turbine speed.

❀ 7-13 Reverse gear Figure 7-14 shows the members involved in producing reverse. Figure 7-15 shows the power flow in reverse. The holding members are:

1. The reverse clutch (1) is engaged, locking the turbine input shaft to the reverse sun gear.
2. The low-reverse band is applied, holding the planet carrier stationary.

The power flow in reverse is shown in Figs. 7-14 and

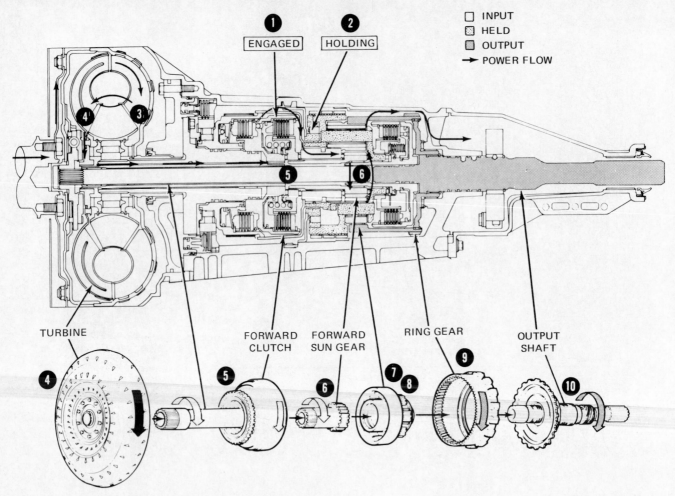

Fig. 7-7 Power flow in first gear. *(Ford Motor Company)*

7-15. The engine drives the converter impeller (3) and turbine (4) clockwise.

1. The turbine shaft drives the reverse-clutch (5) shell and reverse sun gear (5) clockwise.
2. The reverse sun gear drives the long planet pinions (6) counterclockwise.
3. The carrier is stationary, so the long planet pinions drive the ring gear (7) and output shaft (7) counterclockwise.

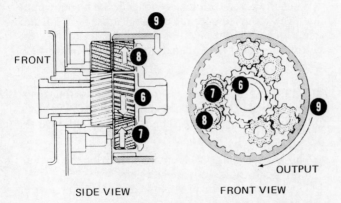

Fig. 7-8 Gear rotation in planetary gearset in first gear. *(Ford Motor Company)*

The short planet pinions and forward sun gear also turn but are not involved in the power flow. The long planet pinions become, in effect, idler gears. Gear reduction is 2:1, with overall torque multiplication as high as 1:4. This is because the power flow is through the torque converter which further multiplies torque.

✿ 7-14 Direct clutch and mechanical input Figure 7-16 shows the elements in the direct clutch which are involved in third (direct) and fourth (overdrive) gears. The direct clutch cylinder or drum (1) is driven mechanically through the direct-drive input shaft rather than by the turbine input shaft. It therefore bypasses the torque converter.

The direct-clutch cylinder (1) is supported by bushing B13 and is splined to the direct-drive input shaft. It always turns at engine speed. But it does not transmit power until the clutch is engaged.

Power to the gearset is through the direct-clutch hub (2), which is splined to the planet carrier (3). Friction plates (4) are splined on the exterior of the hub. Steel plates (5), including a pressure plate, are splined into the direct-clutch cylinder (1). The clutch piston (6) is applied by hydraulic pressure and released by coil springs when the pressure is exhausted.

When the direct clutch is engaged, the direct-drive input shaft will drive the planet carrier clockwise at

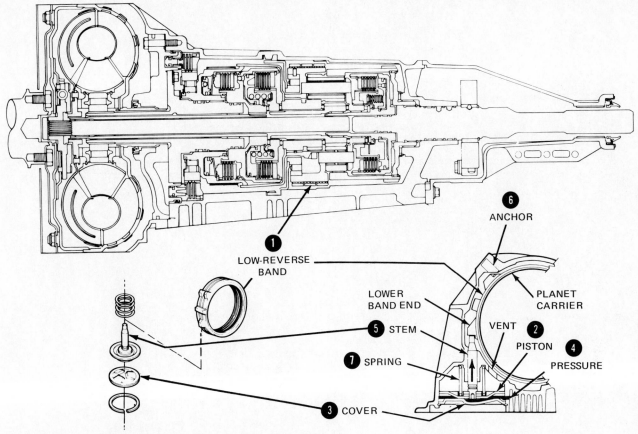

Fig. 7-9 Low-reverse band and servo. *(Ford Motor Company)*

LOW-REVERSE
BAND

6 ANCHOR

PLANET
CARRIER

LOWER
BAND END

5 STEM

VENT

2

4

PISTON

PRESSURE

7 SPRING

3 COVER

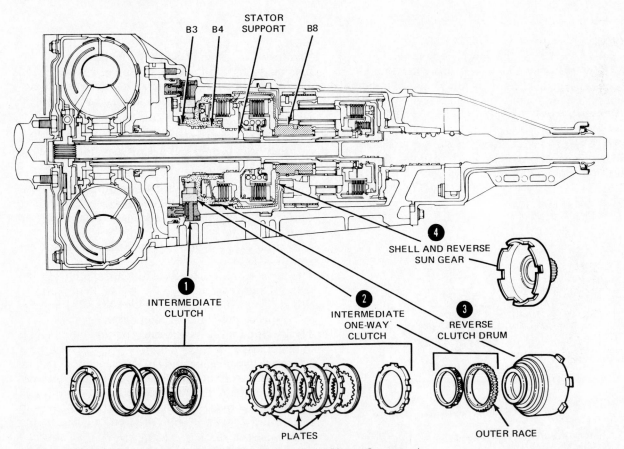

B3 B4 STATOR
SUPPORT B8

SHELL AND REVERSE
SUN GEAR

4

1 INTERMEDIATE
CLUTCH

2 INTERMEDIATE
ONE-WAY
CLUTCH

3 REVERSE
CLUTCH DRUM

PLATES

OUTER RACE

Fig. 7-10 Reverse sun gear and holding members. *(Ford Motor Company)*

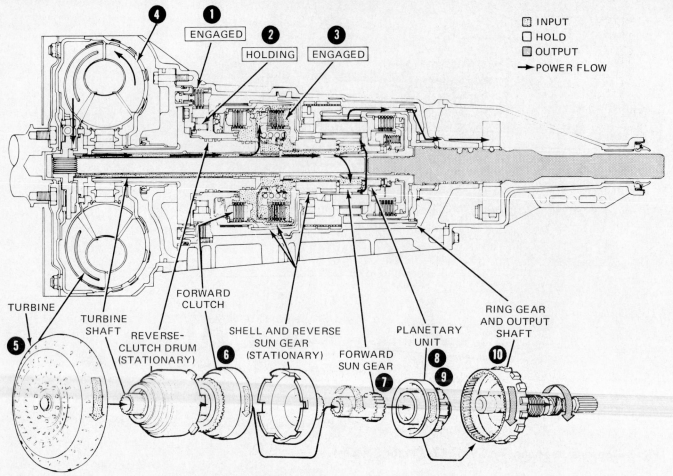

Fig. 7-11 Power flow in second gear. *(Ford Motor Company)*

engine speed, bypassing the torque converter. The planetary (low) one-way clutch overruns when the direct clutch is engaged. The direct clutch handles about 60 percent of the total input torque in direct drive. In overdrive (fourth gear), it handles all the torque.

✿ 7-15 Third gear (direct drive) Figure 7-17 shows the holding members involved in third gear, direct drive. The holding members are:

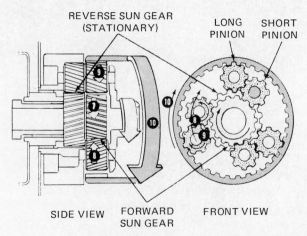

Fig. 7-12 Gear rotation in planetary gearset in second gear. *(Ford Motor Company)*

1. The forward clutch (1 and 6), which locks the turbine input shaft to the forward sun gear
2. The direct clutch (2 and 4), which locks the direct-drive input shaft to the planet-carrier assembly (7)

The power flow is shown in Fig. 7-18. When any two members of a planetary gearset are locked together, the system is locked up and it turns as a unit. Note the following (Fig. 7-17):

1. The converter cover (3) drives the direct-drive shaft and the direct clutch at engine speed.
2. The direct clutch drives the planet carrier at engine speed.
3. The turbine (5) is driven hydraulically at almost engine speed.
4. The turbine shaft drives the forward clutch and the forward sun gear.
5. When two members of the gearset are driven at the same speed, they in effect lock up the gearset.
6. The long pinions drive the ring gear and output shaft in direct drive.
7. There are two torque inputs to the transmission. One is through the direct-drive shaft. The other is through the turbine shaft. The turbine shaft is driven through the torque converter and some slippage occurs in the torque converter. For this reason, less of the torque passes into the turbine shaft than through the direct-drive shaft, where no slippage occurs. The

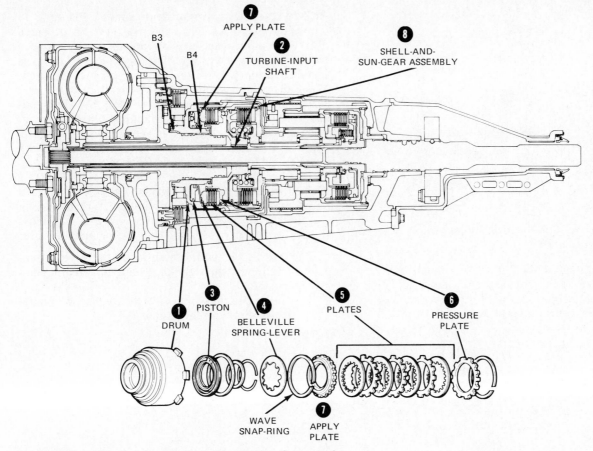

Fig. 7-13 Reverse-clutch assembly. *(Ford Motor Company)*

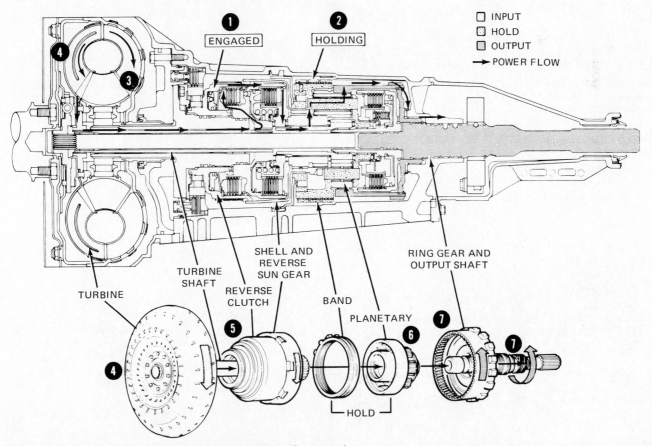

Fig. 7-14 Power flow in reverse gear. *(Ford Motor Company)*

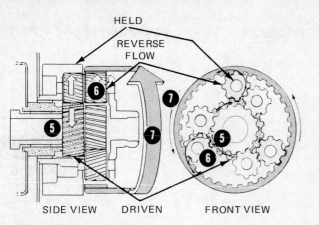

Fig. 7-15 Gear rotation in planetary gearset in reverse. *(Ford Motor Company)*

actual split is about 40-60. About 60 percent of the torque is delivered by the direct-drive shaft, and 40 percent by the turbine shaft. If there were no slippage in the converter, the split would be 50-50.

✿ 7-16 Overdrive band and servo Figure 7-19 shows the overdrive servo and overdrive band, which act, with other transmission members, to produce fourth gear, or overdrive. The overdrive band (1) is installed around the drum (2) of the reverse clutch. When the band is applied, it holds the drum stationary. When this drum is stationary, it holds the shell-and-reverse-sun-gear assembly stationary. The band is applied in overdrive, and also in second gear in range 1. The overdrive servo is constructed and operates the same as the low-reverse servo (Figs. 7-9 and 7-10).

✿ 7-17 Fourth gear ⓓ (overdrive) Figure 7-20 shows the holding members involved in achieving fourth gear, or overdrive. Figures 7-20 and 7-21 show the power flow in overdrive. The holding members are:

1. The overdrive band, which locks the reverse sun gear stationary through the reverse-clutch drum and driving shell
2. The direct clutch, which directly couples the planet carrier assembly to the engine

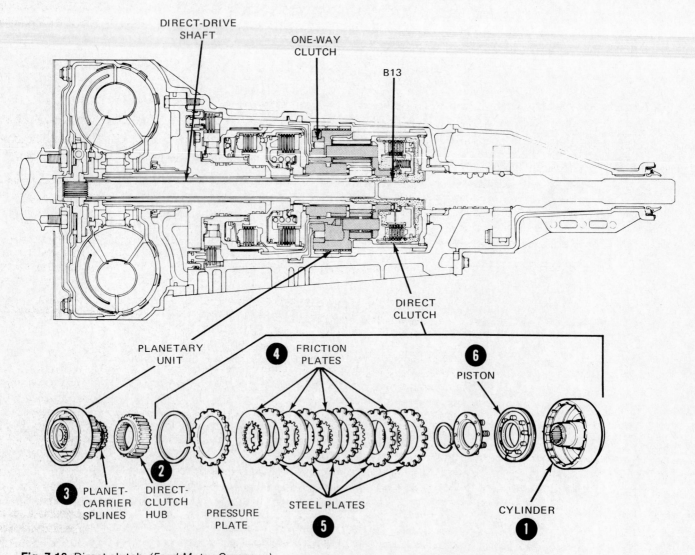

Fig. 7-16 Direct clutch. *(Ford Motor Company)*

90

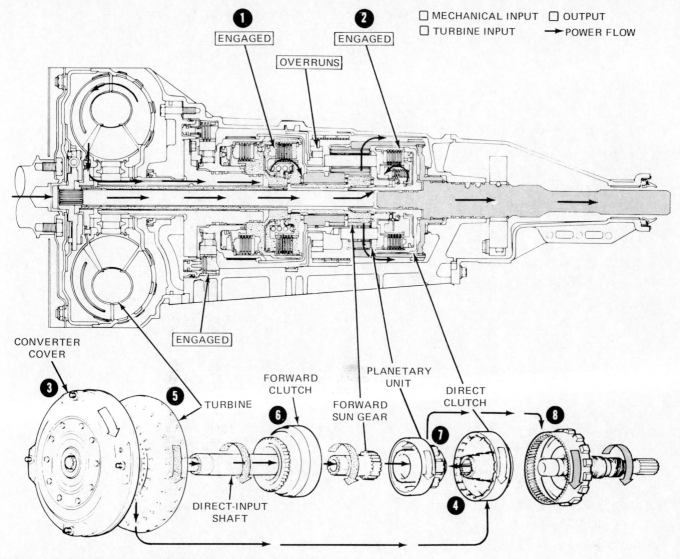

Fig. 7-17 Third gear or direct-drive power flow. *(Ford Motor Company)*

The power flow is as follows:

1. Overdrive is accomplished by holding a sun gear and driving the pinion carrier (Figs. 7-20 and 7-21).
2. The converter cover drives the direct-drive shaft clockwise at engine speed.
3. The direct-drive shaft drives the direct clutch.
4. The direct clutch drives the planet-carrier assembly clockwise at engine speed.
5. The long pinions (6) walk around the stationary reverse sun gear, in a clockwise direction.
6. The ring gear and output shaft are driven at a faster speed by the rotation of the long planet pinions.

In overdrive, the gear ratio is 0.667:1. The input is completely mechanical so there is no slip in the torque converter. The torque converter is completely bypassed. The converter cover has a damper assembly (Fig. 7-3) which smooths and cushions engine shock. This assembly serves the same purpose as the damper springs in the friction disk of clutches used with manual transmissions.

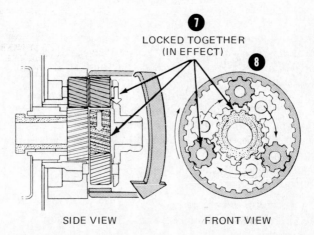

Fig. 7-18 Gear rotation in planetary gearset in direct drive. *(Ford Motor Company)*

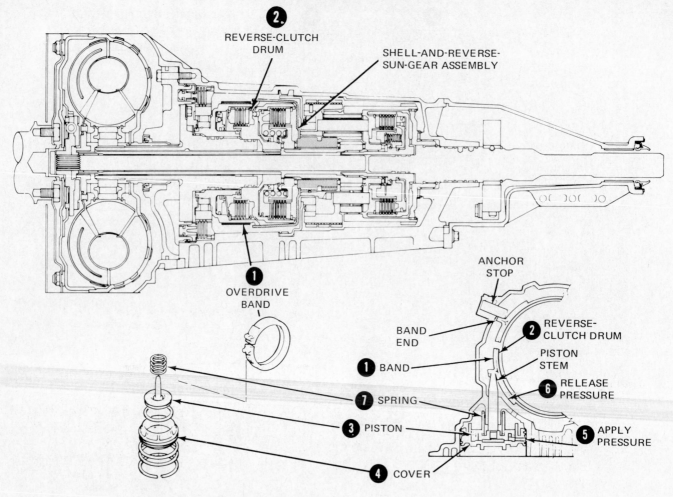

Fig. 7-19 Overdrive band and servo. *(Ford Motor Company)*

✿ 7-18 Summary of clutch and band operation
Figure 7-22 summarizes the operation of the clutches and bands in the various gear positions. This chart can be helpful in trouble-diagnosing a transmission. For example, the direct clutch is engaged in third and overdrive. If the transmission slips in both these gears, the cause could be slipping of the direct-clutch.

✿ 7-19 Hydraulic system Figure 7-23 shows the hydraulic system for the Ford automatic overdrive transmission. Figure 7-24 is an exploded view of the valve body. The parts are numbered the same in the two illustrations. For example, the manual valve is 27 in both. Refer to Figs. 7-23 and 7-24 as each numbered item in the illustrations is listed and described. To help you locate valves in Fig. 7-23, they are in the same positions they occupy in the valve body as shown in Fig. 7-24.

The main components in the automatic overdrive transmission are:

1. The sump, where reserve fluid is kept.
2. The screen, which keeps dirt from entering the system.
3. The oil pump, which pumps fluid through the system.
4. The oil-pressure-booster valve, which changes fluid pressure as throttle opening and drive range or gear ratio changes.
5. The main regulator valve, which regulates main (line) pressure.
6. The converter relief valve, which prevents excessive pressure in the converter.
7. The torque converter.
8. The check valve, which prevents fluid in the converter from draining back when the engine is turned off.
9. The cooler, which removes excess heat from the fluid.
10. The 3-4 accumulator, which cushions the 3-4 upshift.
11 and 12. 1-2 capacity modulator valve and the 1-2 accumulator valve, which operate together to smooth the 1-2 upshift.
13. The 3-4 shuttle valve, which controls the operation of the overdrive servo-regulator valve and the 3-4 accumulator.
14. The overdrive servo-regulator valve, which determines the apply pressure for the overdrive servo on the 4-3 downshift.
15. The 1-2 shift valve, which controls the automatic 1-2 upshift and 2-1 downshift.

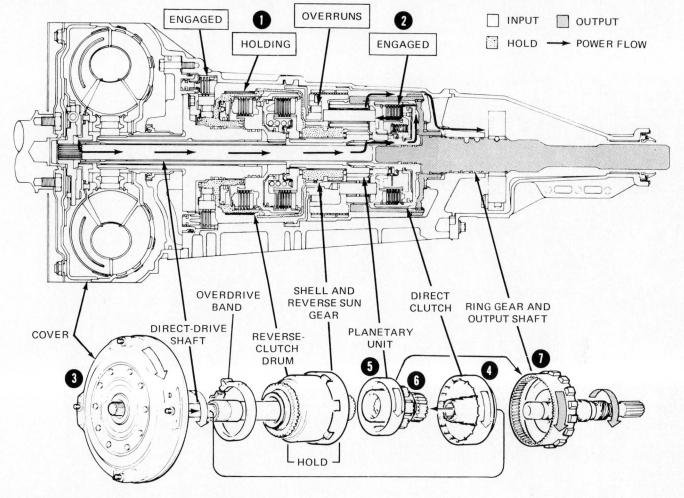

Fig. 7-20 Power flow in overdrive. *(Ford Motor Company)*

16. The throttle-valve (TV) limit valve, which limits throttle pressure acting on the pressure booster valve (4) and on the 2-3 shift valve (28) by regulating TV pressure at wide-open throttle (WOT).
17. Eight ball check valves in the valve body that are used as shuttle valves or one-way valves.
18. The 3-4 shift valve, which controls automatic 3-4 upshift and 4-3 downshift.
19. The 3-4 TV modulator valve, which modulates throttle-valve (TV) pressure acting on the 3-4 shift valve.
20. The orifice-control valve, which prevents the transmission from shifting out of second gear directly into overdrive. It also works with the 2-3 back-out valve (24).
21. The 2-3 capacity modulator valve, which controls back pressure on the 2-3 accumulator to smooth the 2-3 upshift.
22. The 2-3 accumulator, which cushions direct-clutch engagement for a smooth 2-3 upshift and also cushions the forward clutch for a smooth initial engagement.
23. The governor, which supplies a road-speed signal to the hydraulic system.
24. The 2-3 back-out valve, which controls the feed

rate of the direct clutch. It also operates with the orifice-control valve (20) to control the feed rate of the forward clutch.
25. The throttle plunger, which varies spring force on the throttle valve with throttle opening. It also op-

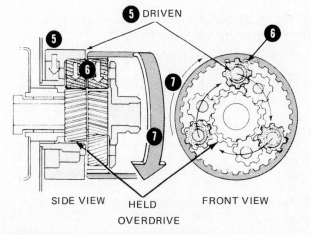

Fig. 7-21 Gear rotation in planetary gearset in overdrive. *(Ford Motor Company)*

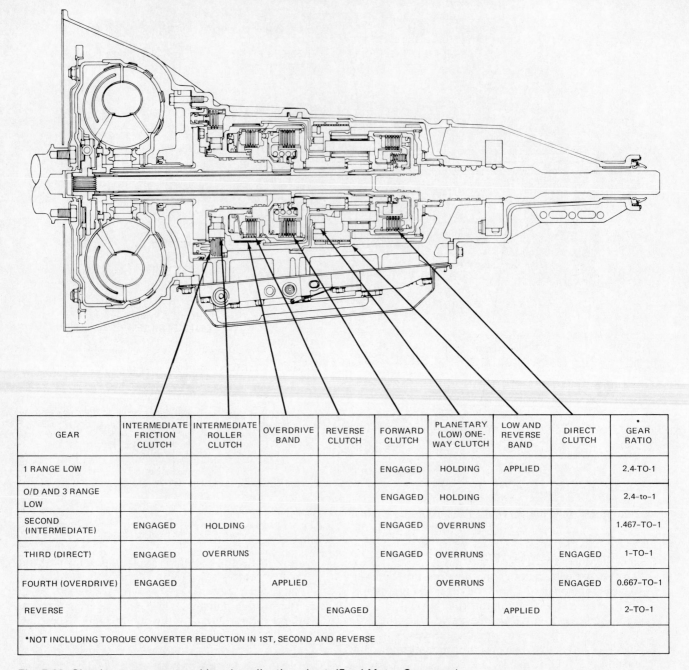

GEAR	INTERMEDIATE FRICTION CLUTCH	INTERMEDIATE ROLLER CLUTCH	OVERDRIVE BAND	REVERSE CLUTCH	FORWARD CLUTCH	PLANETARY (LOW) ONE-WAY CLUTCH	LOW AND REVERSE BAND	DIRECT CLUTCH	*GEAR RATIO
1 RANGE LOW					ENGAGED	HOLDING	APPLIED		2.4-TO-1
O/D AND 3 RANGE LOW					ENGAGED	HOLDING			2.4-to-1
SECOND (INTERMEDIATE)	ENGAGED	HOLDING			ENGAGED	OVERRUNS			1.467-TO-1
THIRD (DIRECT)	ENGAGED	OVERRUNS			ENGAGED	OVERRUNS		ENGAGED	1-TO-1
FOURTH (OVERDRIVE)	ENGAGED		APPLIED			OVERRUNS		ENGAGED	0.667-TO-1
REVERSE				ENGAGED			APPLIED		2-TO-1

*NOT INCLUDING TORQUE CONVERTER REDUCTION IN 1ST, SECOND AND REVERSE

Fig. 7-22 Clutch-engagement and band-application chart. *(Ford Motor Company)*

erates the kickdown or demand downshift system at wide-open throttle.

26. The throttle control valve, which regulates throttle pressure as an engine-load signal to the hydraulic system.

27. The manual valve, which moves with the shift selector and directs control pressure to various passages to engage clutches and apply servos, and to provide automatic functions of the hydraulic system.

28. The 2-3 shift valve, which controls automatic 2-3 upshifts and 3-2 downshifts.

29. The 2-3 TV modulator valve, which modulates throttle-valve (TV) pressure acting on the 2-3 shift valve and the 1-2 shift valve.

30. The 3-4 back-out valve, which smooths the 3-4 upshift if it occurs at closed throttle.

31. The 2-1 scheduling valve, which determines the 2-1 downshift speed when the shift selector is moved to manual low from direct or overdrive range.

32. The low servo-modulator valve, which regulates low-and-reverse servo apply pressure for a smooth 2-1 downshift in manual low range.

33. The TV relief valve, which prevents excess TV limit pressure if the TV limit valve (16) fails to regulate properly.

The following sections describe the conditions in the hydraulic system with the selector in different gear positions and with different gears. The purpose of var-

11. OD-3-1 pressure works the 1-2 accumulator in readiness for the 1-2 upshift.
12. OD-3-1 pressure through the 3-4 shift valve also holds the 3-4 shuttle valve forward.
13 and 14. OD-3-1 line pressure also works on the 1-2 shift valve to oppose governor pressure. Later, when car speed and governor pressure increase, the 1-2 shift valve will move, causing a 1-2 upshift.
15. Line pressure to the 2-3 valve opposes governor pressure.
16. Line pressure holds the 3-4 shift valve out and inactive.
17. Line pressure is applied to the release side of the overdrive servo.

❂ 7-22 Second gear in Ⓓ range

Figures 7-11 and 7-12 show the holding members and power flow in second gear in the Ⓓ (overdrive) range. The vehicle has started to move and increases enough in speed that governor pressure forces an upshift.

1 and 2. Same as in ❂ 7-21.
3. Line boost continues to cause an increase in line pressure. Adding to the increase is throttle-valve (TV) pressure increase due to opening of the throttle valve.
4. The throttle plunger has moved in and the throttle valve is producing a regulated TV pressure.
5. Throttle-valve pressure moves the 2-3 back-out valve out.
6. Throttle-valve pressure flows through the TV limit valve. When the pressure reaches about 85 psi [586 kPa], the limit valve will work to prevent further TV pressure increase.
7. With low TV pressure, the TV modulator valves are not active.
8. Throttle-valve limit pressure moves the 3-4 back-out valve out.
9. The OD-3-1 passage is charged and line pressure is available to the governor and forward clutch.
10. The governor is sending a "road-speed" pressure signal to all valves.
11. Because of TV pressure on the 2-3 back-out valve, forward-clutch apply pressure takes a different route through the valve. Orifice A is bypassed.
12. Clutch apply pressure continues to hold the 2-3 accumulator down and the 3-4 shuttle valve in.
13. OD-3-1 pressure remains effective on the 1-2 accumulator.
14. Governor pressure has moved the 1-2 shift valve in.
15. The OD-3-1 passage is now open to the overdrive servo-regulator valve.
16. OD-3-1 pressure is transmitted through the 1-2 capacity modulator to the INT CL passage. The intermediate clutch engages.
17. The apply pressure, through orifice F, passes into the 1-2 accumulator. It cushions the clutch engagement.
18. Line pressure is applied to the overdrive servo-release side through the OD-3 and OD REL passages.
19. The OD servo-regulator valve produces a regulated

pressure to the apply side of the servo. Pressure (18) keeps the servo released.

❂ 7-23 Third gear in Ⓓ range

Figures 7-17 and 7-18 show the holding members and power flow in third gear. Plate 18 shows the conditions in the hydraulic system. The vehicle has increased enough in speed that the governor pressure has forced an upshift to third gear.

1. High-volume output from the pump is being controlled by the regulator valve.
2. The 3-4 accumulator is up, preparatory to the 3-4 upshift.
3. Throttle-valve limit pressure is increasing line pressure. LINE BOOST is exhausted.
4. The OD-3 and OD-3-1 passages are charged at the manual valve with line pressure.
5. The throttle valve is producing a high TV pressure. This pressure holds the 2-3 back-out valve out.
6. The 3-4 TV modulator valve is regulating modulated TV pressure to the 3-4 shift valve.
7. The TV limit valve is regulating TV limit pressure to 85 psi [586 kPa]. This pressure moves the 3-4 back-out valve out and causes the 2-3 TV modulator valve to regulate.
8, 9, and 10. These are the same as in second gear. Increased governor pressure causes the 2-3 shift valve to move inward. The 2-3 upshift takes place.
11. The overdrive servo release is now pressurized from the forward-clutch apply system.
12. The OD-3 passage pressurizes the direct-clutch (DIR CL) passage.
13. Direct-clutch apply pressure routes through orifice K.
14 and 15. The direct clutch engages. The direct-clutch engagement is cushioned by the 2-3 accumulator.

❂ 7-24 Fourth gear (Ⓓ)

Figures 7-20 and 7-21 show the holding members and power flow in overdrive. Plate 19 shows the conditions in the hydraulic system. The vehicle has increased enough in speed and the throttle is fairly light so that governor pressure has forced the upshift to overdrive gear. To get into overdrive, the overdrive servo must be applied and the forward clutch disengaged. Here is how this happens.

1, 2, and 3. Same as in third (❂ 7-23).
4. The throttle valve is regulating a moderate TV pressure. The 2-3 and 3-4 back-out valves are forced out by TV and TV limit pressure.
5. The governor is regulating at high pressure due to high road speed.
6. Governor pressure forces the 3-4 shift valve in.
7. The pressure apply for the forward clutch is now blocked at the 3-4 shift valve.
8. The 3-4 shift valve opens the forward-clutch apply system to an exhaust passage.
9. The forward clutch disengages along with the top of the 2-3 accumulator.
10. The overdrive servo release is exhausted through

the 2-3 shift valve, 3-4 back-out valve, and 2-3 back-out valve.

11. Pressure at the end of the 3-4 shuttle valve is exhausted so the valve moves out.

12. OD-3-1 pressure is now effective on the overdrive servo-regulator valve. It moves the valve in so that it stops regulating and sends out line pressure.

13. Full line pressure is applied to the servo apply side through orifice D.

14. The 3-4 accumulator cushions the band application and upshift.

15. The direct clutch remains engaged from the 2-3 shift valve through the DIR CL passages.

16. The intermediate clutch also remains engaged but is inactive.

NOTE: In range 3 (overdrive lockout) the overdrive lockout is through the P-3-1-R passage at the manual valve and through the shuttle valve (4) to the P-3-1-R/KD passage. With the orifice-control valve in, this pressure continues to the bore between the 3-4 shift valve and the 3-4 TV modulator valve. Line pressure is then effective on the spring end of the 3-4 shift valve to prevent an upshift to overdrive. It also forces the 3-4 TV modulator valve to shift to its normal (in) position. Otherwise, the hydraulic system is operating essentially the same as shown in Plate 19.

✿ **7-25 Reverse** Figures 7-14 and 7-15 show the holding members and power flow in reverse. Plate 20 shows the conditions in the hydraulic system. The low-reverse band and the reverse clutch are on. Full line pressure is applied to the reverse servo.

1. The main control-pressure system is regulating.

2. The 3-4 accumulator is forced up by line pressure.

3. Line pressure is increased by LINE BOOST and REV boost pressures acting on the booster valve.

4. At the manual valve, the REV, P-1-R, and P-3-1-R passages are charged by line pressure.

5 and 6. When the throttle is opened to move the car, the TV passage will be charged. Throttle-valve limit pressure will further increase line pressure by acting on the oil-pressure booster valve. Maximum line pressure results in reverse because three pressures act on the booster valve.

7. The low servo-modulator valve is forced outward by line pressure. This opens the REV passage to the REV servo.

8. The reverse servo applies with full line pressure on it.

9. The reverse clutch is engaged with full line pressure from the REV passage. Orifice I delays the clutch engagement slightly so that the band will always apply first. This smooths engagement.

10, 11, and 12. Other passages are charged but have no effect on reverse operation. The governor is inactive in reverse.

Chapter 7 review questions

Select the *one* correct, best, or most probable answer to each question. Then check your answers against the correct answers given at the end of the book.

1. The Ford overdrive automatic transmission has:
   a. three pinions
   b. four pinions
   c. five pinions
   d. six pinions

2. The Ford overdrive automatic transmission, counting the pinions, has:
   a. nine gears
   b. eight gears
   c. seven gears
   d. six gears

3. The Ford overdrive automatic transmission has:
   a. three clutches
   b. four clutches
   c. five clutches
   d. six clutches

4. The symbol Ⓓ means:
   a. downshift
   b. overdrive
   c. third gear
   d. none of the above

5. Mechanic A says the short pinions and the long pinions are carried on separate pinion carriers. Mechanic B says the short and long pinions are carried on a single pinion carrier. Who is right?
   a. A only
   b. B only
   c. both A and B
   d. neither A nor B.

6. The two input shafts to the planetary gearset are the:
   a. direct-drive shaft and indirect-drive shaft
   b. sun-gear shaft and planet-carrier shaft
   c. turbine shaft and input shaft
   d. turbine shaft and direct-drive shaft

7. In first gear in Ⓓ or 3, the holding members are the:
   a. forward clutch and reverse clutch
   b. forward clutch and planetary one-way clutch
   c. forward clutch and forward band
   d. first-gear clutch and band

8. In manual low gear (1), the carrier is prevented from turning by use of the:
   a. forward band
   b. low-and-reverse band
   c. first clutch
   d. low clutch

9. In second gear in ⒟ or 3, the holding members are the:
   a. intermediate, forward, and one-way clutches
   b. intermediate clutch and forward band
   c. number 2 band and forward clutch
   d. forward clutch and number 2 clutch

10. In reverse, the holding members are the:
    a. reverse and intermediate clutches
    b. reverse, intermediate, and forward clutches
    c. reverse clutch and low-reverse band
    d. reverse clutch and the two bands

11. In third gear, the holding members are the:
    a. forward, intermediate, and direct clutches
    b. forward clutch and forward band
    c. forward band and intermediate clutch
    d. forward clutch and direct clutch

12. In ⒟, the holding members are the:
    a. overdrive band and overdrive clutch
    b. drive clutch and band
    c. direct clutch and overdrive band
    d. direct clutch and overdrive clutch

13. The purpose of the governor is to:
    a. prevent excessive engine speed
    b. supply a road-speed signal to the hydraulic system
    c. control the line pressure
    d. back up the manual-valve action

14. The purpose of the accumulators is to:
    a. accumulate enough pressure to assure a quick shift
    b. absorb excessive pressures
    c. cushion shifts
    d. firm up band applications

15. *TV pressure* stands for:
    a. pressure to watch television
    b. throttle-valve pressure
    c. top-value pressure
    d. none of the above

16. The manual valve has:
    a. four positions
    b. five positions
    c. six positions
    d. seven positions

17. Mechanic A says that in first gear in ⒟, line pressure is applied to the release side of the overdrive servo. Mechanic B says line pressure is applied to the forward clutch. Who is right?
    a. A only
    b. B only
    c. both A and B
    d. neither A nor B

18. Mechanic A says that upshifts are produced by the difference between TV and governor pressures. Mechanic B says upshifts are produced by movement of the manual valve. Who is right?
    a. A only
    b. B only
    c. both A and B
    d. neither A nor B

19. Which of these statements is correct?
    a. To go into overdrive, the overdrive servo must release and the forward clutch must engage.
    b. To go into overdrive, the overdrive servo must apply and the forward clutch must disengage.
    c. To go into overdrive, both bands must apply and all clutches must disengage.
    d. To go into overdrive, both bands must release and the forward and overdrive clutches must engage.

20. Which of these statements is correct?
    a. Maximum line pressure results in reverse because three pressures act on the booster valve.
    b. In reverse, the REV passage is charged by line pressure at the manual valve.
    c. In reverse, full line pressure is applied to the reverse servo and the reverse clutch.
    d. All of the above.

GENERAL MOTORS 200C AUTOMATIC TRANSMISSION

After studying this chapter, you should be able to:

1. Describe the construction of the compound planetary gearset and explain how it produces three forward speeds and reverse.
2. Name the holding members in action in all gear positions and describe their effects on the gearset.
3. List six major valves in the transmission and explain their purpose and how they work.
4. Explain the purpose of the accumulators.
5. Describe the construction of the torque-converter clutch and explain its purpose and how it works.
6. Describe the actions in the transmission as it shifts up from first to second to direct, in drive.
7. Explain what happens when the selector lever is moved to L, and from L to 2.
8. Discuss what happens when the selector lever is moved to R.

8-1 General Motors 200C applications This chapter describes the construction and operation of the General Motors type 200C automatic transmission. This transmission is widely used in General Motors cars, including Chevrolet, Pontiac, Oldsmobile, and Buick. The 200C automatic transmission has a compound planetary gearset, three multiple-disk clutches, a one-way clutch, and a band. These provide three forward speeds and reverse. The 200C automatic transmission uses an oil cooler as explained in ❂ 3-15. Servicing the 200C transmission is described in Chap. 11.

The 200 and 200C automatic transmissions are made in various versions. For example, if you compared the 200 used in the Chevette, which is a small car, with the 200 used in the Caprice or Impala, which are larger cars, you would find several basic differences. The transmission used in the larger cars has a much larger torque converter—11¾ inches [298 mm] in diameter—as contrasted with the torque converter for the smaller car—9 inches [229 mm] in diameter. Also, the transmission for the larger cars has four more plates in the clutches and four more pinion gears in the planetary gearset.

8-2 Type 200C automatic transmission The 200C transmission was introduced in 1980. It is essen-

tially the same as the earlier 200 transmission. However, the 200C transmission has an added converter-clutch assembly. When the car reaches cruising speed on the highway, the clutch locks up the torque converter. This prevents any slippage through the converter. The *C* following the 200 identifies the transmission as having a torque-converter clutch.

Figure 8-1 is a sectional view of the 200C automatic transmission. Note the locations of the torque-converter clutch, the torque converter, the oil pump, the intermediate band, the main valve body, the three multiple-disk clutches, and the governor.

❂ 8-3 Hydraulic circuits of the 200 and 200C automatic transmissions Figures 8-2 and 8-3 show the hydraulic circuits for the earlier 200 and the later 200C automatic transmissions. Note that the lower parts of these illustrations are the same. However, at the top of the two illustrations, there is a difference. In the 200C (Fig. 8-3) the intermediate servo has been moved to the right and down. This makes room in the circuit diagram for the added actuator-apply valve, solenoid, and governor pressure switch, which control the converter clutch.

The converter clutch and its controls are described in ❂ 4-23 and 4-24. Both the 200C and the 350C use converter clutches and controls that are similar in con-

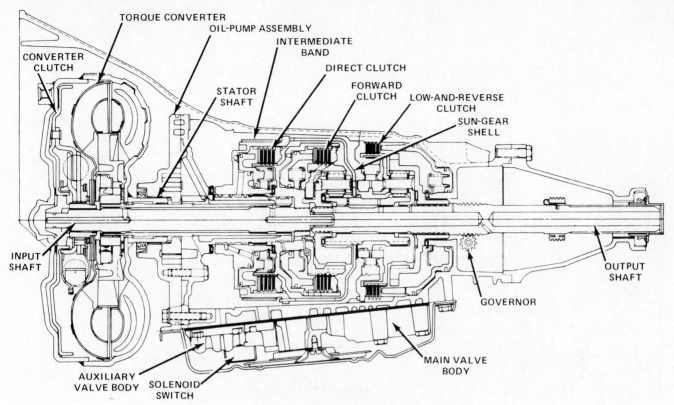

Fig. 8-1 Sectional view of the type 200C automatic transmission. This transmission has a torque-converter clutch. *(Oldsmobile Division of General Motors Corporation)*

struction and operation. The basic difference is that the actuator-apply valve is in the oil-pump cover in the 200C. The valve is in the auxiliary valve body in the 350C. Figure 4-21 shows the torque converter and clutch disassembled.

Following sections describe the operation of the valves, clutches, and band in the various ranges and gears in the 200C automatic transmission.

✿ 8-4 Clutches and band
Figure 8-1 is a sectional view of the 200C automatic transmission, showing the locations of the clutches and band. From left to right they are: converter clutch, pump, intermediate band, direct clutch, forward clutch, one-way clutch, and low-and-reverse clutch. The valves and servo in the hydraulic system control the clutches and band. Figure 8-4 shows which are on and off in the various driving modes.

✿ 8-5 200C valves
Figure 8-3 shows all the valves in the hydraulic system of the 200C automatic transmission. The various hydraulic components of the transmission are described below, starting from the lower left in Fig. 8-3 and moving up.

1. **1-2 accumulator piston** The accumulator piston eases the upshift from first to second in the drive range (selector lever at D).

2. **1-2 accumulator valve** The 1-2 accumulator valve directs drive oil to the 1-2 accumulator valve. The drive oil comes from the manual valve, in the center of the hydraulic system (Fig. 8-3).

3. **Direct-clutch exhaust** The direct-clutch exhaust operates to exhaust oil from the direct clutch when governor pressure is reduced enough (car speed has fallen enough) to call for a downshift.

4. **Intermediate booster** When the manual lever is moved to intermediate, or range 2, intermediate oil pressure acts on the intermediate booster valve.

5. **Pressure-regulator valve** The pressure-regulator valve regulates line pressure. In the intermediate, low, and reverse ranges, line-boost pressure is directed to the top of the pressure-regulator valve. This causes the pressure-regulator valve to increase the drive pressure. Additional pressure is needed to hold the intermediate band and the low-and-reverse clutch.

6. **Converter-clutch valves** The converter-clutch valves, when actuated, cause oil to fill the cavity in the torque converter so that the pressure-plate assembly is forced against the converter housing cover. This locks the converter, as explained in ✿ 4-23 and 4-24.

7. **Intermediate servo** Now, continuing in a clockwise direction, the intermediate servo (upper right in Fig. 8-3) applies the band. This occurs when the intermediate servo is actuated.

8. **2-3 Shift valve** The 2-3 shift valve and 2-3 throttle valve control the 2-3 upshifts and 3-2 downshifts in the drive range. Governor pressure works at the left end of the 2-3 shift valve. Increasing car speed means increasing governor pressure. This acts to promote movement of the 2-3 shift valve and therefore an upshift. Opposing this action is the 2-3 throttle valve.

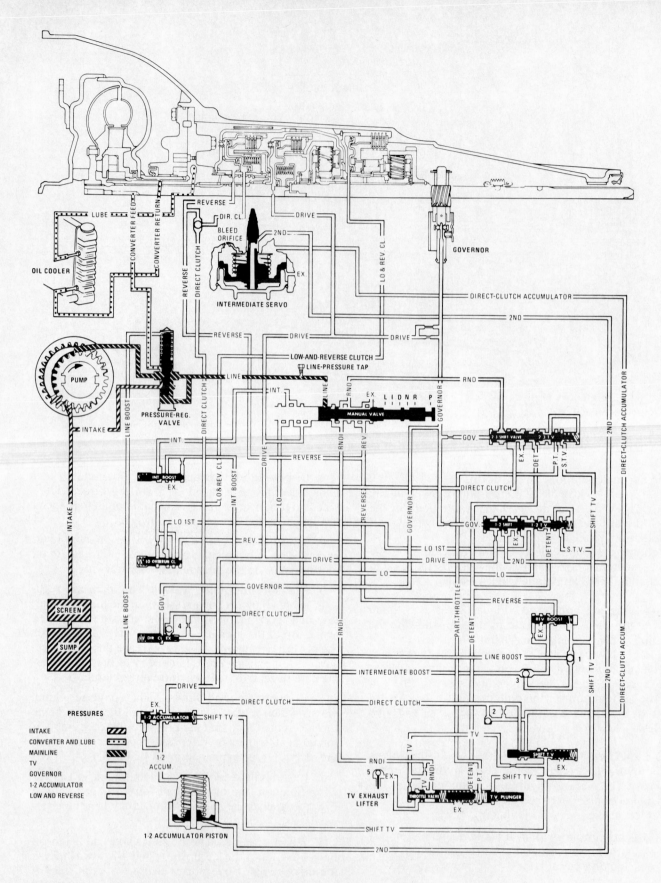

Fig. 8-2 Hydraulic system for the type 200 automatic transmission. This is the same as the hydraulic system for the type 200C automatic transmission with the torque-converter clutch except that it does not have the clutch. *(Buick Motor Division of General Motors Corporation)*

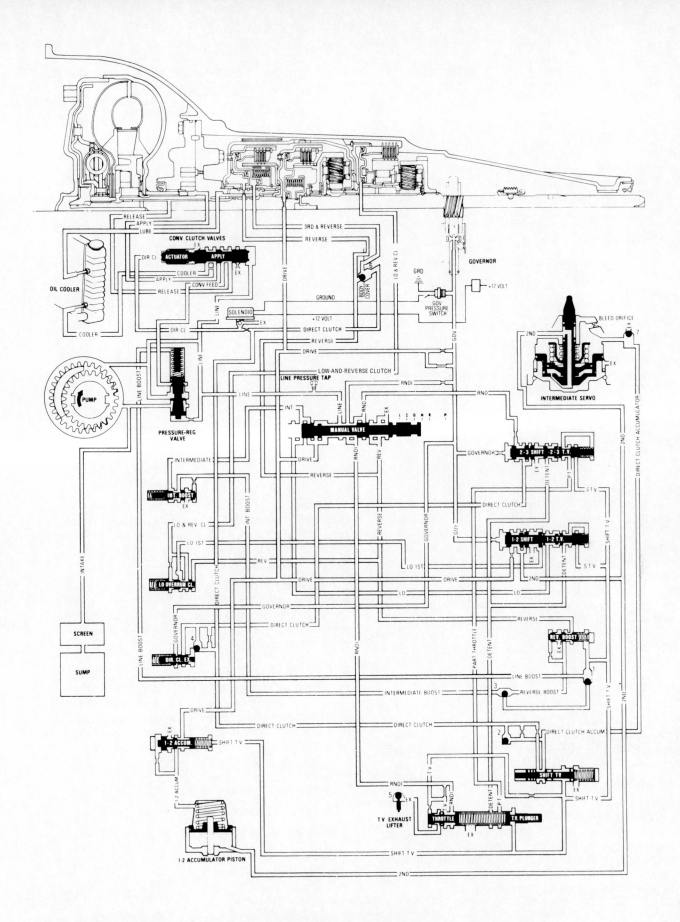

Fig. 8-3 Hydraulic system for the type 200C automatic transmission, with a torque-converter clutch. *(Oldsmobile Division of General Motors Corporation)*

	INTERMEDIATE BAND	DIRECT CLUTCH	FORWARD CLUTCH	ROLLER CLUTCH	LOW-AND-REVERSE CLUTCH
Park (engine on)	R	D	D		D
N	R	D	D		D
D1			E	L	
D2	A		E		
D3		E	E		
Part-throttle 3-2 downshift	A		E		
Detent downshift	A		E		
Intermediate range			E		
Low			E		E
R		E			E

R—released; A—applied; E—engaged; D—disengaged; L—locked

Fig. 8-4 Clutch-engagement and band-application chart for the 200C automatic transmission.

The 2-3 throttle valve is linked by a cable to the throttle-valve plunger (at the lower right in Fig. 8-3). As the throttle is opened, the plunger moves. This increases the shift-throttle-valve pressure to the 2-3 throttle valve. If the throttle is only slightly open, there will be little pressure on the 2-3 throttle valve. This means that the governor pressure does not have to go very high to produce an upshift. The 2-3 upshift will take place at fairly low car speed. If the throttle is opened fairly wide, the throttle-valve pressure will be higher. Therefore car speed must go higher before 2-3 upshift can take place.

9. **1-2 Shift valve** The 1-2 shift and the 1-2 throttle valve control upshifts and downshifts between first and second in the drive range. The action is similar to that of the 2-3 shift valve and 2-3 throttle valve. The difference is that the 1-2 shifts take place at a lower governor pressure.

10. **Reverse-boost valve** The reverse-boost valve operates when reverse pressure is directed to it from the manual valve (when the manual valve is shifted to reverse). When reverse pressure is applied to the reverse-boost valve, it sends line-boost pressure to the top of the pressure-regulator valve. This causes the line pressure to increase. The low-and-reverse clutch needs a higher pressure to hold when the transmission is in reverse.

11. **Shift throttle valve** The shift throttle valve operates as the transmission shifts to direct drive in the drive range. It directs direct-clutch pressure to the intermediate servo. This pressure acts to release the intermediate band. At the same time, it smooths the engagement of the direct clutch so that the shift is made without noticeable feel.

12. **Governor** The governor is similar to a valve in operation. It receives drive oil pressure and reduces it more or less depending on car speed. The governor-regulated oil coming from the governor is called *governor oil* or *governor pressure*.

❋ **8-6 200C hydraulic system** The difference between the older type 200 and the new 200C automatic-transmission hydraulic systems is shown in Figs. 8-2 and 8-3. Following sections describe the actions that take place in the nine operating conditions of the 200C transmission:

Drive range—first gear

Drive range—second gear

Drive range—third gear

Drive range—third gear, converter locked up

Part-throttle 3-2 downshift

Detent downshift

Intermediate range

Low range

Reverse

The hydraulic circuits for some of these operating conditions are shown in color in the *Color Plates of Automatic-Transmission Hydraulic Circuits,* sixth edition.

✿ 8-7 Drive range, first gear In drive range, first gear, both planetary gearsets are in reduction, so there is maximum gear reduction in the transmission. The forward clutch is engaged and the one-way or roller clutch is holding. Plate 21 shows the conditions in the hydraulic system. When the manual valve is moved to Ⓓ (drive), line pressure enters the drive passage. Drive oil then flows to the:

Forward clutch

Governor assembly

1-2 shift valve

1-2 accumulator valve

Drive oil flows to the forward clutch so it engages. Drive oil flowing to the governor is regulated to a varying pressure, called governor pressure, depending on car speed. Governor pressure increases with car speed. This increasing pressure acts against the 1-2 and 2-3 shift valves and the direct-clutch exhaust valve. Drive oil also flows to the 1-2 accumulator valve and is regulated to another variable pressure called the *1-2 accumulator pressure*. This pressure enters the 1-2 accumulator piston so that it acts as a cushion for the band apply on the 1-2 upshift.

✿ 8-8 Drive range, second gear In drive range, second gear, the front planetary gearset is in reduction. The forward clutch and intermediate band are on. The manual valve is in D (drive). As car speed and governor pressure increase, the force of the governor pressure acts on the 1-2 shift valve to overcome the pressure of the 1-2 throttle valve and throttle-valve-spring force. This allows the 1-2 shift valve to move and direct drive oil into the second oil passage. This oil is called *second (2nd) oil*.

The second oil from the 1-2 shift valve is directed to the:

Intermediate servo

1-2 accumulator piston

As the second oil enters the intermediate servo, it is actuated. Its piston moves up and causes the intermediate band to apply. This puts the front planetary gearset in reduction. However, the rear planetary gearset is locked and has no effect on the operation, so there is gear reduction through the transmission. During the shifting action, second oil enters the 1-2 accumulator piston against 1-2 accumulator pressure and the accumulator spring. This causes a more gradual buildup of pressure on the intermediate servo during the 1-2 upshift. The result is a smooth and not a sudden band application.

✿ 8-9 Drive range, third gear In drive range, third gear, the front planetary gearset is locked, so there is

direct drive through the transmission. Conditions in the hydraulic system are as shown in Plate 22.

The forward clutch remains engaged as in first and second gears. At the same time, the direct clutch engages. This is caused by the following. As car speed and governor pressure continue to increase, the governor pressure acting on the 2-3 shift valve overcomes the force of the 2-3 throttle-valve spring and throttle-valve oil pressure. This causes the 2-3 shift valve to move and open the oil passage to the direct clutch. The oil, as it leaves the 2-3 shift valve, is called the *RND oil* in Plate 22. It is also referred to as the *direct-clutch oil*.

Direct-clutch oil from the 2-3 shift valve is directed to the following:

Direct-clutch exhaust check ball (4)

Direct-clutch exhaust valve

Direct clutch

Direct-clutch accumulator exhaust check ball (2)

Shift throttle-valve valve

Intermediate servo

Direct-clutch oil from the 2-3 shift valve flows to the direct-clutch exhaust check ball (4), seating the ball. Now the oil flows through an orifice to the inner area of the direct-clutch piston, engaging the direct clutch.

At the same time, direct-clutch oil flows past the direct-clutch accumulator exhaust check ball (2) and becomes direct-clutch accumulator oil. This direct-clutch accumulator oil flows into the release side of the intermediate servo. The pressure of this oil, combined with the cushion spring force, moves the servo piston against the second oil and acts as a 2-3 accumulator for a smooth intermediate-band release and direct-clutch engagement.

✿ 8-10 Drive range, third gear, converter locked-up Plate 22 shows the conditions in the hydraulic system before the converter clutch has locked up the converter. The only difference after lockup is that the converter clutch has been activated. When car speed reaches the operating value, the governor pressure is great enough to close the governor pressure switch. This connects the solenoid to the battery so it shuts off the exhaust line from the direct clutch. Direct-clutch oil now enters the converter cavity and causes the pressure plate to move to the left (in Plate 22) and lock up the converter.

✿ 8-11 Part-throttle 3-2 downshift A part-throttle 3-2 downshift will occur if the throttle pedal is pushed down far enough. As the throttle pedal is pushed down, it causes the throttle-valve plunger to move. If the throttle-valve plunger is moved far enough, it will allow shift throttle-valve oil to enter the part throttle (PT) passage. This oil is called *part-throttle* or *PT oil*.

Part-throttle oil from the throttle-valve plunger is directed to the 2-3 throttle valve. The PT oil and the 2-3 spring move the 2-3 shift valve against governor pressure. This shuts off the RND oil to the direct-clutch passage. Direct-clutch oil is exhausted into the reverse passage. The direct-clutch oil is forced through an orifice, or is regulated by the direct-clutch exhaust valve, according to governor pressure and vehicle speed. This slows the disengagement of the direct clutch.

At the same time, direct-clutch accumulator oil from the intermediate servo flows through the open shift (throttle-valve) valve. It also exhausts through the same route as the direct-clutch oil. This allows second oil on the apply side of the servo to move the servo piston as the direct-clutch accumulator oil leaves. The result is a smooth band application at the same time the direct clutch disengages.

❂ 8-12 Detent downshift Detent downshift occurs when the throttle is opened wide. This produces a downshift from 3 to 2, or from 2 to 1, depending on the car speed. The action is caused by the forward clutch being on and the intermediate band applying or releasing. When the throttle is fully opened, the throttle-valve plunger is moved so that shift throttle-valve oil can pass through and enter the detent passage. This oil is then called *detent oil*. It is routed to the following:

2-3 throttle valve

1-2 throttle valve

Detent oil flows to the 2-3 throttle valve. Detent oil, part-throttle pressure, and the 2-3 spring close the 2-3 shift valve against governor-oil pressure. This opens the direct-clutch oil passage so that it exhausts in the reverse passage.

At high car speed, governor oil acting on the direct-clutch exhaust valve moves the valve against spring force. Exhausting direct-clutch oil will move the direct-clutch check ball (4) off its seat for quick direct-clutch exhaust and a fast clutch disengagement.

When the throttle is wide open, high throttle-valve pressure positions the shift throttle-valve valve against its spring. This forces the exhausting direct-clutch accumulator oil from the intermediate servo to seat the direct-clutch-accumulator exhaust check ball (2). Then the oil can flow through an orifice, which slightly slows the application of the intermediate band. This produces a smooth 3-2 shift at high speeds. At high speeds, the direct-clutch exhaust valve and the shift throttle-valve valve control the rate at which the direct clutch disengages and the intermediate band applies for a smooth 3-2 downshift.

A detent 2-1 downshift will occur at lower speeds. Detent oil and the 1-2 spring acting on the 1-2 throttle valve will close the 1-2 shift valve. This shifts the transmission into first gear.

❂ 8-13 Intermediate range When the selector is moved to the intermediate range (2), the intermediate band applies and the forward clutch engages. This is the same condition that occurs in second gear, drive range shifts from first to second (**❂ 8-8**). But the action in the hydraulic system producing this condition is dif-

ferent. For example, when the selector lever is moved from D to 2 (intermediate range), RND oil will exhaust at the manual valve and intermediate-range oil from the manual valve will flow to the intermediate-boost valve.

This causes the intermediate-boost valve to increase the pressure. The oil from the boost valve, called *intermediate-boost oil,* flows to the intermediate reverse check ball (3). The ball seats against the reverse-boost passage. Intermediate-boost oil then flows to the shift throttle-valve boost check ball (1) and seats it against the shift throttle-valve passage. Boost oil then flows to the spring end of the pressure-regulator valve to cause the valve to increase line pressure.

Shift throttle-valve pressure then is great enough to seat the check ball (1) against intermediate-boost oil. The oil then flows to the boost passage, which varies the line pressure, according to throttle opening. Because RND oil is exhausted, the transmission will shift to second gear, regardless of vehicle speed. The RND oil is the feed for direct-clutch oil in third gear. With the transmission in second gear, intermediate range, it cannot upshift to third gear regardless of car speed.

❂ 8-14 Low range When the selector lever is moved to L, or low, the forward clutch and low-and-reverse clutch engage. This provides a solid drive through the transmission so that engine braking takes place. Low is used for maximum downhill braking and for pulling heavy loads. Low-range oil is regulated by the intermediate-boost valve and the pressure regulator to provide high pressure. The oil is directed from the manual valve to the 1-2 throttle valve. Low oil pressure and the 1-2 throttle-valve spring acting on the 1-2 valve will close the 1-2 shift valve.

Note that this action can take place only if the car is moving slowly—below about 30 mph [48 km/h].

❂ 8-15 Reverse When the selector lever is moved to R (reverse), the direct clutch and the low-and-reverse clutches engage. The manual valve is moved to the position which allows line pressure to enter three passages, as shown in Plate 23:

Reverse

RNDI (reverse, neutral, drive, and intermediate)

RND (reverse, neutral, and drive)

Reverse oil from the manual valve is directed to the 2-3 shift valve and the 2-3 throttle valve, where it enters the direct-clutch passage. Reverse oil also flows through an orifice to the outer area of the direct clutch to provide additional apply area for reverse. Reverse oil enters the low-and-reverse clutch passage at the low overrun clutch valve. This engages the low-and-reverse clutch. Reverse oil also flows to the reverse-boost valve and exits at a regulated pressure.

The regulated oil flows to the intermediate reverse check ball (3), seating it in the intermediate-boost passage. It then flows to the shift throttle-valve check ball (1), seating it in the shift throttle-valve passage, and to the spring end of the pressure-regulator valve. This boosts the reverse pressure for increased holding pressure of the two clutches.

RNDI oil from the manual valve flows to the throttle valve and is regulated to throttle-valve pressure. Throttle-valve oil pressure flows through the shift throttle valve and is limited by it to a lower pressure.

Oil from the shift throttle-valve valve is called *shift TV oil* and is directed to the throttle-valve plunger, the 1-2 accumulator valve, the reverse-boost valve, and the 1-2 and 2-3 throttle valves. Shift throttle-valve oil

on the reverse-boost valve boosts the reverse line pressure to as much as 240 psi [1660 kPa], depending on throttle opening.

RND oil from the manual valve is directed to the 2-3 shift valve. Regardless of the position of the 2-3 shift valve, the inner area of the direct clutch will be filled.

To sum up, in reverse the hydraulic system operates the various valves to prevent any other action but a shift to reverse. The hydraulic system also produces a considerable boost in reverse-oil pressure.

Chapter 8 review questions

Select the *one* correct, best, or most probable answer to each question. Then check your answers against the correct answers given at the end of the book.

1. The 200C automatic transmission includes a compound planetary gearset, a roller or one-way clutch, and
   a. two multiple-disk clutches and two bands
   b. three multiple-disk clutches and two bands
   c. three multiple-disk clutches and one band
   d. four multiple-disk clutches and no bands

2. Mechanic A says that the difference between the 200 and the 200C automatic transmissions is that the 200C has an extra clutch. Mechanic B says that the 200C has a torque-converter lockup feature. Who is right?
   a. A only
   b. B only
   c. both A and B
   d. neither A nor B

3. The three devices that control the torque-converter clutch are the actuator-apply valve, the solenoid, and the:
   a. governor
   b. manual valve
   c. 2-3 shift valve
   d. governor-pressure switch

4. Mechanic A says the 1-2 accumulator piston smooths the 1-2 upshift. Mechanic B says the servo piston acts as a 2-3 accumulator to smooth the 2-3 upshift. Who is right?
   a. A only
   b. B only
   c. both A and B
   d. neither A nor B

5. Opposing the governor pressure working on the 2-3 shift valve is the:
   a. 2-3 throttle valve
   b. manual valve
   c. regulator valve
   d. clutch valve

6. In drive range, first gear, the following are holding:
   a. forward and reverse clutches
   b. roller clutch and band
   c. forward and intermediate clutches
   d. forward clutch and roller clutch

7. In drive range, second gear, the following are holding:
   a. forward and reverse clutches
   b. forward clutch and intermediate band
   c. forward and intermediate clutches
   d. forward clutch and roller clutch

8. In drive range, third gear, the following are holding:
   a. forward and direct clutches
   b. direct and roller clutches
   c. direct clutch and intermediate band
   d. direct and reverse clutches

9. During a detent downshift to second, the following hold:
   a. direct and forward clutches
   b. forward clutch and intermediate band
   c. forward and roller clutches
   d. forward and reverse clutches

10. When the selector lever is moved to R, the following hold:
    a. direct and intermediate clutches
    b. forward and direct clutches
    c. low-and-reverse clutch and intermediate band
    d. low-and-reverse and direct clutches

AUTOMATIC-TRANSMISSION SERVICE

After studying this chapter, and with proper instruction and equipment, you should be able to:

1. Discuss trouble diagnosis of automatic transmissions and explain the general procedure for determining trouble causes.
2. Explain how to check fluid level.
3. Describe how to examine the condition of the fluid.
4. Discuss the reasons why transmissions can fail.
5. Explain the purpose of the trouble-diagnosis charts.
6. Describe how to road-test a transmission and what you should learn from the test.
7. List and explain nine transmission-service precautions.
8. Discuss the importance of cleanliness when servicing transmissions.

9-1 Introduction to automatic-transmission service This chapter covers the general maintenance procedures on automatic transmissions. Then the fundamentals of trouble diagnosis and of removing, overhauling, and reinstalling transmissions are covered. Following chapters describe in detail the specific procedures on the various models of automatic transmissions in use in modern automotive vehicles.

9-2 Normal automatic-transmission maintenance The level of the automatic-transmission fluid (also called *ATF* or *oil*) should be checked every time the engine oil is changed. In addition, many car manufacturers recommend changing the transmission fluid and filter at periodic intervals. The length of the intervals depends on how the car is used.

For example, Chevrolet recommends changing the fluid and filter every 100,000 miles [160,000 km] for normal service. For severe service, as in taxis, in vehicles hauling trailers, in police vehicles, in stop-and-go city driving, or in delivery service, Chevrolet recommends changing the fluid and filter every 15,000 miles [24,000 km].

Linkages and bands may require relatively frequent adjustment if the vehicle is in severe service. General adjustment procedures are covered in following sections. Later chapters, on specific automatic transmis-

sions, cover the procedures step by step as recommended by the transmission manufacturer.

9-3 Checking the fluid Check the fluid level in the transmission, and also the condition of the fluid. This should be done at every engine-oil change.

1. **Checking fluid level** Clean dirt from around the dipstick cap. Pull the dipstick, wipe it, reinsert it, and pull it out again. Note the level of the fluid on the dipstick. The level will vary under normal operating conditions as much as ¾ inch [19 mm] from cold to hot. For example, as the temperature of the fluid goes up from 60° F [16° C] to 180° F [82° C], the level of the fluid will rise as much as ¾ inch [19 mm]. This is the reason that many dipsticks are marked to indicate proper levels at different temperatures (Fig. 9-1).

Do not add too much fluid. Normally, not more than one pint [0.5 L] should be required. Too much fluid will cause foaming. Foaming fluid cannot operate clutches and bands effectively. They will slip and probably burn. This could result in an expensive transmission overhaul.

You can get a general idea of the fluid temperature by cautiously touching the transmission end of the dipstick to find out if it feels cool, warm, or hot. If the fluid feels cool, the fluid level on the dipstick should be on the low side of the dipstick. If the fluid feels warm, or

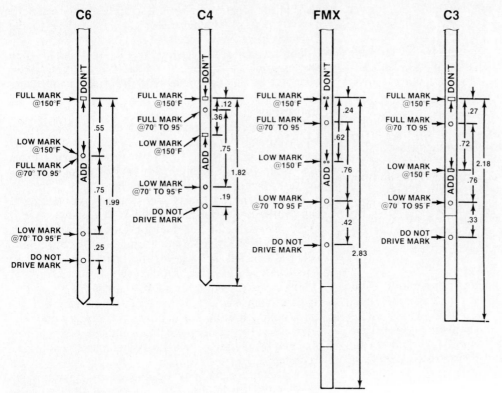

Fig. 9-1 Dipsticks used in various automatic transmissions.

hot (too hot to hold), the fluid level on the dipstick should be on the high side. Specific instructions for the various automatic transmissions are given in later chapters.

2. **Condition of the fluid**   The transmission fluid is normally red in color. If it is brown or black and has a burned odor, the transmission may be in trouble. Bands and clutch plates may have overheated and burned. In such a case, particles of friction material from the bands and clutch plates have probably circulated through the transmission and oil cooler. These particles can build up and cause valves in the valve body to hang. Shifts will be noisy or will not take place. Servos and clutches can malfunction. The transmission shifts may be jerky, or may not take place. Or the transmission may slip.

Use a piece of absorbent white paper such as facial tissue to wipe the dipstick. Examine the stain for evidence of solids (specks on the paper) or for evidence of antifreeze leakage (gum or varnish on the dipstick).

NOTE: In the General Motors 125 automatic transaxle (Chap. 18), the COLD reading on the dipstick is above the FULL mark.

If the fluid is dark in color, or if you find specks on the paper or gum or varnish on the dipstick, remove the transmission-oil pan and look for further evidence of trouble. If you find contamination in the oil pan, it is added proof of transmission trouble. The transmission must be removed from the car for overhaul. Overhaul of specific transmission models is covered in later chapters.

When the fluid is contaminated, the oil cooler and

lines must be flushed out. Also, the torque converter must be flushed out or replaced. Most manufacturers state that a torque converter cannot be flushed out to remove all debris. It must be replaced with a new or remanufactured torque converter.

3. **Engine coolant in transmission fluid**   If engine coolant has leaked into the transmission fluid, the transmission must be removed for complete overhaul. This includes cleaning and replacement of seals, composition-faced clutch plates, nylon washers, and speedometer and governor gears. All these parts can be affected by coolant. The converter should be flushed out. The cooler in the engine radiator must be repaired or replaced. Then the cooler lines must be flushed out.

4. **Checking for transmission leaks**   If the transmission fluid is low, suspect leakage. Some fluid might be lost through the vent if it has foamed. The fluid is red and this helps in finding any place where a leak occurs. Here is one procedure.

a. Clean the suspected area with solvent to remove all traces of fluid.
b. Remove the converter shield, if present, to expose as much of the converter as possible.
c. Spray the cleaned area with a spray can of white foot powder. This will show up the red fluid at the leak points.
d. Start and run the engine at high idle.
e. If the leak does not show up immediately, have the owner bring the car back the next day or later. Then recheck the suspected area.
f. Repair any leak points found by replacing gaskets or

seals, tightening attached bolts, or replacing porous castings.

❋9-4 Linkage and band adjustments There are basically two linkage adjustments, the shift-linkage adjustment and the throttle- or kickdown-linkage adjustment.

1. Shift-linkage adjustment   This is a critical adjustment. When the transmission shift lever is moved into any position, the manual valve in the transmission valve body must be exactly centered in its selected position. If the manual valve is off, delays in shifting and slipping while operating can occur. Slipping of clutch plates or bands can soon damage the transmission so seriously that an overhaul will be required (❋9-3, item 2). Adjustment procedures vary with different car and transmission models. Figure 9-2 shows a typical adjustment.

2. Throttle- or kickdown-linkage adjustment   The throttle or kickdown linkage causes the transmission to shift down if the throttle is opened wide (within a certain speed range). Basically, the adjustment is correct if the throttle linkage causes the kickdown linkage to produce the downshift when the throttle is opened to the full-throttle-stop position.

3. Band adjustments   Some transmission models require periodic band adjustments if the transmission is used in severe service. Also, bands are always adjusted when the transmission is overhauled.

❋9-5 Causes of automatic-transmission failure The usual causes of automatic-transmission trouble are abuse and neglect. Most often these are overloading the transmission and not checking transmission fluid and adding or changing it if necessary.

Extremes of heat and cold are also hard on transmissions. Long periods of idling in stop-and-go traffic can overheat the transmission fluid if the transmission is left in a driving range. This continues to whip the

fluid in the torque converter and adds to the heat. The remedy is to shift to neutral or park if the car is to idle for more than a few seconds.

The car, and transmission, may be overloaded when the car pulls a trailer, or when it is carrying extra weight. Revving the engine with the brakes applied and the transmission in a driving range also overloads the transmission.

For mountain driving, either with a heavy load or when pulling a trailer, use the 2 (second) or 1 (low) selector-lever position when on upgrades requiring a heavy throttle for ½ mile [0.8 km] or more. This reduces transmission and converter heating.

Working the transmission hard and overloading it can overheat the transmission fluid. The heat can eventually cause the fluid to deteriorate. Gum and varnish may form in the transmission. These could cause poor valve action and slippage of the bands and clutches. This can lead to further trouble (❋9-3, item 2).

Neglect is the other abuse that can damage an automatic transmission. If the transmission is being worked hard (pulling a trailer, or in police, taxicab, or door-to-door delivery service) the fluid and filter should be changed frequently. The bands should also be adjusted frequently.

Another problem may be caused by operation in consistently low temperatures. For example, Ford recommends that the fluid be changed every 7500 miles [12,000 km] if the vehicle is operated for more than 60 days in temperatures consistently below 10° F [−12° C] and with short trips of less than 10 miles [16 km].

Even in normal, conservative operation, some manufacturers recommend changing the transmission fluid and filter periodically. For example, Chevrolet recommends changing these every 100,000 miles [160,000 km].

Careful: When towing a rear-wheel drive car with automatic transmission, lift the rear of the car or remove the drive shaft. The car can be towed with the rear wheels on the ground and the drive shaft connected for short distances at low speed. But the transmission is lubricated only when the engine is running. To guard against transmission damage, lift the rear end or remove the drive shaft. For front-wheel-drive cars, lift the front end.

❋9-6 Trouble diagnosis Trouble diagnosis is more than following a series of steps in an attempt to find the solution to a problem. It is a way of looking at systems that are not working right. Here are the basic rules:

1. Know the system. This means you should know how the parts go together, how they work together as a system, and what happens if some part goes bad or the parts fail to work together as they should.

2. Know the history of the system. How old is the system? What sort of treatment has it had? What is its service history? Has it been serviced before for the same problem? The answers to these questions might save you a lot of time.

3. Know the history of the condition. Did it start all at

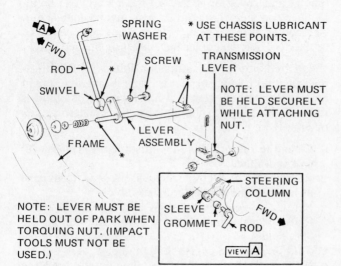

Fig. 9-2 A typical shift or manual-valve linkage adjustment. *(Chevrolet Motor Division of General Motors Corporation)*

once? Or did the trouble come on gradually? Was it related to some other condition, such as an accident or a previous service problem?

4. Know the odds. Some troubles happen more often than others. Be aware of what can happen frequently and what happens but rarely. A trouble such as the engine cranks normally but does not start is more likely to be caused by a loose wire than by a broken timing chain.

5. Don't cure the symptom and leave the cause. Charging a run-down battery may be a temporary "fix." But if the trouble is really a defective alternator, you have not eliminated the problem.

6. Be sure you have fixed the basic condition that caused the trouble.

7. Be sure you get all the information (items 1, 2, and 3) you can from the customer. This information may greatly simplify your search for the cause of a trouble.

❀ **9-7 Automatic-transmission trouble-diagnosis charts** Trouble-diagnosis charts take different forms. Each manufacturer works out a chart that fits the transmission models it makes. The following chart lists three typical transmission troubles and several possible causes of each. This is not a working chart. Do not use it to diagnose troubles on specific automatic transmissions. The chart is included here to introduce you to trouble-diagnosis charts and how to use them.

NOTE: Be sure the engine is in good condition before making any tests of the transmission. If the engine lacks power, the transmission probably will not work properly.

There are other possible troubles, such as no braking in low 1 or low 2, no drive in reverse, rough shifts, and failure to hold in park. Each trouble can be caused by certain specific conditions. When you are trouble-diagnosing a transmission, your job is to first verify that the trouble exists. Road-test the car if necessary.

Automatic-Transmission Trouble-Diagnosis Chart

TROUBLE	POSSIBLE CAUSE
No drive in drive range.	This could be caused by low oil pressure (low oil, defective pump, screen or line plugged), incorrect manual-valve-linkage adjustment, defective clutch or band, or defective overrunning clutch.
Does not shift automatically.	Throttle-valve cable misadjusted, vacuum modulator defective, governor not working properly, 1-2 or 2-3 shift valves sticking, band or clutch defective, low oil pressure (low oil, defective pump, screen or line plugged).
Transmission slips.	Low oil pressure, defective band or clutch, misadjusted linkage, vacuum modulator defective.

Then consider the various possible conditions that could cause the trouble. If the transmission has been recently overhauled, perhaps some part was left out or not assembled properly. If the transmission is older and has had hard service, suspect worn bands, clutches, and bearings. The purpose of trouble-diagnosis charts is to help you pinpoint the cause of trouble quickly.

Some troubles might be fixed with a minor repair or adjustment. Other troubles mean removing the transmission for a complete overhaul.

❀ **9-8 Diagnostic test procedures** Most automatic-transmission troubles require some testing to verify the trouble and also to check oil pressures under different operating conditions. The tests can be made on the road, on a chassis dynamometer (❀ 1-15), or in the shop with the driving wheels either on or off the floor. Some manufacturers recommend one procedure, other manufacturers another. Before any test is made, the fluid level and condition should be checked and the control linkages checked and adjusted if necessary.

The oil pressure in the hydraulic system is usually checked during the test procedure. This requires a pressure gauge with a long enough hose that it can be connected to the transmission and read from the driver's seat. Test procedures for some transmissions also require a tachometer to measure engine rpm because some tests are made at specific engine speeds. An engine that lacks power can make a transmission perform improperly even though nothing is wrong with it.

The purpose of the diagnostic test is to operate the car at various speeds and check each gear position for slipping, or incorrect shifting. Note the oil pressures and whether the shifts are harsh or spongy. Note also the speeds and throttle positions at which the shifts take place.

1. Flare-up Slipping or flare-up (engine speedup or runaway on a shift) usually means clutch, band, or overrunning-clutch trouble. Flare-up is a sudden surge or increase in engine speed before or during a shift. You can usually determine which band or clutch is slipping by pinpointing the gear position in which the slipping occurs. Figure 9-3 is a sample chart showing the elements in use at each position of the selector lever.

2. Torque converter Poor acceleration could be caused by burned engine valves or by a clogged exhaust system. It could also be caused by a slipping stator clutch in the torque converter. To check this out, accelerate the engine in N. If it reaches high rpm normally, the trouble could be that the stator clutch is slipping. Check for poor performance in R and in D.

If the one-way clutch is locked up, the result will be limited high vehicle speed and high engine rpm. The engine and torque converter may overheat. The overheating may actually cause the converter to turn blue. If this happens, the torque converter must be replaced. Also, a locked-up or slipping one-way clutch requires a new torque converter. Repairs on the one-way clutch cannot be made.

LEVER POSITION	GEAR RATIO	START SAFETY	PARKING SPRAG	CLUTCHES				BANDS	
				FRONT	REAR	OVER-RUNNING	LOCKUP	FRONT (KICKDOWN)	REAR (LOW-REVERSE)
P—park		X	X						
R—reverse	2.21			X					X
N—neutral		X							
D—drive									
First	2.45				X	X			
Second	1.45				X			X	
Direct	1.00			X	X		X		
2—second									
First	2.45				X	X			
Second	1.45				X			X	
1—low (first)	2.45				X				X

Fig. 9-3 Elements in transmission used in each selector-lever position. (*Chrysler Corporation*)

⚙ 9-9 Stall test The stall test checks the holding ability of the torque-converter stator clutch and the transmission clutches. It is made by applying the car brakes, shifting to different selector-lever positions, and opening the throttle wide to check the maximum engine speed or rpm (or the oil pressure, on some transmissions).

A tachometer is required to accurately measure engine speed. The throttle should not be held wide open for more than a few seconds. Also, the front wheels should be blocked in addition to applying the car brakes. Be sure no one is standing in front or in back of the car! It could lurch forward or backward as engine speed reaches maximum, in spite of the brakes and blocks.

Not all manufacturers recommend the stall test. Following chapters, which discuss servicing of specific automatic transmissions, describe in detail the procedures recommended for each transmission type.

Automatic-transmission service work

⚙ 9-10 Transmission-service precautions Cleanliness is very important during automatic-transmission service. The slightest piece of dirt or lint can cause valves to hang. This can prevent normal clutch or band operation and cause serious transmission damage. Here are other precautions.

1. Never mix parts between transmissions. Keep all parts belonging to a transmission in one place. Use pans to prevent scattering of parts. Muffin pans are handy for small parts.
2. Clean the outside of the transmission thoroughly before starting to disassemble it. Plug all openings and use a steam cleaner if it is available. The bench area, tools, your hands, and all parts must be kept clean. Do not allow dust blowing in from outside or from other areas of the shop to settle on transmission parts. This could cause trouble later.
3. Before installing screws in aluminum parts, dip the screws in transmission fluid to lubricate them. Lubrication will prevent galling of the aluminum threads and prevent threads from seizing.
4. If the threads in an aluminum part are stripped, repair can be made with a threaded insert (Fig. 9-4). To repair a defective thread with a Heli-Coil insert, drill the hole and tap it with a special Heli-Coil tap. Then, install the Heli-Coil, which brings the hole back to its original thread size.
5. Use special care and special tools to protect and prevent damage to seals during installation and assembly. The slightest flaw in a seal or sealing surface can result in an oil leak.
6. Aluminum castings are relatively soft and can be easily nicked, scratched, or burred. Use special care in handling them.
7. Discard all old O rings, gaskets, and oil seals. Use new ones on reassembly.
8. During disassembly, clean and inspect all transmission parts (⚙ 9-11).
9. After you disassemble a unit, such as a clutch, reassemble it before disassembling another unit. This avoids mixing parts, such as plates or snap rings, between the units.

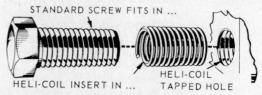

Fig. 9-4 Installation of a Heli-Coil, which is a type of threaded insert. (*Chrysler Corporation*)

10. Do not overexpand snap or retainer rings when removing or replacing them.
11. During reassembly, lubricate all internal parts with transmission fluid.
12. Always use seal-protecting tools during reassembly to prevent damage to seals.
13. If you must replace oil-cooler lines (between the transmission and the cooler in the engine radiator) use only double-wrapped and brazed steel tubing. Copper or aluminum tubing could soon fail from fatigue due to car vibration.

✿ 9-11 Cleaning and inspecting transmission parts After disassembly, clean and inspect all parts. Wash all metal parts in solvent and blow them dry with compressed air. Do not clean clutch faces with solvent. Make sure all fluid passages are clean. Check small passages with a small wire as from a tag. Inspect parts as follows:

1. Check linkages and pivot points for wear. If plastic grommets require replacement, cut or pry them out and snap new grommets into place with pliers. New grommets are required if linkages must be disconnected. Ford supplies a special tool to remove and install grommets (✿ 13-5).
2. Check bearing and thrust faces for wear and scoring.
3. Check mating surfaces of castings and end plates for burrs that could cause poor seating and oil leaks. Burrs may be removed by laying a piece of crocus cloth on a flat surface such as a piece of plate glass. Then lay the part on the crocus cloth and move it back and forth. This process is called lapping. It will bring the metal surfaces to a smooth finish.
4. Check for damaged grooves or lands where O rings seat. Irregularities can cause serious oil leaks.
5. Check castings for cracks and sand holes and for damaged threads. Thread repairs can often be made with a threaded insert (Fig. 9-4).
6. Check gear teeth for chipping, scoring, or wear.
7. Make sure the valves are free of burrs and that the shoulders are sharp and square. Burrs can be removed from bores by honing. Valves must slide freely in the bores.
8. Inspect the composition clutch plates for damaged surfaces and loose facings. If you can remove flakes of a facing with your thumbnail, replace the plates. However, discoloration is normal and not a sign of failure.
9. Inspect steel clutch plates for scored surfaces.
10. Inspect springs for distortion and collapsed coils.
11. Check the bushings and test for wear by inserting the shaft into the bushing and noting the amount of looseness.
12. If the transmission has internal damage to clutches or other parts, or if it shows evidence that foreign material has been circulating in the oil, thoroughly clean all passages and parts that come into contact with the oil. Also, flush out the oil lines and oil cooler. In addition, a new torque converter probably will be required. Some manufacturers have supplied instructions for flushing out the torque converter. But other manufacturers reject this process, so a new torque converter is required.
13. Be sure the lubrication holes in the turbine shaft are open.

✿ 9-12 Tools for automatic-transmission service You need a lift or a jack and safety stands to get the car up so you can work underneath it. Your major concern is to have a means of *safely* supporting the car so you can work under it without danger.

In addition, you need a set of basic hand tools and common shop equipment. Some special tools also are needed for disassembling clutches and adjusting bands. These tools and their use are described in later chapters.

The shop should have a parts-cleaner tank large enough to hold the transmission parts and an oil-drain pan large enough to catch the fluid flowing from the torque converter and transmission. A *must* is a transmission jack to lower the transmission from the car when it is detached from the engine (Fig. 9-5). The transmission jack must also be used to lift the transmission into position to reinstall it.

A clean workbench and work area are essential. The fit of valves in the valve body is very close. A trace of dust in the hydraulic system can cause the valves to malfunction. This can lead to clutch and band burning and serious transmission trouble.

✿ 9-13 Supplies for automatic-transmission service In addition to solvent for cleaning the transmission metal or hard parts, the proper type of automatic-transmission fluid is needed. Never substitute another type of fluid for the one specified. This can lead to transmission failure.

Manufacturers supply overhaul kits which contain the basic seals, gaskets, O rings, and other parts that should be replaced during an overhaul. Some kits include clutch packs and bands. Filters and vacuum modulators also are needed. Filters are always replaced during overhaul. Vacuum modulators may require replacement since they have a flexible diaphragm that can tear or leak.

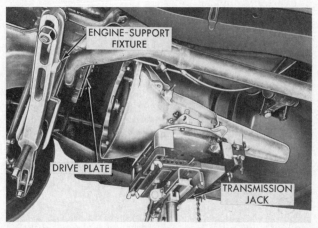

Fig. 9-5 Using a transmission jack to raise or lower the transmission to or from the car. *(Chrysler Corporation)*

⚙9-14 Removing and installing an automatic transmission The removal and installation procedure varies with different cars and transmissions. In general, here are the items you may have to disconnect or remove. First, open the hood and disconnect the battery ground cable. This eliminates the possibility of causing an electrical short-circuit. Other parts that may have to be removed or disconnected include the following:

Starting motor
Backup-light wire
Neutral-park starting switch
Vacuum line to modulator
Oil lines to oil cooler
Shift linkages
Drive shaft
Torque-converter flexplate bolts
Housing-to-engine bolts

There may be additional items to be disconnected or removed. In following chapters, specific removal procedures are described.

Installation includes lifting the transmission into position and reattaching it. Then all leads, lines, and parts removed when the transmission was taken off the vehicle must be attached or installed.

⚙9-15 Overhauling an automatic transmission Overhaul procedures vary considerably from transmission model to model. These procedures also vary according to whether you are replacing one defective component or performing a complete overhaul. In the complete overhaul, you will discard all old gaskets, oil seals, O rings, metal sealing rings, clutch friction and steel disks, filters, and modulators. You will also inspect and replace, if needed, such parts as bands, bushings, pumps, gears, governor, linkage, and converter.

Following chapters describe in detail the servicing of specific automatic transmissions.

Chapter 9 review questions

Select the *one* correct, best, or most probable answer to each question. Then check your answers against the correct answers given at the end of the book.

1. Mechanic A says you can check the fluid level only if the transmission is hot. Mechanic B says you can check fluid level either hot or cold. Who is right?
   a. A only
   b. B only
   c. both A and B
   d. neither A nor B
2. Mechanic A says that if the transmission fluid is brown or black, the clutch plates probably have overheated and burned. Mechanic B says the bands have overheated and burned. Who is right?
   a. A only
   b. B only
   c. both A and B
   d. neither A nor B
3. Which statement is correct?
   a. bands never need adjustment.
   b. bands need adjustment in severe service.
   c. if a band is out of adjustment, replace the band; do not adjust it.
   d. most bands cannot be adjusted.
4. Mechanic A says the purpose of the throttle or kickdown linkage is to cause the throttle to kick back when the transmission upshifts. Mechanic B says the linkage causes the transmission to downshift if the throttle is wide open. Who is right?
   a. A only
   b. B only
   c. both A and B
   d. neither A nor B
5. Severe service that requires periodic changing of the fluid and filter includes:
   a. taxi service
   b. trailer pulling
   c. stop-and-go delivery service
   d. all of these
6. The purpose of the road test is to check for slipping, flare-up, or improper shifting with the selector lever in:
   a. direct drive
   b. second
   c. reverse
   d. all of these
7. Mechanic A says that if the fluid is black and has particles of dirt in it, most manufacturers recommend that the torque converter be discarded. Mechanic B says you can always flush out the converter and use it again. Who is right?
   a. A only
   b. B only
   c. both A and B
   d. neither A nor B
8. The shift linkage is connected between the:
   a. selector lever and throttle
   b. throttle lever and kickdown lever
   c. selector lever and manual valve
   d. manual valve and throttle valve
9. Probably the most important precaution to observe in automatic transmission work is:
   a. don't mix parts.
   b. keep everything clean.
   c. replace all seals.
   d. handle castings with care.
10. Important supply items to have on hand in the automatic-transmission service shop include:
    a. solvent
    b. the proper transmission fluid
    c. thread-repair kits
    d. all of these

GENERAL MOTORS 350C AUTOMATIC-TRANSMISSION SERVICE

After studying this chapter, and with proper instruction and equipment, you should be able to:

1. Explain how to check transmission-fluid level and add fluid if necessary.
2. Describe the procedure for changing the fluid.
3. Explain how to use the trouble-diagnosis chart in the chapter.
4. Discuss the converter-clutch trouble-diagnosis procedure and how to modify a governor to test the system.
5. Explain how to remove and install the transmission.
6. Discuss the procedure of disassembling and assembling the transmission.
7. Explain how to make adjustments to the transmission.
8. Perform the services listed above.

 10-1 Introduction to servicing the GM 350C automatic transmission This chapter describes the trouble diagnosis, checks, adjustments, removal, overhaul, and installation of the General Motors type 350C automatic transmission. Chapter 4 describes the construction and operation of this transmission.

This chapter also describes the service procedures on the type 350C transmission. Chapter 9 covers general servicing procedures on all automatic transmissions. The procedures described in that chapter also apply to the type 350C automatic transmission. However, this chapter provides step-by-step instructions on the type 350C automatic transmission.

✿ 10-2 Type 350C transmission variations The type 350C transmission is used on many cars with a wide variety of engines. Some minor modifications of the transmission are made so that it will operate properly with the engine with which it is used. These modifications have no effect on the service procedures, but they do give the characteristics the transmission needs for engine torque and power curves.

For example, the transmission used on some six-cylinder engines has fewer plates in the clutches than the transmission used on many eight-cylinder engines. Fewer plates are required to handle the lower torque

of the smaller engine. Also, springs and valves may be changed so as to provide different shift points. However, the service procedures are essentially the same for all models of the type 350C Turbo Hydra-Matic transmission.

NOTE: The instructions in this chapter also apply to the earlier type 350, with the exception that the torque-converter clutch information does not apply. The earlier 350 does not have this clutch.

✿ 10-3 Checking and adding transmission fluid Checking the fluid level and condition is covered in **✿** 9-3. Use only Dexron II transmission fluid or the equivalent. Using the wrong type of fluid could damage the transmission. Fluid level should be checked at every engine-oil change. If oil is needed, it should be added as follows. Every 100,000 miles [160,000 km] the oil pan should be drained and the strainer cleaned. Then fresh oil should be added, as explained later. If the car is used in severe service (**✿** 9-2), the oil pan should be drained and the strainer cleaned every 15,000 miles [24,000 km].

Fluid level can be checked with the engine cold or hot. But it is better to check the fluid level with the engine and transmission hot (180° F) [82.2° C]. Both methods are described below.

1. **Checking fluid level and adding fluid with transmission hot** Proceed as follows:

a. Block the car wheels and apply the parking brake. Start the engine. *Do not race the engine!* Move the selector lever through each range. If the parking brake releases, hold the car with the foot brake. Reapply the parking brake when you return the selector lever to park.

b. Immediately check the fluid level with the selector lever in park. The engine should be running at slow idle and the car should be on a level surface. The fluid level should be at the FULL mark on the dipstick (Fig. 10-1).

c. If the fluid level is low, add only enough to bring it up to FULL. *Do not overfill!* Overfilling will cause foaming and loss of fluid through the vent pipe. If too much fluid is lost in this manner, complete loss of drive through the transmission could occur.

d. Note the condition of the fluid and its level (❈ 9-3).

2. **Checking fluid level and adding fluid with transmission cold** Proceed as you did for checking with the transmission hot, with this exception: With the transmission cold (80° F [26.7° C] or less), the fluid level should be ¼ inch [6.35 mm] below the ADD mark on the dipstick. If it is lower, add fluid to bring the fluid level up to ¼ inch [6.35 mm] below the ADD mark, no more. When the transmission heats up, the fluid will expand and rise to the FULL mark. *Do not overfill!*

3. **Draining the oil pan and cleaning the strainer** The oil pan should be drained and the strainer cleaned every 100,000 miles [160,000 km] or 15,000 miles [24,000 km] in severe city service.

a. Raise the car on a lift and remove the oil pan and gasket. Throw the gasket away. Drain the oil from the pan. Clean the pan with solvent and dry it thoroughly with clean compressed air.

b. Remove the strainer assembly and clean it. Discard the oil-strainer-to-valve-body gasket.

c. Install a new gasket on the strainer and reinstall it.

d. Install a new gasket on the oil pan and install the oil pan. Tighten attaching bolts to 13 pound-feet [17.6 N·m].

e. Lower the car and add 1½ quarts [1.42 L] of transmission fluid.

f. Block the car wheels, set the parking brake, and start the engine. *Do not race the engine.* Move the selector lever through each range.

g. Immediately check the fluid level with the selector lever in P (park) and with engine idling. The car must be on a level surface.

h. Add fluid to bring the fluid level to ¼ inch [6.35 mm] below the ADD mark. *Do not overfill!*

❈ **10-4 Towing instructions** Usually, a car with rear-wheel drive may be safely towed on its rear wheels with the selector lever in neutral for short distances at speeds no greater than 35 mph [56 km/h]. But there are exceptions: If the transmission is not operating properly, if higher speeds are necessary, or if the car must be towed more than 50 miles [80 km], the *drive shaft must be disconnected*.

As an alternative, the car can be towed on its front wheels. Then, the steering wheel must be secured to keep the front wheels in the straight-ahead position.

❈ **10-5 Adjustments** Only two adjustments are required with the transmission in the car: linkage to the selector lever and linkage to the accelerator (for the detent downshift). The adjustments are different for different cars. Always refer to the appropriate shop manual for the car you are servicing. There, you will find the instructions that apply. Figures 10-2 to 10-5 show typical adjustment procedures for the shift linkage to the selector lever and for the detent or downshift cable.

❈ **10-6 Trouble diagnosis** Although the automatic transmission is complex and packs many moving parts into a small space, causes of troubles are usually easy to spot. A specific trouble can have only certain specific causes.

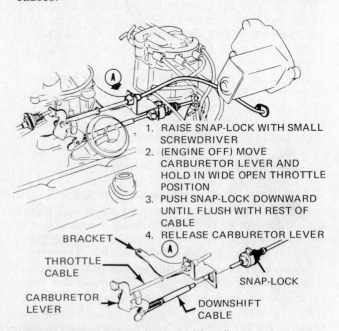

1. RAISE SNAP-LOCK WITH SMALL SCREWDRIVER
2. (ENGINE OFF) MOVE CARBURETOR LEVER AND HOLD IN WIDE OPEN THROTTLE POSITION
3. PUSH SNAP-LOCK DOWNWARD UNTIL FLUSH WITH REST OF CABLE
4. RELEASE CARBURETOR LEVER

BRACKET
THROTTLE CABLE
CARBURETOR LEVER
SNAP-LOCK
DOWNSHIFT CABLE

Fig. 10-2 Downshift or detent-cable adjustment on some car models. (*Oldsmobile Division of General Motors Corporation*)

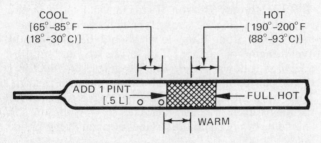

COOL [65°–85° F (18°–30° C)]
HOT [190°–200° F (88°–93° C)]

ADD 1 PINT [.5 L]
FULL HOT
WARM

NOTE: DO NOT OVERFILL. IT TAKES ONLY 1 PINT TO RAISE LEVEL FROM "ADD" TO "FULL" WITH A HOT TRANSMISSION.

Fig. 10-1 Dipstick for 350C automatic transmission. (*Chevrolet Motor Division of General Motors Corporation*)

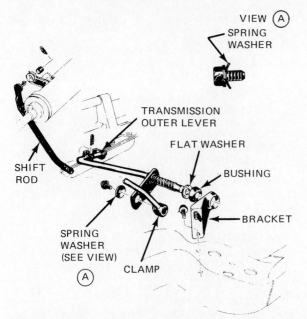

VIEW Ⓐ
SPRING WASHER

TRANSMISSION OUTER LEVER

FLAT WASHER

BUSHING

BRACKET

SHIFT ROD

SPRING WASHER (SEE VIEW)

CLAMP

Ⓐ

SHIFT-ROD ADJUSTMENT

1. WITH SHIFT-ROD CLAMP SCREW LOOSENED, SET TRANSMISSION OUTER LEVER IN NEUTRAL POSITION.
2. HOLD UPPER SHIFT LEVER AGAINST NEUTRAL POSITION STOP IN UPPER STEERING COLUMN (DO NOT RAISE LEVER).
3. TIGHTEN SCREW IN CLAMP ON LOWER END OF SHIFT ROD TO SPECIFIED TORQUE.
4. CHECK OPERATION:
   A. WITH KEY IN "RUN" POSITION AND TRANSMISSION IN "REVERSE," BE SURE THAT KEY CANNOT BE REMOVED AND THAT STEERING WHEEL IS NOT LOCKED.
   B. WITH KEY IN "LOCK" POSITION AND SHIFT LEVER IN "PARK," BE SURE THAT KEY CAN BE REMOVED, THAT STEERING WHEEL IS LOCKED, AND THAT THE TRANSMISSION REMAINS IN PARK WHEN THE STEERING IS LOCKED.
   C. WITH BRAKES FIRMLY APPLIED, CHECK TO MAKE SURE THAT THE STARTER WILL NOT WORK IN ANY SHIFT-LEVER POSITION EXCEPT NEUTRAL AND PARK.

Fig. 10-3 Adjustment of column shift linkage to manual valve on some car models. (*Oldsmobile Division of General Motors Corporation*)

Manufacturers who use the Type 350C automatic transmission recommend that oil-pressure checks be made with the transmission in operation to help locate causes of trouble. The procedures recommended, and the pressures specified under different operating conditions, vary from manufacturer to manufacturer. Always refer to the shop manual that covers the car and transmission you are checking. A typical diagnosis procedure follows.

The chart in Fig. 10-6 shows various troubles (across the top) and possible causes (to the left). This chart applies to both the earlier type 350 transmissions and the later type 350C transmission. Figure 10-7 is another chart that shows possible troubles with the torque-converter clutch and how to diagnose them. Figure 10-8 shows how to prepare test governors for three transmissions using the converter clutch. Figure 10-9 shows, in chart form, which components are holding during

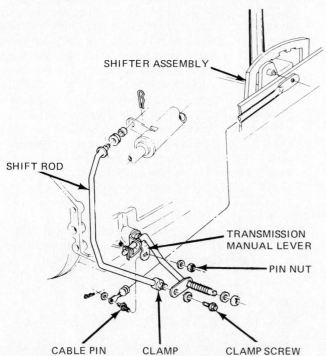

SHIFTER ASSEMBLY

SHIFT ROD

TRANSMISSION MANUAL LEVER

PIN NUT

CABLE PIN   CLAMP   CLAMP SCREW

SHIFT-CABLE ADJUSTMENT

1. LOOSEN SHIFT-ROD CLAMP SCREW, LOOSEN PIN IN TRANSMISSION MANUAL LEVER.
2. PLACE SHIFT LEVER IN "PARK" POSITION. PLACE TRANSMISSION MANUAL LEVER IN "PARK" POSITION AND IGNITION KEY IN LOCK POSITION.
3. TIGHTEN CABLE PIN NUT.
4. ROTATE THE TRANSMISSION MANUAL LEVER FULLY AGAINST THE "PARK" STOP, THEN RELEASE THE LEVER.
5. PULL SHIFT ROD DOWN AGAINST LOCK STOP TO ELIMINATE LASH AND TIGHTEN CLAMP SCREW.
6. CHECK OPERATION:
   A. MOVE SHIFT HANDLE INTO EACH GEAR POSITION AND SEE THAT TRANSMISSION MANUAL LEVER IS ALSO IN DETENT POSITION.
   B. WITH KEY IN "RUN" POSITION AND TRANSMISSION IN "REVERSE," BE SURE THAT KEY CANNOT BE REMOVED AND THAT STEERING WHEEL IS NOT LOCKED.
   C. WITH KEY IN "LOCK" POSITION, AND TRANSMISSION IN "PARK," BE SURE THAT KEY CAN BE REMOVED AND THAT STEERING WHEEL IS LOCKED.
   D. ENGINE MUST START IN PARK AND NEUTRAL.
   E. WITH BRAKES FIRMLY APPLIED, CHECK TO MAKE SURE THAT THE STARTER WILL NOT WORK IN ANY SHIFT-LEVER POSITION EXCEPT NEUTRAL AND PARK.

Fig. 10-4 Adjustment of console linkage to manual valve on some car models. (*Oldsmobile Division of General Motors Corporation*)

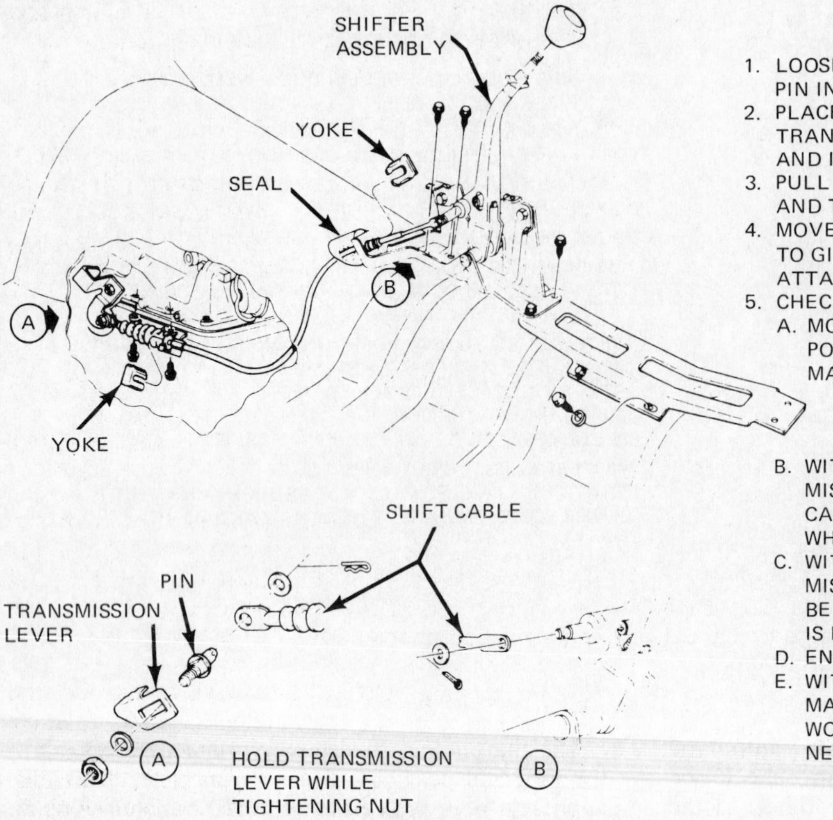
SHIFTER ASSEMBLY
YOKE
SEAL
A
B
YOKE
SHIFT CABLE
PIN
TRANSMISSION LEVER
A
HOLD TRANSMISSION LEVER WHILE TIGHTENING NUT
B
SHIFT-CABLE ADJUSTMENT
1. LOOSEN SHIFT-ROD CLAMP SCREW. LOOSEN PIN IN TRANSMISSION MANUAL LEVER.
2. PLACE SHIFT LEVER IN "P" POSITION. PLACE TRANSMISSION MANUAL LEVER IN "P" POSITION AND IGNITION KEY IN LOCK POSITION.
3. PULL SHIFT ROD LIGHTLY AGAINST LOCK STOP AND TIGHTEN CLAMP SCREW.
4. MOVE PIN IN MANUAL TRANSMISSION LEVER TO GIVE "FREE PIN" FIT AND TIGHTEN ATTACHING NUT.
5. CHECK OPERATION:
A. MOVE SHIFT HANDLE INTO EACH GEAR POSITION AND SEE THAT TRANSMISSION MANUAL LEVER IS ALSO IN DETENT POSITION.
B. WITH KEY IN "RUN" POSITION AND TRANSMISSION IN "REVERSE," BE SURE THAT KEY CANNOT BE REMOVED AND THAT STEERING WHEEL IS NOT LOCKED.
C. WITH KEY IN "LOCK" POSITION AND TRANSMISSION IN "PARK," BE SURE THAT KEY CAN BE REMOVED AND THAT STEERING WHEEL IS LOCKED.
D. ENGINE MUST START IN PARK AND NEUTRAL.
E. WITH BRAKES FIRMLY APPLIED, CHECK TO MAKE SURE THAT THE STARTER WILL NOT WORK IN ANY SHIFT-LEVER POSITION EXCEPT NEUTRAL AND PARK.

Fig. 10-5 Adjustment of console linkage to manual valve on some cars using a shift cable. (*Oldsmobile Division of General Motors Corporation*)

the various operating modes of the transmission. A typical diagnosis procedure follows.

1. Check and correct the oil level (see ✽ 10-3).
2. Check and correct the detent-cable adjustment.
3. Check the vacuum line and fittings for leakage.
4. Check and adjust the manual linkage as necessary.
5. Shop-test and road-test the car, as follows. Install oil-pressure gauges by connecting them to the transmission pressure taps. Gauges must have long hoses so they can be read from the driver's seat. Then test the car in the shop and on the road, using all selector-lever positions. Note the transmission operation and oil pressures under different conditions at three taps—line pressure, 1-2 clutch (intermediate) pressure, and 2-3 clutch (direct) pressure (Fig. 10-10). Refer to the manufacturer's shop manual for the specified pressures that you should read while making the tests. Note that items 3 to 8 along the top of the chart in Fig. 10-6 refer to high or low pressures—line, 1-2 clutch, and 2-3 clutch. Therefore the chart will help you find causes of incorrect pressures.

NOTE: If the engine is not performing satisfactorily, it should be tuned or serviced as necessary to restore normal operation. Poor engine performance can cause faulty transmission shifts and improper operation.

CAUTION: When road-testing a car, obey all traffic laws. Drive safely. Use a chassis dynamometer (✽ 1-15) if it is available so that you can make all the tests in the shop and will not have to drive the car on the road.

6. Refer to the trouble-diagnosis chart (Fig. 10-6) to determine possible causes of troubles that show up during the tests. In most cases, the chart will help you locate the causes so that you can make the necessary repairs. Sometimes, it will be necessary to remove the transmission from the car for disassembly so that defective parts can be replaced. However, other troubles may be solved by replacement of parts without removal of the transmission.
7. Refer to the detailed diagnostic procedure in the chart in Fig. 10-7 to locate the cause if the converter clutch does not engage, engages improperly, or is engaged at all times. The chart also explains how to check the thermal vacuum valve and the low vacuum switch.

✽ 10-7 On-the-car repairs Some parts can be removed and replaced without removing the transmission from the car. These include:

Oil pan and gasket

Oil screen

Valve-body assembly

Direct-clutch accumulator and servo (after valve body is removed)

Manual control linkage and parking pawl

Extension housing and seal

Vacuum modulator

Speedometer drive gear

Transmission-control-spark switch

Governor assembly

Intermediate-clutch accumulator

Governor pressure switch

Electrical connector

Solenoid and check ball (350C)

Auxiliary valve body and valve (350C)

These parts, with the exception of the last six, require the same procedure used during complete disassembly and reassembly of the transmission. Therefore, these removal-and-replacement procedures are explained on later pages where complete transmission disassembly and reassembly are described. Governor-assembly, intermediate-clutch-accumulator, governor-pressure-switch, electrical-connector, solenoid, and auxiliary-valve-body and valve removal and replacement require different procedures if the transmission is not removed from the car. These procedures are described next.

❋ **10-8 Governor replacement** The governor assembly can be removed and replaced with the transmission in the car, as follows:

1. Raise the car on a lift and disconnect the speedometer cable at the transmission.
2. Remove the governor cover retainer and cover. Do not damage the O-ring seal.
3. Remove the governor. Check weights and valve to make sure they move freely. If they do not, repair or replace the parts as necessary.
4. To install the governor, slip it into place and install the cover, using a brass drift punch around the outside flange of the cover. Do not distort the cover. Be sure the O-ring seal is not cut or damaged.
5. Install the retainer and connect the speedometer cable. Lower the car from the lift and check the fluid level.

❋ **10-9 Intermediate-clutch-accumulator replacement** The intermediate-clutch accumulator can be removed and replaced with the transmission in the car, as follows:

1. Remove the two transmission oil-pan bolts below the intermediate-clutch cover. Install the accumulator-cover remover in place of the bolts removed (Fig. 10-11).
2. Press in on the tool handle and use an awl to remove the retaining ring.
3. Remove the cover O-ring seal, spring, and intermediate-clutch accumulator.
4. To install put the intermediate-clutch accumulator into place. Lubricate the rings. Rotating the piston slightly will help start the rings into the piston bore.

5. Put the spring, O-ring seal, and cover into place. Press in on the cover and install the retaining ring.
6. Remove the tool and install the oil-pan bolts.

❋ **10-10 Governor pressure switch** The governor pressure switch is used on the 200C and 350C transmissions, which have the converter clutch. It can be removed by removing the oil pan, disconnecting the lead from the governor pressure switch, and using a $1^1/_{16}$-inch oil-sending-unit socket to remove the switch. Install the switch by reversing the procedure.

❋ **10-11 Electrical connector** The electrical connector can be removed by removing the oil pan and disconnecting the wires from the connector. The connector can then be removed by compressing the fingers on the connector sleeve (Fig. 10-12).

❋ **10-12 Solenoid** On the 350C (and 250C also), the solenoid can be removed by removing the oil pan and filter and disconnecting the solenoid wire clip and wires. Then remove the two attaching bolts and the solenoid (Fig. 10-13).

❋ **10-13 Auxiliary valve body and valve** On the 350C, the apply valve is installed in an auxiliary valve body. On the 200C, the apply valve is installed in the front pump. To remove the 200C apply valve, the transmission must be removed from the car. But the apply valve and body can be removed from the 350C by removing the solenoid (❋ 10-12) and the two bolts holding the auxiliary valve body (Fig. 10-13). Then the valve can be removed from the auxiliary valve body by removing the retainer pin and retainer.

Transmission Overhaul

❋ **10-14 Transmission removal and installation** Individual variations in engine-compartment arrangements require different removal and installation procedures. Always check the shop manual for the car you are working on for the specific instructions that apply. The following is a typical removal-and-replacement procedure.

Before raising the car, disconnect the negative battery cable and release the parking brake. Then proceed as follows:

1. Raise the car on a lift and scribe marks on the drive shaft and flange so that the drive shaft can be reinstalled in its original position. Then remove the drive shaft.
2. Disconnect the shift-control linkage.
3. Support the transmission with a transmission jack.
4. Disconnect the rear mount from the frame cross member.
5. Remove two bolts at each end of the frame cross member and remove the cross member.
6. Remove the oil-cooler lines, vacuum-modulator line, speedometer cable, and detent cable at the transmission.
7. Remove the pan from under the converter. Mark

(*Text continued on page 124*)

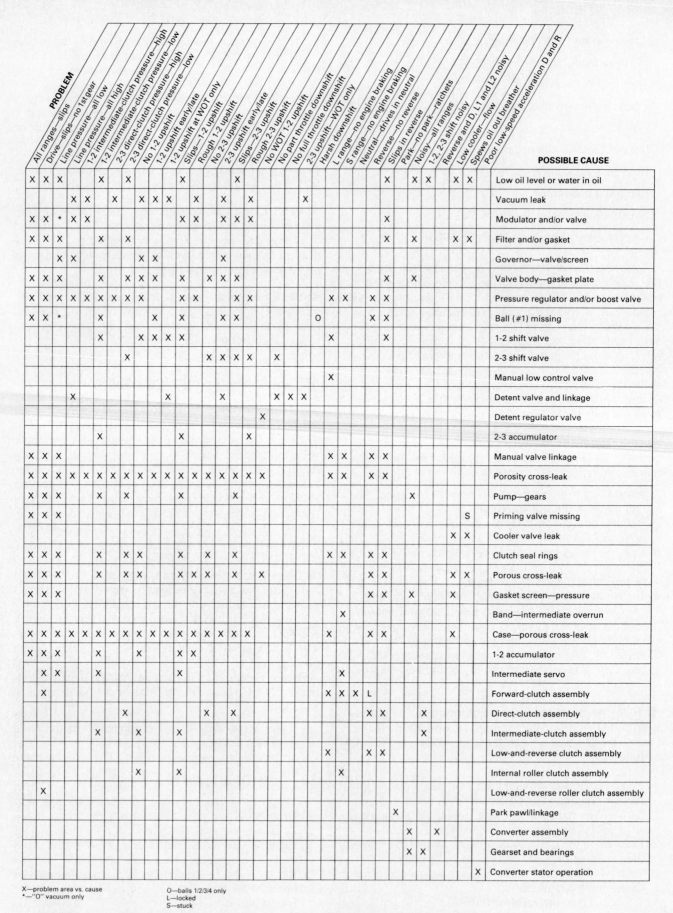

Fig. 10-6 Trouble-diagnosis chart for the 350 automatic transmission. (*Oldsmobile Division of General Motors Corporation*)

TORQUE CONVERTER CLUTCH DIAGNOSIS (GAS)

THE ENGINE MUST BE RUNNING PROPERLY AND ALL ADJUSTMENTS SET TO SPECIFICATIONS BEFORE ROAD TESTING OR ATTEMPTING TO DIAGNOSE THE TRANSMISSION. POOR ENGINE PERFORMANCE CAN RESULT IN ROUGH SHIFTING OR OTHER MALFUNCTIONS.

CONVERTER CLUTCH DOES NOT APPLY

ENGINE AT NORMAL OPERATING TEMPERATURE. CHECK TRANSMISSION OIL LEVEL AND CORRECT AS NECESSARY. CHECK MANUAL LINKAGE ADJUSTMENT AND CORRECT AS REQUIRED. CHECK GAUGES FUSE. WITH IGNITION ON, CHECK CRUISE AND T.C.C. BRAKE SWITCH FOR 12V AT BOTH BLU LT AND PNK D/BLK WIRE TERMINALS. (BRAKE PEDAL RELEASED.)

- **12V AT ONE TERMINAL ONLY** — CHECK BRAKE SWITCH ADJUSTMENT. IF OK, REPLACE BRAKE SWITCH.

- **12V NOT PRESENT AT EITHER TERMINAL** — CHECK AND REPAIR OPEN IN PNK D/BLK WIRE FROM CRUISE AND T.C.C. BRAKE SWITCH TO 39 SPLICE.

- **12V AT BOTH TERMINALS** — ROAD TEST CAR
  - CONVERTER CLUTCH APPLIES, RETURN CAR TO OWNER.
  - CONVERTER CLUTCH DOES NOT APPLY — DISCONNECT ELECTRIC CONNECTOR AT TRANSMISSION CASE AND CHECK FEMALE CONNECTOR FOR 12 VOLTS USING VOLT/OHM METER. (2000 RPM IN NEUTRAL.)

NO VOLTAGE AT CONNECTOR — CHECK FOR 12V AT BOTH LOW VACUUM SWITCH TERMINALS

- **12V AT ONE TERMINAL ONLY** — CHECK FOR VACUUM AT LOW VACUUM SWITCH
  - VACUUM PRESENT, CHECK LOW VACUUM SWITCH OPERATION. (SEE LOW VACUUM SWITCH CHECK.) — REPLACE IF NECESSARY
  - NO VACUUM, CHECK VACUUM HOSES AND REPAIR AS REQUIRED. IF VACUUM HOSES OK, CHECK THERMAL VACUUM VALVE OPERATION IF EQUIPPED. (SEE THERMAL VACUUM VALVE CHECK.)

- **12V AT BOTH TERMINALS** — REPAIR OPEN IN WIRE FROM LOW VACUUM SWITCH TO TRANSMISSION CASE CONNECTOR

- **12V NOT PRESENT AT EITHER TERMINAL** — CHECK FOR OPEN IN WIRING FROM LOW VACUUM SWITCH TO CRUISE AND T.C.C. BRAKE SWITCH. REPAIR AS REQUIRED.

12 VOLTS PRESENT AT CONNECTOR — USING A 12-VOLT TEST LIGHT, CONNECT POSITIVE LEAD TO FEMALE CONNECTOR AND GROUND LEAD TO MALE CONNECTOR.

START ENGINE IN "PARK" AND MAINTAIN 2000 RPM

- **TEST LIGHT "ON"** — GOVERNOR SWITCH OR WIRING INTERNALLY SHORTED TO GROUND — CHECK WIRING AND/OR REPLACE GOVERNOR SWITCH

- **TEST LIGHT "OFF"** — RAISE REAR WHEELS AND RUN CAR IN DRIVE RANGE UNTIL 3D GEAR UPSHIFT IS OBTAINED. MAINTAIN 2000 RPM.

(A)

Fig. 10-7 Torque-converter-clutch trouble-diagnosis chart. (*Oldsmobile Division of General Motors Corporation*) (Figure 10-7 continued on pages 122 and 123.)

Fig. 10-7 TORQUE CONVERTER CLUTCH DIAGNOSIS (GAS) (CONTINUED)

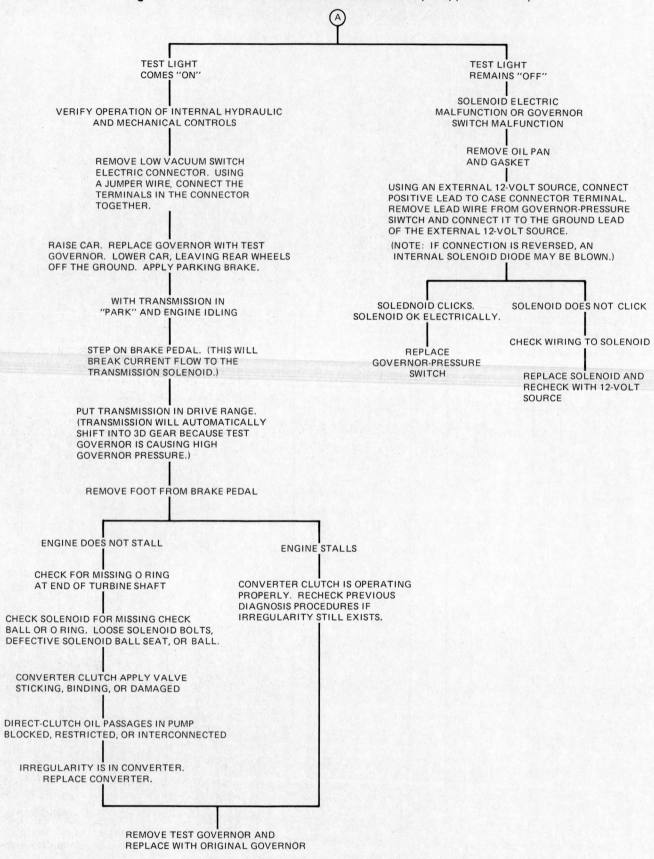

(A)

TEST LIGHT COMES "ON"

VERIFY OPERATION OF INTERNAL HYDRAULIC AND MECHANICAL CONTROLS

REMOVE LOW VACUUM SWITCH ELECTRIC CONNECTOR. USING A JUMPER WIRE, CONNECT THE TERMINALS IN THE CONNECTOR TOGETHER.

RAISE CAR. REPLACE GOVERNOR WITH TEST GOVERNOR. LOWER CAR, LEAVING REAR WHEELS OFF THE GROUND. APPLY PARKING BRAKE.

WITH TRANSMISSION IN "PARK" AND ENGINE IDLING

STEP ON BRAKE PEDAL. (THIS WILL BREAK CURRENT FLOW TO THE TRANSMISSION SOLENOID.)

PUT TRANSMISSION IN DRIVE RANGE. (TRANSMISSION WILL AUTOMATICALLY SHIFT INTO 3D GEAR BECAUSE TEST GOVERNOR IS CAUSING HIGH GOVERNOR PRESSURE.)

REMOVE FOOT FROM BRAKE PEDAL

ENGINE DOES NOT STALL

CHECK FOR MISSING O RING AT END OF TURBINE SHAFT

CHECK SOLENOID FOR MISSING CHECK BALL OR O RING, LOOSE SOLENOID BOLTS, DEFECTIVE SOLENOID BALL SEAT, OR BALL.

CONVERTER CLUTCH APPLY VALVE STICKING, BINDING, OR DAMAGED

DIRECT-CLUTCH OIL PASSAGES IN PUMP BLOCKED, RESTRICTED, OR INTERCONNECTED

IRREGULARITY IS IN CONVERTER. REPLACE CONVERTER.

ENGINE STALLS

CONVERTER CLUTCH IS OPERATING PROPERLY. RECHECK PREVIOUS DIAGNOSIS PROCEDURES IF IRREGULARITY STILL EXISTS.

TEST LIGHT REMAINS "OFF"

SOLENOID ELECTRIC MALFUNCTION OR GOVERNOR SWITCH MALFUNCTION

REMOVE OIL PAN AND GASKET

USING AN EXTERNAL 12-VOLT SOURCE, CONNECT POSITIVE LEAD TO CASE CONNECTOR TERMINAL. REMOVE LEAD WIRE FROM GOVERNOR-PRESSURE SIWTCH AND CONNECT IT TO THE GROUND LEAD OF THE EXTERNAL 12-VOLT SOURCE.

(NOTE: IF CONNECTION IS REVERSED, AN INTERNAL SOLENOID DIODE MAY BE BLOWN.)

SOLEDNOID CLICKS. SOLENOID OK ELECTRICALLY.

REPLACE GOVERNOR-PRESSURE SWITCH

SOLENOID DOES NOT CLICK

CHECK WIRING TO SOLENOID

REPLACE SOLENOID AND RECHECK WITH 12-VOLT SOURCE

REMOVE TEST GOVERNOR AND REPLACE WITH ORIGINAL GOVERNOR

Fig. 10-7 TORQUE CONVERTER CLUTCH DIAGNOSIS (GAS) (CONTINUED)

```
┌─────────────────────────────┐        ┌─────────────────────────────┐
│   CONVERTER CLUTCH APPLIES   │        │   CONVERTER CLUTCH APPLIES   │
│         ERRATICALLY          │        │  AT VERY LOW OR VERY HIGH    │
│                              │        │        3D GEAR SPEEDS        │
└─────────────────────────────┘        └─────────────────────────────┘
```

VACUUM HOSE LEAK.
VACUUM SWITCH MALFUNCTION.
RELEASE OIL-EXHAUST ORIFICE AT
PUMP BLOCKED OR RESTRICTED.
TURBINE SHAFT O RING DAMAGED
OR MISSING.
SOLENOID BOLTS LOOSE.
GOVERNOR-PRESSURE SWITCH
MALFUNCTION.

GOVERNOR SWITCH MALFUNCTION.
GOVERNOR MALFUNCTION.
HIGH LINE PRESSURES.
CONVERTER CLUTCH VALVE
STICKING OR BINDING.
SOLENOID MALFUNCTION.

```
┌─────────────────────────────┐        ┌─────────────────────────────┐
│   CONVERTER CLUTCH APPLIED   │        │   CONVERTER CLUTCH APPLIED   │
│  IN ALL RANGES—ENGINE STALLS │        │   AT ALL TIMES IN 3D GEAR    │
│ WHEN TRANSMISSION IS PUT INTO│        │                              │
│            GEAR              │        └─────────────────────────────┘
└─────────────────────────────┘
```

CONVERTER CLUTCH APPLY VALVE
STUCK IN THE APPLY POSITION.

GOVERNOR-PRESSURE SWITCH
SHORTED TO GROUND. GROUND
WIRE FROM SOLENOID SHORTED
TO CASE.

THERMAL VACUUM VALVE CHECK

1. DISCONNECT VACUUM HOSE AT LOW VACUUM SWITCH.
2. ATTACH VACUUM GAUGE TO HOSE.
3. START ENGINE AND CHECK VACUUM IN "PARK."
   A. WITH ENGINE COLD, ENGINE COOLANT TEMPERATURE
      BELOW $120°$F 3.8 L (VIN A), $130°$F 5.0 L
      (VIN H), $150°$F 4.3 L (VIN S), VACUUM AT
      IDLE AND AT 2000 RPM SHOULD READ ZERO.
   B. WITH ENGINE WARM, ENGINE COOLANT TEMPERATURE
      ABOVE $120°$F 3.8 L (VIN A), $130°$F 5.0 L (VIN H), $150°$F
      4.3 L (VIN S), VACUUM AT IDLE SHOULD BE ZERO WHILE
      VACUUM AT 2000 RPM SHOULD BE 10 INCHES MINIMUM.

LOW VACUUM SWITCH CHECK

1. DISCONNECT VACUUM HOSE AND ELECTRICAL CONNECTOR FROM VACUUM SWITCH.
2. ATTACH ONE LEAD OF A TEST LIGHT TO EITHER TERMINAL OF THE VACUUM SWITCH.
3. GROUND THE REMAINING TERMINAL OF THE VACUUM SWITCH.
4. ATTACH THE REMAINING LEAD OF THE TEST LIGHT TO THE HOT (+12V) SIDE OF THE
   VACUUM SWITCH CONNECTOR.
5. ATTACH A HAND VACUUM PUMP WITH A GAUGE TO THE VACUUM PORT OF THE
   VACUUM SWITCH.
6. TURN CAR IGNITION ON.
7. IF USING A SELF-POWERED TEST LIGHT, CONNECT ONE LEAD OF THE TEST LIGHT TO
   EITHER TERMINAL OF THE VACUUM SWITCH AND CONNECT THE REMAINING LEAD
   OF THE TEST LIGHT TO THE OTHER TERMINAL OF THE VACUUM SWITCH. CAR IGNITION
   DOES NOT HAVE TO BE TURNED ON WHEN USING THIS TYPE OF TEST LIGHT.
8. ACTUATE HAND VACUUM PUMP. TEST LIGHT SHOULD BE OFF AND REMAIN OFF
   UNTIL VACUUM GAUGE READS SPECIFIED VACUUM FOR SWITCH BEING CHECKED.
9. DECREASE VACUUM SLOWLY. TEST LIGHT SHOULD REMAIN ON UNTIL VACUUM
   DROPS TO SPECIFIED READING.
10. IF THE VACUUM SWITCH DOES NOT TURN THE TEST LIGHT ON AND OFF AT THE
    SPECIFIED VALUES, THE VACUUM SWITCH IS MALFUNCTIONING.

250C AND 350C TEST GOVERNOR
1. OBTAIN KNOWN GOOD GOVERNOR FOR TEST PURPOSES.
2. CUT TWO LENGTHS OF 3/8-INCH O.D. RUBBER HOSE, 3/8-INCH LONG.
3. INSERT ONE PIECE OF HOSE UNDER EACH WEIGHT OF THE GOVERNOR.

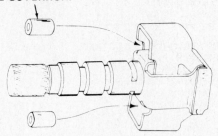

200C TEST GOVERNOR
1. OBTAIN KNOWN GOOD GOVERNOR FOR TEST PURPOSES.
2. CUT TWO LENGTHS OF 3/8-INCH O.D. RUBBER HOSE, 3/8-INCH LONG. INSERT HOSE BETWEEN GOVERNOR SHAFT AND GOVERNOR WEIGHTS AS SHOWN.

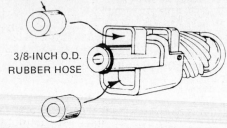

3/8-INCH O.D. RUBBER HOSE

Fig. 10-8 How to prepare test governor to make the test called for in the chart in Fig. 10-7. (*Oldsmobile Division of General Motors Corporation*)

the converter and flywheel so that they can be re-attached in their original position. Then remove the converter-to-flywheel bolts.

8. Loosen the exhaust-pipe-to-manifold bolts about ¼ inch [6.35 mm] (if required). Lower the transmission until the jack is barely supporting it.

Careful: Watch the clearance of all engine components while the rear of the transmission is being low-ered. Be careful that the ignition distributor is not forced against the dash and damaged.

9. Remove the transmission-to-engine mounting bolts and remove the oil-filler tube from the transmission.
10. Raise the transmission to its normal position, support the engine with a jack, and slide the transmission rearward from the engine. Then lower the transmission from the vehicle.

CAUTION: Use a converter-holding tool, or keep the rear of the transmission lower than the front. Otherwise, the converter is likely to slide off and drop to the floor. This could injure you or someone nearby and damage the torque converter.

11. Installation is essentially the reverse of removal except for the following added steps: Before installing the flex-plate-to-converter bolts, make sure that the attaching lugs on the converter are flush with the flex plate and that the converter rotates freely by hand in this position. Then, hand-start all three bolts and tighten them finger tight before torquing to specifications. This ensures proper alignment of the converter (Fig. 10-14). After installation is complete, remove the car from the lift and check the linkage for proper adjustment. Check the fluid level (❂ 10-3) and add fluid as necessary.

NOTE: Some transmissions may require shims at the rear transmission support (Fig. 10-15) to correct the drive-shaft angle. When removing a transmission, always count and save any shims so you can put them back when reinstalling the transmission. If shims are not reinstalled, the drive-shaft angle will be wrong. This can cause rapid wear of the universal joints in the drive shaft.

❂ **10-15 Transmission-service precautions** Cleanliness is very important during transmission service. The tiniest, almost invisible piece of lint from a shop

RANGE	GEAR	INTERMEDIATE CLUTCH	DIRECT CLUTCH	FORWARD CLUTCH	LOW-AND-REVERSE CLUTCH	INTERMEDIATE OVERRUN ROLLER CLUTCH	LOW-AND-REVERSE ROLLER CLUTCH	INTERMEDIATE OVERRUN BAND
Park, neutral		Off	Off	Off	Off	Ineffective	Ineffective	Off
Drive	First	Off	Off	On	Off	Locked	Locked	Off
	Second	On	Off	On	Off	Locked	Overrunning	Off
	Third	On	On	On	Off	Overrunning	Overrunning	Off
Second	First	Off	Off	On	Off	Locked	Locked	Off
	Second	On	Off	On	Off	Locked	Overrunning	On
Low	First	Off	Off	On	On	Locked	Locked	Off
Reverse		Off	On	Off	On	Ineffective	Ineffective	Off

Fig. 10-9 Clutch-engagement and band-application chart. (*Oldsmobile Division of General Motors Corporation*)

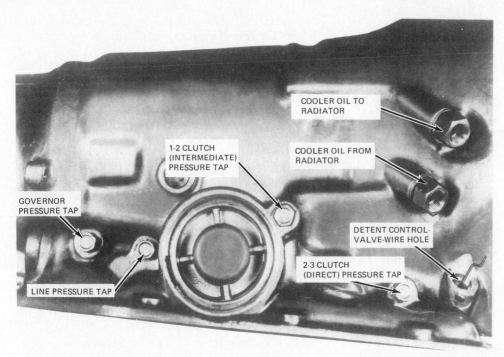

towel or a small piece of dirt can cause a valve to hang and prevent normal transmission action. Under some conditions, this could cause severe damage to the transmission. See ✿ 9-10, which discusses cleanliness and other transmission-service precautions.

✿ 10-16 Cleaning and inspecting transmission parts

After disassembly, clean and inspect all parts, as follows: Wash all metal parts in solvent and blow them dry with compressed air. Do not use cleaning solvents that could damage rubber seals or clutch facings. Make sure all oil passages are clean. Check small passages with small wire such as tag wire. ✿ 9-11 discusses special checks of transmission parts to be made during disassembly.

✿ 10-17 Transmission disassembly

None of the moving parts requires forcing when disassembling or reassembling the transmission. However, bushing removal and replacement require driving or pressing. The cases might fit tightly, and you can loosen them with a rawhide or plastic mallet. Never use a hard hammer.

As a first step, the transmission must be placed in a holding fixture (Fig. 10-16). Proceed as follows:

1. Turn the transmission so that the oil pan is up. If you have used a torque-converter holding tool, remove it. Then slip off the torque converter.
2. Remove the vacuum-modulator-assembly attaching bolt and retainer. Now, remove the vacuum-modulator assembly, O-ring seal, and modulator

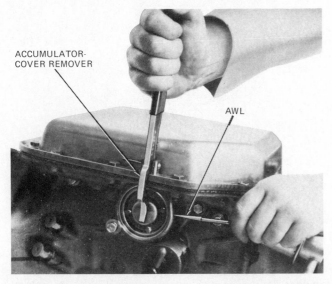

Fig. 10-11 Using a special tool to remove the intermediate-clutch-accumulator ring.

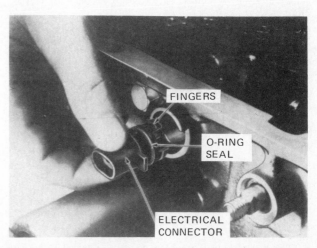

Fig. 10-12 Removing electrical connector from 350C automatic transmission. (*Buick Motor Division of General Motors Corporation*)

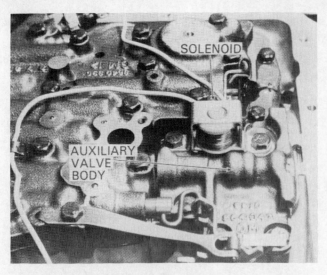

Fig. 10-13 Location of solenoid and auxiliary valve body in the 350C automatic transmission. (*Buick Motor Division of General Motors Corporation*)

valve (Fig. 10-17). The vacuum modulator is checked as explained in ✿ 10-26.

3. Remove the extension housing by taking out the four attaching bolts. Remove the square-cut seal ring. Use a screwdriver to remove the extension-housing lip seal from the housing (Fig. 10-18). If it is necessary to replace the housing bushing, use a screwdriver to collapse it (Fig. 10-19). Clean the housing carefully and inspect it for any damage. Then install a new housing bushing and a new housing lip seal with special tools. The extension housing is now ready to be reinstalled on the transmission case. Set the extension housing aside until the rest of the transmission has been disassembled and reassembled.

4. To remove the yoke seal from the end of the output shaft, install the special tool, as shown in Fig. 10-20, to keep the seal from cocking. Then tap the seal off with a screwdriver. Next, take off the speedometer drive gear by depressing the retaining clip. The gear can be slid off the shaft with the clip depressed. Remove the clip.

5. Next, remove the governor by first prying off the governor retaining wire with a screwdriver. Remove the governor cover and O-ring seal from the case. Use a screwdriver to pry between the cover flange and the case. But be extremely careful to avoid denting or bending the cover or gouging the transmission case. The dimple in the end of the cover provides the proper end play for the governor. If the cover is bent or damaged in any way, it must be replaced. Remove the O-ring seal from the cover. Now, the governor assembly can be withdrawn from the case (Fig. 10-21).

6. Remove the oil pan, strainer, and valve body. First, take out 13 attaching-screw-and-washer assemblies and lift off the oil pan and gasket. Then remove the two screws attaching the oil-pump strainer to the valve body. Take the oil-pump-strainer gasket off the valve body.

Next, remove the detent-spring-and-roller assembly from the valve body (Fig. 10-22). Remove the bolts attaching the valve body to the case. Now, lift the valve body from the case, carefully guiding the manual-valve link from the range-selector inner lever. Remove the detent-control-valve link from

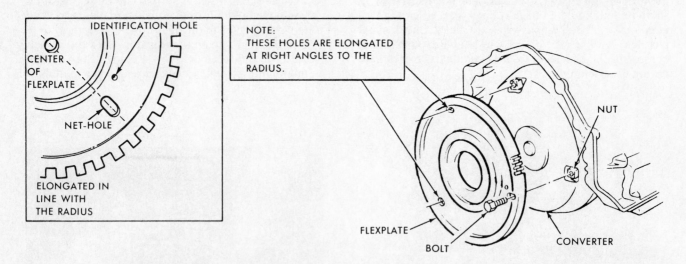

1. MOVE CONVERTER FORWARD TO CONTACT ATTACHING SURFACE ON FLEXPLATE PRIOR TO TIGHTENING BOLTS.

2. ALIGN SLOT IN FLEXPLATE THAT HAS AN IDENTIFICATION HOLE NEAR IT, WITH AN ATTACHMENT HOLE IN CONVERTER. INSTALL BOLT AND NUT AND TIGHTEN TO SPECIFIED TORQUE. TIGHTEN ALL REMAINING BOLTS TO SPECIFIED TORQUE AS THEY ARE INSTALLED.

Fig. 10-14 Attachment holes and bolts for attaching the converter flex plate to the converter. (*Chevrolet Motor Division of General Motors Corporation*)

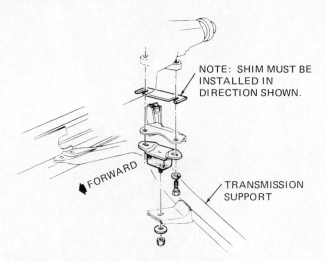

Fig. 10-15 Some cars have shims under the rear transmission support. These shims raise or lower the rear of the transmission as necessary to correct the drive-line angle. (*Chevrolet Motor Division of General Motors Corporation*)

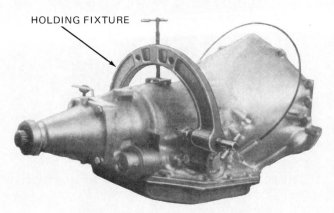

Fig. 10-16 Transmission in a holding fixture. (*Chevrolet Motor Division of General Motors Corporation*)

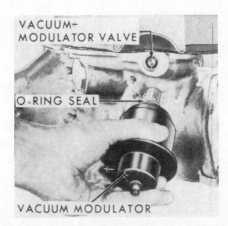

Fig. 10-17 Removing the vacuum-modulator assembly. (*Buick Motor Division of General Motors Corporation*)

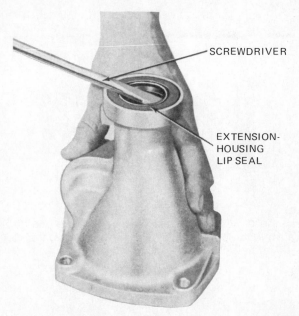

Fig. 10-18 Removing the extension-housing lip seal with a screwdriver. (*Buick Motor Division of General Motors Corporation*)

the detent actuating lever. Lay the valve body aside on a clean surface in preparation for further disassembly.

Remove from the case the following: valve-body-to-spacer-plate gasket, spacer-support-plate bolts, and the spacer support plate (Fig. 10-23). Then remove the valve-body spacer plate and gasket (Fig. 10-24).

7. From the case, remove the oil-pump pressure screen from the pressure hole (Fig. 10-25) and the

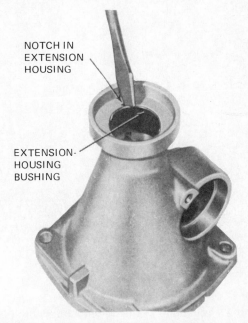

Fig. 10-19 Removing the extension-housing bushing with a screwdriver. (*Buick Motor Division of General Motors Corporation*)

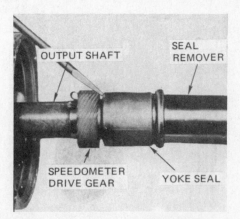

Fig. 10-20 Removing the yoke seal from the end of the transmission output shaft. (*Oldsmobile Division of General Motors Corporation*)

Fig. 10-21 Removing the governor assembly from the case. (*Buick Motor Division of General Motors Corporation*)

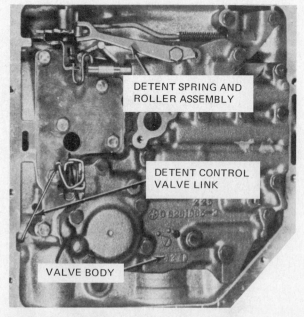

Fig. 10-22 Locations of the detent-spring-and-roller assembly and the detent-control-valve link on the valve body. (*Buick Motor Division of General Motors Corporation*)

Fig. 10-23 Removing the spacer support plate. (*Buick Motor Division of General Motors Corporation*)

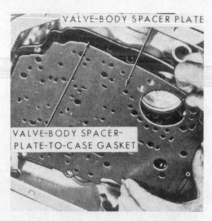

Fig. 10-24 Removing the valve-body spacer plate and gasket. (*Buick Motor Division of General Motors Corporation*)

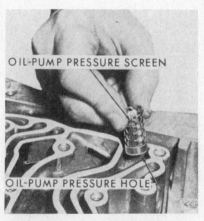

Fig. 10-25 Removing the oil-pump pressure screen. (*Buick Motor Division of General Motors Corporation*)

Fig. 10-26 Removing the governor screens. (*Buick Motor Division of General Motors Corporation*)

Fig. 10-28 Removing the manual-shaft-to-case retainer. (*Buick Motor Division of General Motors Corporation*)

governor screens (Fig. 10-26). Clean the screens and lay them aside in a clean, safe place. Remove the four check balls from the case face. Their locations are shown in Fig. 10-27.

8. Next, remove the manual shaft, inner lever, and parking pawl. First, take off the manual-control-valve-link retainer from the range-selector inner lever. Then use a screwdriver (Fig. 10-28) to remove the manual-shaft-to-case retainer. Next, use a wrench (Fig. 10-29) to remove the jam nut holding the range-selector inner lever to the manual shaft. Now, remove the manual shaft from the case. Remove the range-selector inner lever and parking-pawl actuating rod. If the manual-shaft-to-case lip seal is damaged, remove it with a screwdriver.

If it is necessary to remove the parking pawl or shaft, proceed as follows: Remove the parking-lock bracket. Then remove the retaining plug, parking-pawl shaft, parking pawl, and disengaging spring. These parts are shown in Fig. 10-30.

Fig. 10-29 Removing the jam nut from the manual shaft. (*Buick Motor Division of General Motors Corporation*)

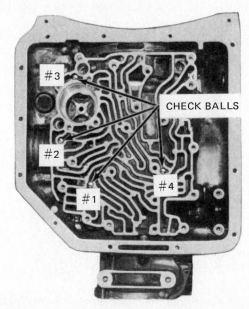

Fig. 10-27 Locations of the check balls in case. (*Buick Motor Division of General Motors Corporation*)

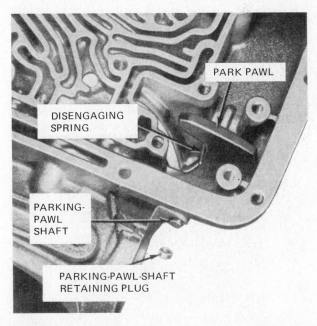

Fig. 10-30 Locations of the parking pawl, parking-pawl shaft, parking-pawl-shaft retaining plug, and disengaging spring. (*Buick Motor Division of General Motors Corporation*)

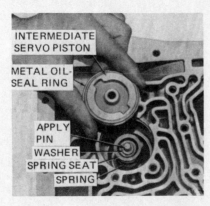

Fig. 10-31 Removing the intermediate-servo piston and metal oil-seal ring. (*Buick Motor Division of General Motors Corporation*)

9. Remove the intermediate-servo piston and metal oil-seal ring (Fig. 10-31). Then remove the washer, spring seat, and apply pin. Now, check to see if a new long or short apply pin is required on reassembly. Do this by using a special apply-pin gauge and straightedge, as shown in Fig. 10-32. Press down on the gauge and note whether the upper end of the gauge is above or below the straightedge. There are two replacement apply pins: long and short. If the gauge is below the straightedge, the long pin should be used; if above, the short pin should be used. Selecting the proper length of pin is equivalent to adjusting the band. Make a note of which pin is required so that it can be installed on reassembly.

10. Check the gear-train end play by attaching a dial indicator on the case. Set the finger against the end of the turbine shaft and move it forward and back.

11. Remove the pump as follows: With the transmission turned pump-end up, remove the eight pump-attaching bolts and washer-type seals. Discard the seals. Install two threaded slide hammers into the threaded holes in the pump body and tighten the jam nuts. Use the slide hammers to pull the pump up and out of the case (Fig. 10-33). Remove and discard the pump-assembly-to-case gasket. Lay the pump aside on a clean surface for later disassembly.

12. Remove the intermediate-clutch plates and band as follows: First take out the intermediate-clutch

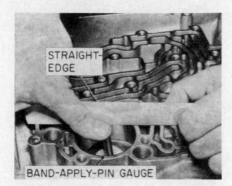

Fig. 10-32 Using a special gauge and straightedge to determine which apply pin to use on reassembly (*Buick Motor Division of General Motors Corporation*)

Fig. 10-33 Removing the pump assembly with slide hammers. (*Buick Motor Division of General Motors Corporation*)

cushion spring. Then remove the intermediate-clutch plates, as shown in Fig. 10-34. Inspect the plates as explained in ✿ 9-11. Next, remove the intermediate-clutch pressure plate and the overrun band.

13. Remove the forward- and direct-clutch assemblies from the case (Fig. 10-35). Lay them aside in a clean place to await further disassembly.

14. Remove the front input-ring-gear thrust washer, and the input ring gear. Check the bushing in the ring gear. If it is worn or galled, use a special tool to remove and replace it. Thread the tool onto the drive handle and use the tool, as shown in Fig. 10-36, to remove the bushing. The same tool is used to install the new bushing.

15. To remove the output-carrier assembly, first remove the input-ring-gear-to-output-carrier thrust washer. Then remove and discard the output-carrier-to-output-shaft snap ring. Now the output-carrier assembly will slide out, as shown in Fig. 10-37. Lay it aside on a clean surface to await further disassembly.

16. Next, remove the sun-gear-and-drive-shell assembly by pulling it out.

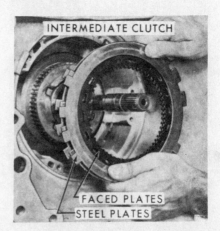

Fig. 10-34 Removing the intermediate-clutch plates. The faced plates are also known as *composition plates*. (*Buick Motor Division of General Motors Corporation*)

Fig. 10-35 Removing the forward- and direct-clutch assemblies. (*Buick Motor Division of General Motors Corporation*)

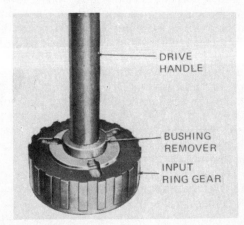

DRIVE HANDLE

BUSHING REMOVER

INPUT RING GEAR

Fig. 10-36 Removing the bushing from the input ring gear with special tools. (*Buick Motor Division of General Motors Corporation*)

OUTPUT-CARRIER ASSEMBLY

Fig. 10-37 Removing the output-carrier assembly. (*Buick Motor Division of General Motors Corporation*)

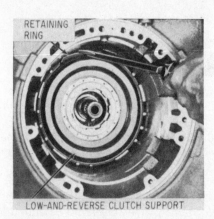

RETAINING RING

LOW-AND-REVERSE CLUTCH SUPPORT

Fig. 10-38 Removing the low-and-reverse-clutch-support retaining ring. (*Buick Motor Division of General Motors Corporation*)

17. Remove the low-and-reverse-clutch-support assembly by first prying out the retaining ring, as shown in Fig. 10-38. Now, grasp the output shaft and pull up until the low-and-reverse-roller-clutch-support assembly clears the low-and-reverse-clutch-support retainer spring. The support assembly will then slide out. Now, use pliers to remove the retainer spring.

18. Reach into the case and remove the low-and-reverse-clutch plates and the reaction carrier. Inspect the clutch plates, as explained in ✿ 9-11. If the reaction-carrier bushing is worn or galled, replace it with the special tools.

19. Remove the output-ring-gear-and-shaft assembly from the case. Remove the output-ring-gear-to-case needle bearing from either the case or the output shaft. Then take the tanged thrust washer from the assembly. Next, take off the output-ring-gear-to-output-shaft snap ring and discard it. The snap-ring location is shown in Fig. 10-39. Slip the output shaft from the output ring gear.

Remove the output-ring-gear-to-case needle-bearing assembly.

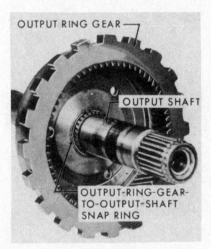

OUTPUT RING GEAR

OUTPUT SHAFT

OUTPUT-RING-GEAR-TO-OUTPUT-SHAFT SNAP RING

Fig. 10-39 Location of the output-ring-gear-to-output-shaft snap ring. (*Buick Motor Division of General Motors Corporation*)

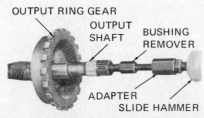

Fig. 10-40 Removing the output-shaft bushing. (*Buick Motor Division of General Motors Corporation*)

Check the output-shaft bushing. If it is worn or galled, replace it with special tools. The procedure is to assemble the bushing remover onto the adapter and then assemble both to the slide hammer (Fig. 10-40). Thread the assembly into the bushing and clamp the slide hammer in a vise. Then grasp the output shaft to remove the bushing. Assemble the bushing installer onto the drive handle and press the new bushing into place 0.140 inch [3.56 mm] below the end surface of the output shaft.

20. To remove the low-and-reverse-clutch piston, the springs behind the piston must be compressed by installing a clutch-spring compressor (Fig. 10-41). Then the piston retaining ring and spring retainer and springs can be removed. Figure 10-42 shows in sectional view how the tool is installed to compress the springs. Now, the low-and-reverse-clutch-piston assembly can be removed. Use compressed air to aid in the removal of the piston assembly. Apply compressed air in the passage shown pointed out by a pencil in Fig. 10-43. The low-and-reverse-clutch-piston assembly has three seals: the outer seal, center seal, and inner seal. All three seals should be removed and discarded.

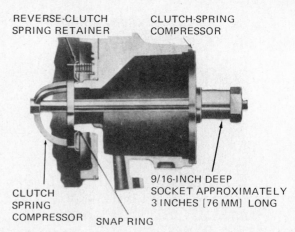

Fig. 10-42 Sectional view of the clutch-spring compressor compressing the springs. (*Oldsmobile Division of General Motors Corporation*)

21. If the case bushing is worn or damaged, remove it by using the special removing tool installed on the drive handle. The bushing is driven inward. The same tools are then used to drive the new bushing into place. The bushing should be pressed in to 0.195 inch [4.95 mm] below the chamfered edge of the case. Make sure that the split on the bushing is opposite the notch on the case.

22. Removal and installation of the intermediate-clutch accumulator are discussed in ✦ 10-9. The tool used is shown in Fig. 10-11. A disassembled view of the intermediate-clutch accumulator with retaining ring, cover and seal, spring, and piston is shown in Fig. 10-44.

✦ **10-18 Oil-pump-assembly service** The oil-pump assembly is shown in exploded view in Fig. 10-45. Note

Fig. 10-41 Using a clutch-spring compressor to compress the low-and-reverse-clutch-piston springs. (*Chevrolet Motor Division of General Motors Corporation*)

Fig. 10-43 Pencil points to the passage at which compressed air is applied to aid in removal of the low-and-reverse-clutch piston. (*Buick Motor Division of General Motors Corporation*)

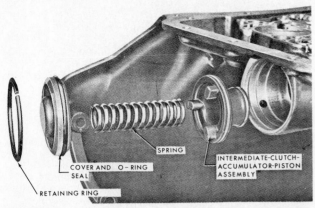

Fig. 10-44 Disassembled view of the intermediate-clutch accumulator. (*Buick Motor Division of General Motors Corporation*)

that the assembly includes the intermediate-clutch-piston assembly and related parts.

Disassembly To disassemble the oil-pump assembly, put it on the bench with the shaft through a hole in the bench.

1. Remove the five pump-cover-to-pump-body bolts. This permits removal of the spring retainer, springs, and intermediate-clutch-piston assembly, all shown in Fig. 10-45. The intermediate-clutch piston has two seals: an inner seal and an outer seal. Both seals should be removed and discarded so that new seals can be installed.
2. There are five oil-seal rings on the pump hub, as shown in Fig. 10-45. Remove the rings by lifting them up and off the hub. The selective thrust washer,

shown in Fig. 10-45, or the thrust bearing and shim, can now be removed. Note that this thrust washer comes in various thicknesses. The correct thickness must be selected to produce the proper end play in the transmission. The thrust bearing may be used alone, or with one or two 0.017-inch [0.43-mm] shims.

3. Lift the pump-cover-and-stator-shaft assembly from the pump body. Remove the pump drive and driven gears from the pump body. These are shown in Fig. 10-45. Remove and discard the large seal ring.
4. Turn the pump body over on two wood blocks to prevent damage to the surface, and remove the pump-to-converter-hub lip seal. Turn the pump body inner face up, and remove the priming valve and spring. These are shown in Fig. 10-45.
5. Then remove the cooler bypass-valve seat, check ball, and spring. Two methods for removal are given. In one method, a bolt extractor is used. In the other, the bypass passage is filled with grease and a special tool is driven into the passage to force out the valve seat, check ball, and spring, as shown in Fig. 10-46.

Inspection of parts Wash all parts in cleaning solvent and blow out all oil passages. Do not use rags to dry parts. Traces of lint can cause valves to hang and cause faulty transmission performance. Check all parts for nicks, scoring, or other damage. If the pump-body bushing is worn or galled, replace the bushing or pump body, according to recommendations in the shop manual.

With parts clean and dry, install the pump gears and check the clearance between the pump-body face and the gear faces with a straightedge and thickness gauges. The clearance should be between 0.0005 and 0.0015

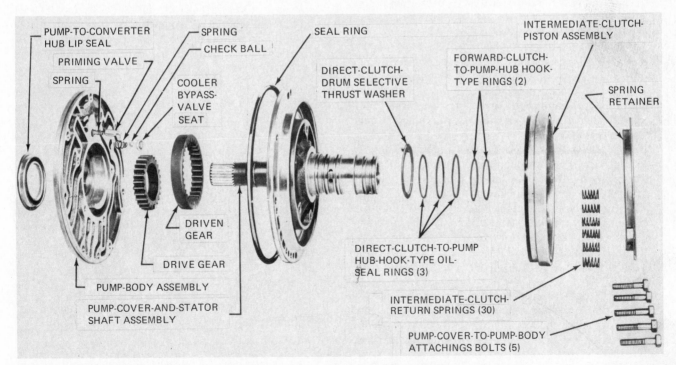

Fig. 10-45 Exploded view of the oil-pump assembly. (*Chevrolet Motor Division of General Motors Corporation*)

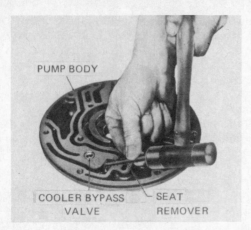

Fig. 10-46 Removing the cooler bypass-valve seat with grease and an extractor. (*Buick Motor Division of General Motors Corporation*)

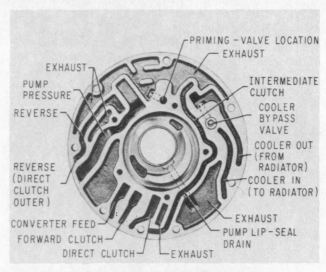

Fig. 10-47 Pump-body oil passages. (*Chevrolet Motor Division of General Motors Corporation*)

inch [0.013 and 0.043 mm]. If the clearance is excessive, the gears and body should be replaced.

Reassembly

1. If the pump-to-converter-hub lip seal has been removed, replace it by using a special driver tool.
2. Turn the gear body over and install the gears. Make sure that the driving tang on the driving gear is up and that the marks on the gears, where present, are up and align.
3. Install the priming valve and spring, cooler bypass-valve seat, check ball, and spring. These are all shown in Fig. 10-45. When installing the valve seat, tap it down with a soft hammer or brass drift punch until the seat is flush to 0.010 inch [0.25 mm] below the surface.
4. To the hub on the pump cover, install five hook-type oil-seal rings. These are shown in Fig. 10-45.
5. Install the inner seal and outer seal on the intermediate-clutch piston. Now install the intermediate-clutch-piston assembly in the pump cover, being careful not to damage the seals. Put the intermediate-clutch-spring retainer and springs into place and install five bolts.
6. Bring the pump-cover-and-stator-shaft assembly down onto the pump body. Be sure to align them properly. Figures 10-47 and 10-48 show the various passages in the faces of the pump body and cover. One convenient alignment procedure is to align the priming valve in the pump body with the priming-valve cavity in the pump cover (Figs. 10-47 and 10-48). Install the square-cut seal ring. Tighten the five attaching bolts to 18 lb-ft [24 N-m] of torque.

✿ 10-19 Direct-clutch service The direct-clutch assembly, with overrun-clutch parts, is shown in Fig. 10-49. To remove the overrun parts, pry out the retaining ring with a screwdriver. The retainer, outer race, and roller-clutch assembly can then be lifted off. Next, the direct-clutch assembly itself can be disassembled. It is shown in exploded view in Fig. 10-50.

Disassembly

1. Start by removing the clutch-drum-to-forward-clutch-housing special thrust washer.
2. Use a screwdriver to remove the retaining ring, pressure plate, and clutch plates.
3. Use the special tool, as shown in Fig. 10-51, to compress the springs. Then remove the retaining ring, spring retainer, clutch return springs, and direct-clutch-piston assembly.

Inspection Check clutch plates for signs of burning, wear, or scoring. Check springs for collapsed coils or distortion. Check the piston and clutch housing for wear and scores. Make sure all oil passages are open and that the ball check works freely. Make sure roller-clutch inner and outer races are free of scratches, indentations, or other signs of wear. The roller cage

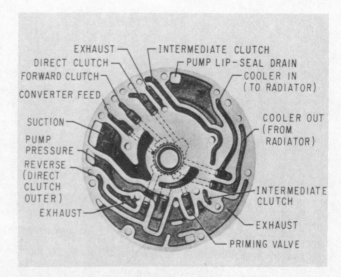

Fig. 10-48 Pump-cover oil passages. (*Chevrolet Motor Division of General Motors Corporation*)

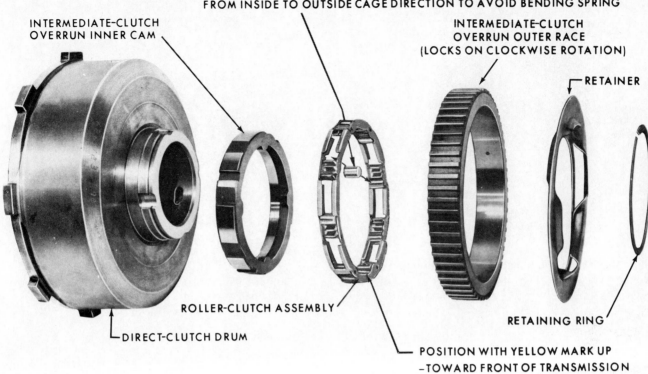

INTERMEDIATE-CLUTCH OVERRUN INNER CAM

INTERMEDIATE-CLUTCH OVERRUN OUTER RACE (LOCKS ON CLOCKWISE ROTATION)

RETAINER

ROLLER-CLUTCH ASSEMBLY

DIRECT-CLUTCH DRUM

RETAINING RING

POSITION WITH YELLOW MARK UP —TOWARD FRONT OF TRANSMISSION

Fig. 10-49 Direct-clutch assembly with intermediate-overrun-clutch parts. (*Chevrolet Motor Division of General Motors Corporation*)

should be free of excessive wear, and roller springs should be in good condition.

Reassembly

1. Install new inner and outer seals on the direct-clutch piston.
2. Install the direct-clutch-piston center seal on the drum with the lip facing upward.

3. Install the direct-clutch piston in the drum using a piece of 0.020-inch [0.51-mm] piano wire crimped into copper tubing, as shown in Fig. 10-52, to help get the seal into place.
4. Install the clutch return springs and put the spring retainer in position. Compress springs with a special tool, as shown in Fig. 10-51, and install the retaining ring.

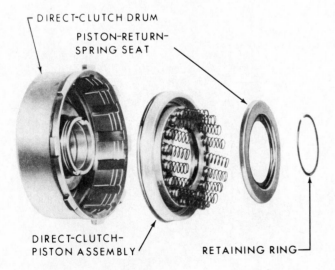

DIRECT-CLUTCH DRUM

PISTON-RETURN-SPRING SEAT

DIRECT-CLUTCH-PISTON ASSEMBLY

RETAINING RING

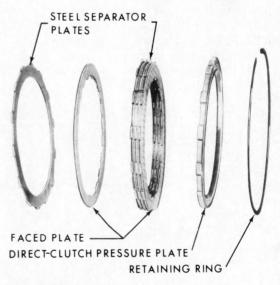

STEEL SEPARATOR PLATES

FACED PLATE

DIRECT-CLUTCH PRESSURE PLATE

RETAINING RING

Fig. 10-50 Exploded view of a direct-clutch assembly. (*Chevrolet Motor Division of General Motors Corporation*)

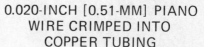
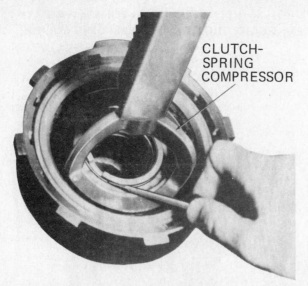

Fig. 10-51 Using a clutch-spring compressor to compress the springs so that the retaining ring can be removed. (*Chevrolet Motor Division of General Motors Corporation*)

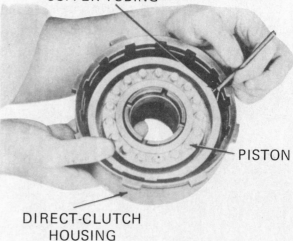

Fig. 10-52 Using a piece of 0.020-inch [0.51-mm] piano wire to install the direct-clutch piston in the drum. (*Chevrolet Motor Division of General Motors Corporation*)

5. Soak the clutch plates in transmission fluid for at least 30 minutes and install them, alternating faced and steel plates.
6. Install the pressure plate and retaining ring.
7. Install the roller-clutch assembly, outer race, retainer, and retainer ring (all shown in Fig. 10-49). Make sure that the outer race can freewheel in a counterclockwise direction only.

✿ 10-20 Forward-clutch service   Figure 10-53 shows the forward-clutch assembly in exploded view.

Disassembly

1. Remove the pressure-plate-to-drum retaining ring.
2. Remove the pressure plate, clutch plates, and cushion spring. These are all shown in Fig. 10-53.

3. Use the special compressing tool, as shown in Fig. 10-51, to compress springs so that the retaining ring can be removed. Now, the spring retainer, springs, and forward-clutch-piston assembly can be removed.
4. Remove the piston inner and outer seals.

Inspection   Check clutch plates for signs of wear, burning, or scoring. Check springs for collapsed coils or signs of distortion. Check the piston and clutch drum for signs of wear or other damage. Make sure oil passages are open. Inspect the input shaft for damaged splines, worn bushing journals, cracks, or other damage, and make sure the oil passages are open. Check the ball-check exhaust to make sure it is free.

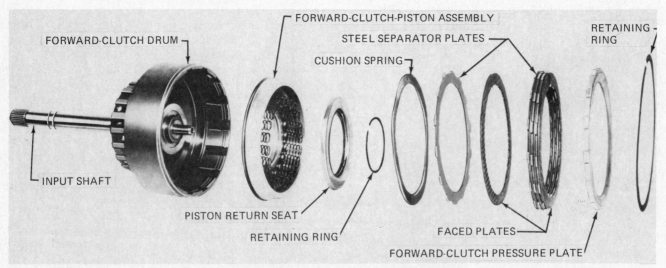

Fig. 10-53 Exploded view of the forward-clutch assembly. (*Chevrolet Motor Division of General Motors Corporation*)

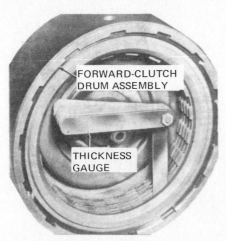

Fig. 10-54 Measuring clearance between the pressure plates and the nearest clutch plate. (*Buick Motor Division of General Motors Corporation*)

Reassembly

1. Install the inner and outer seals on the piston. Then install the forward-clutch-piston assembly in the clutch drum, using a piece of 0.020-inch [0.51-mm] piano wire crimped into copper tubing. This procedure is similar to the procedure for installing the direct-clutch piston, shown in Fig. 10-52.
2. Install the 21 springs and the spring retainer. Compress the springs with a special tool, as shown in Fig. 10-51, and install the retaining ring.
3. Soak the clutch plates in transmission oil for at least 30 minutes. Then install the cushion spring, clutch plates, forward-clutch pressure plate, and retaining ring.
4. Use a thickness gauge to check the clearance between the pressure plate and the clutch plate, as shown in Fig. 10-54. If the clearance is less than 0.0105 inch [0.263 mm], a thinner pressure plate should be used. If the clearance is greater than 0.082 inch [2.08 mm], a thicker pressure plate should be used. The correct clearance is between 0.0105 and 0.082 inch [0.263 and 2.08 mm]. Pressure plates of three thicknesses are available.

✿ 10-21 Sun-gear and drive-shell service The sun gear and drive shell, along with the other components of the planetary-gear train, are shown in exploded view in Fig. 10-55.

1. Remove the sun-gear-to-drive-shell rear retaining ring. Lift off the rear flat-steel thrust washer. Then remove the front retaining ring.
2. Check the gear and shell for wear.
3. On reassembly, use new retaining rings.

NOTE: If sun-gear bushings require replacement, they may be removed with a cape chisel. Then the new bushings are pressed in with the special tool required.

✿ 10-22 Low-and-reverse-roller-clutch service This assembly is shown in exploded view in Fig. 10-56.

The parts are separated by removing the thrust washer and retaining ring. Inspect the roller races and rollers for wear, scratches, or other damage. Check the springs for distortion.

When reassembling, be sure that the roller-clutch assembly is installed in the support with the oil holes to the rear of the transmission. Make sure that the inner race freewheels in a clockwise direction only.

✿ 10-23 Governor service The governor is serviced by replacing it completely, with the exception of the driven gear and springs, which can be replaced. The reason is that the parts are selectively fitted and each assembly is calibrated during manufacture to give the required performance. If driven-gear replacement is required, or if foreign matter has gotten into the governor so that it does not perform properly, the governor can be disassembled, as follows.

Refer to Fig. 10-57. Cut off one end of each governor weight pin and remove the pins, thrust cap, weights, and springs. Remove the valve from the governor sleeve.

Wash all parts in cleaning solvent and air-dry. Blow out all passages. Inspect the sleeve and valve for nicks, burrs, or other damage. Check the governor sleeve for free operation in the bore of the transmission case. The governor valve should slide freely in the bore of the sleeve. Springs and weights should be in good condition. If the driven gear is damaged, it may be replaced as follows.

A special service package is available, containing a new driven gear, two weight retaining pins, and a gear-retainer split pin. To replace the driven gear, drive out the old split pin with a small punch. Support the sleeve on $3/16$-inch [4.76-mm] plates installed in the exhaust slots of the sleeve, and press the gear out of the sleeve in an arbor press. Clean the governor sleeve. Press the new driven gear into the sleeve until it is nearly seated. Remove any chips that may have been shaved off the gear, and press the gear on in until it bottoms on the shoulder. Drill a new pinhole 90° from the old one. Install the split retaining pin. Wash the sleeve off to remove any chips or dirt.

Reassemble the governor by installing the valve in the sleeve, large-land end first. Install the weights, springs, and thrust cap and secure with new pins. Crimp both ends of the pins to keep them from falling out. Check for free operation of weights and valve in the sleeve.

✿ 10-24 Valve-body service The valve body is shown disassembled in Fig. 4-9. While disassembling the body, use care to avoid dirt and damage to finished surfaces. The various valves and springs are held in place by retaining pins. The direct-clutch-accumulator piston is held in place by a retaining ring. Cleanliness is essential in working on the valve body and valves. The slightest piece of lint or dirt can cause a valve to hang, and this could cause the transmission to malfunction or fail. Identify all springs as they are removed so that they can be reinstalled in the proper positions.

Handle the valves with care to avoid damaging the

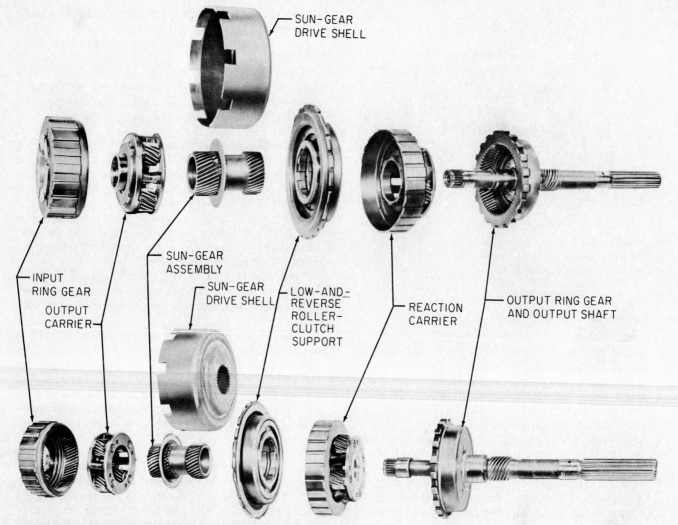

Fig. 10-55 Planetary-gear train in exploded view. Note that there are two views of each part. (*Chevrolet Motor Division of General Motors Corporation*)

operating surfaces. Any dent or burr can cause the valve to hang.

Wash all parts in clean solvent and air-dry them. Make sure all oil passages in the valve body are open. Follow Fig. 4-9 carefully when reinstalling the valves.

✷ 10-25 Transmission reassembly Cleanliness is very important when handling any transmission parts. Hands, tools, and work area must be clean. If work is stopped before reassembly is complete, cover all openings with clean lint-free cloths.

During reassembly, lubricate all bushings with transmission oil. Coat thrust washers on both sides with petroleum jelly.

Use all new seals and gaskets. Do not reuse old seals or gaskets.

Tighten all parts evenly and in the proper sequence when installing screws or bolts. Use new retaining rings, where required. Be careful not to overstress retaining rings. This could cause them to loosen in service and result in transmission failure. The reassembly procedure follows.

1. Install the low-and-reverse-clutch piston, with new seals in place, with the notch in the piston adjacent to the parking pawl.
2. Install 17 piston return springs and the spring retainer. Use the compressor tool shown in Figs. 10-41 and 10-42 to compress the springs so that the retaining ring can be installed.
3. Install the output ring gear on the output shaft. Install the retaining ring (Fig. 10-39). Install the reaction-carrier-to-output-ring-gear thrust washer into the output-ring-gear support. This thrust washer has three tangs.
4. Put the output-ring-gear-to-case needle bearing into position and install the output-shaft assembly in the case.
5. Install the reaction-carrier assembly.
6. Oil and install the low-and-reverse-clutch plates. Start with a steel plate and alternate with faced plates. Install the retainer spring (Fig. 10-58).
7. Install the low-and-reverse-clutch-support assembly and retaining ring.

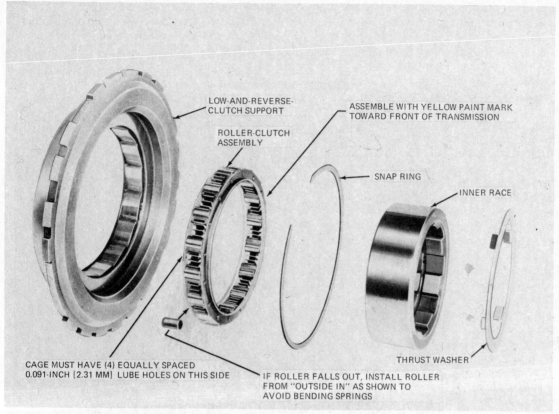

Fig. 10-56 Exploded view of the low-and-reverse-clutch assembly. (*Chevrolet Motor Division of General Motors Corporation*)

Labels in figure:
LOW-AND-REVERSE-CLUTCH SUPPORT
ROLLER-CLUTCH ASSEMBLY
ASSEMBLE WITH YELLOW PAINT MARK TOWARD FRONT OF TRANSMISSION
SNAP RING
INNER RACE
THRUST WASHER
CAGE MUST HAVE (4) EQUALLY SPACED 0.091-INCH [2.31 MM] LUBE HOLES ON THIS SIDE
IF ROLLER FALLS OUT, INSTALL ROLLER FROM "OUTSIDE IN" AS SHOWN TO AVOID BENDING SPRINGS

NOTE: The splines in the inner race of the roller clutch must align with the splines on the reaction carrier.

8. Install the low-and-reverse-roller-clutch-inner-race-to-sun-gear-shell thrust washer. Install the retaining ring, as shown in Fig. 10-38.
9. Install the sun-gear-and-drive-shell assembly.
10. Install the output-carrier assembly (Fig. 10-37). Secure with a retainer ring. Install the thrust washer.
11. Install the input ring gear and the thrust washer.
12. Install the direct-clutch assembly and special thrust washer to the forward-clutch assembly.

There are two designs here. One design uses a thrust washer between the forward and direct clutches. The second design uses a Torrington needle bearing. When replacing the first design, use the new design parts. These include a new direct clutch, forward-clutch housing, and needle bearing.

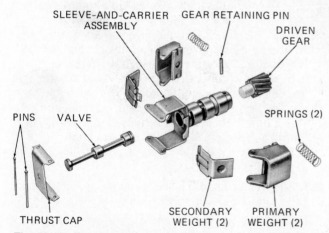

Fig. 10-57 Disassembled view of the governor. (*Chevrolet Motor Division of General Motors Corporation*)

Labels in figure:
SLEEVE-AND-CARRIER ASSEMBLY
GEAR RETAINING PIN
DRIVEN GEAR
PINS
VALVE
SPRINGS (2)
THRUST CAP
SECONDARY WEIGHT (2)
PRIMARY WEIGHT (2)

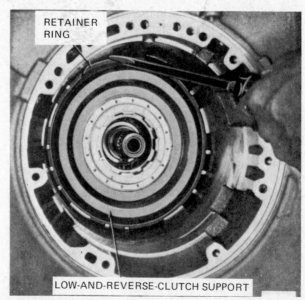

Fig. 10-58 Installing retainer spring. (*Chevrolet Motor Division of General Motors Corporation*)

Labels in figure:
RETAINER RING
LOW-AND-REVERSE-CLUTCH SUPPORT

13. Install the clutch assemblies in the case, as shown in Fig. 10-35. Make sure that the forward-clutch face plates are positioned over the input ring gear and that the tangs on the direct-clutch housing are installed into the slots on the sun-gear drive shell.
14. Install the intermediate-overrun band.
15. Install the intermediate-clutch pressure plate.
16. Soak the intermediate-clutch plates in transmission oil at least 30 minutes and install, starting with a face plate and alternating steel and face plates (Fig. 10-34). The notch in the steel reaction plates is installed toward the selector-lever inner bracket.
17. Install the intermediate-clutch cushion spring.
18. The pump is installed next, but during this procedure, the end play of the input shaft must be checked. If the end play is incorrect, the pump must be removed so that a selective thrust washer of the proper thickness can be installed. The procedure is as follows:

Install a selective-fit thrust washer or thrust bearing alone or with one or two 0.017-inch [0.43-mm] shims, an oil-pump gasket, and the oil pump. Use two guide pins in two opposing holes in the case to align the oil pump as it is brought into position. Install and tighten the pump-to-case bolts. Mount a dial indicator, as shown in Fig. 10-59. Push in on the input shaft and set the dial indicator to zero. Pull out on the input shaft to read the end play. The end play should be between 0.032 and 0.064 inch [0.81 and 1.63 mm]. If it is not, then a different selective-fit thrust washer should be used. This means removal and reinstallation of the pump. Selective-fit washers are available in three thicknesses: 0.066, 0.083, and 0.100 inch [1.68, 2.11, and 2.54 mm].
19. After the proper selective-fit thrust washer is determined, the complete pump-installation procedure is as follows:

a. Install a new pump-to-case gasket.
b. Install the correct thrust washer or bearing and shims. Cover both sides of the thrust washer with petroleum jelly.

c. Install a new square-cut oil-seal ring.
d. Install the pump, using two guide pins to align it. Attach with bolts, using new washer-type seals.

Careful: Check the rotation of the input shaft as the pump is pulled down into place. If the input shaft cannot be rotated freely, the direct- and forward-clutch housings have not been properly installed. The clutch plates are not indexing. Remove the necessary parts to secure proper indexing. Otherwise, parts will break as the pump is pulled down into place.

20. Put the drive-gear retainer clip into the hole in the output shaft, align the slot in the speedometer drive gear with a retainer clip, and install the gear on the shaft. Then install the yoke seal.
21. Install the extension-housing-to-case square-cut seal ring, and attach the extension housing to the case with attaching bolts. Torque to specifications.
22. Install the parking pawl, with the tooth toward the inside of the case. Then install the disengaging spring on the pawl and slide the shaft into place (Fig. 10-30). Drive the retainer plug flush to 0.010 inch [0.25 mm] below the case, using a ⅜-inch [9.53-mm] rod. Stake the plug in three places.
23. Install the park-lock bracket and torque the bolts to specifications.
24. With the actuating rod attached to the range-selector inner lever, put the assembly into place.
25. Install the manual shaft through the case and the range-selector inner lever. Install the retaining nut on the manual shaft and torque to specifications. Install the manual-shaft-to-case spacer clip.
26. Next, install the intermediate-servo piston, washer, spring seat, and apply pin (Fig. 10-31). How to check the band apply pin for correct length was explained earlier (Fig. 10-32). When installing the piston, use a new metal oil-seal ring.
27. Install the four check balls in the transmission-case pockets, as shown in Fig. 10-27.

NOTE: If check balls are missing, complete transmission failure may occur.

28. Install the oil-pump pressure screen and governor screens in the case, as shown in Figs. 10-25 and 10-26.
29. Install the valve-body spacer-plate-to-case gasket and spacer plate, as shown in Fig. 10-24.
30. Install the valve-body-to-spacer-plate gasket and spacer support plate (Figs. 10-23 and 10-24). Torque the support-plate bolts to specifications.
31. Connect the detent-control-valve link to the actuating lever (Fig. 10-22).
32. Install the valve body. Connect the manual-control-valve link to the range-selector inner lever. Torque bolts in random sequence to specifications. Leave a bolt loose for the detent-spring-and-roller assembly (Fig. 10-22).

Careful: When handling the valve body, be careful that the retainer pins do not fall out.

33. Install the manual-shaft-to-case retainer ring and detent-spring-and-roller assembly, as shown in Fig. 10-25.

DIAL INDICATOR

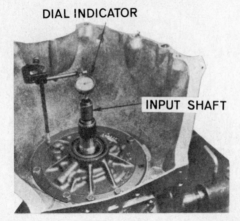

INPUT SHAFT

Fig. 10-59 Dial indicator in place to check input-shaft end play. (*Oldsmobile Division of General Motors Corporation*)

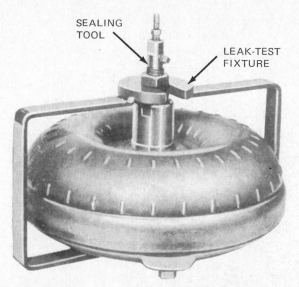

Fig. 10-60 Using the leak-test fixture to check the converter for leaks. (*Oldsmobile Division of General Motors Corporation*)

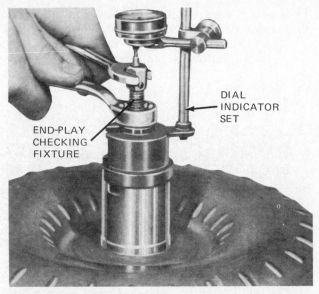

Fig. 10-61 Using a dial indicator and end-play checking fixture to check converter end play. (*Oldsmobile Division of General Motors Corporation*)

34. Install the strainer-assembly gasket and strainer.
35. Install the new gasket and oil pan.

NOTE: The intermediate-clutch-accumulator assembly must be installed before the oil pan is installed (✿ 10-9).

36. Install the governor assembly, cover, seal, and retainer wire (Fig. 10-21).
37. Install the vacuum-modulator valve and retainer clip (Fig. 10-17). Lubricate the O-ring seal to prevent damaging it. Torque the clip bolt to specifications.
38. Before installing the converter, check for leaks by installing the sealing tool, as shown in Fig. 10-60. Fill the converter with air to 80 psi [552 kPa], and submerge it in water to check for leaks.
39. Check converter end play with special tools and dial indicator, as shown in Fig. 10-61. Tighten the hex nut, and zero the dial indicator. Loosen the hex nut and note indicator reading. This is the end clearance. It should not exceed 0.050 inch [1.27 mm].

40. Converter leakage or excessive end play requires replacement of the assembly.
41. Refer to ✿ 10-14 for the transmission-installation procedure.

✿ 10-26 Vacuum-modulator check Turn the vacuum modulator so that the vacuum side points down. If oil comes out, the diaphragm is defective and the modulator must be replaced. Make a bellows-comparison check with a vacuum modulator known to be good, as shown in Fig. 10-62. Using the gauge as shown, install the two modulators on either end of the gauge.

Hold the modulators horizontal, and push them toward each other until either modulator sleeve end just touches the line in the center of the gauge. The gap between the other modulator sleeve end and the line should be $1/16$ inch [1.59 mm] or less. If it is more, the modulator being checked should be discarded. To make checking a modulator easier, a tool is available which has a white strip at its center rather than gauge lines.

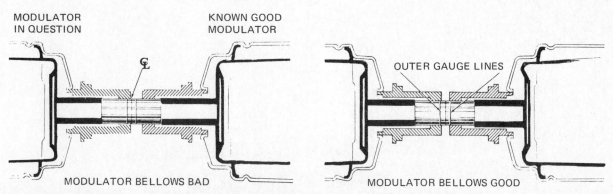

Fig. 10-62 Checking a possible defective modulator bellows against a modulator known to be good. (*Chevrolet Motor Division of General Motors Corporation*)

If the gauge shows white when the modulators are pushed together, the modulator being checked can be reinstalled. But if a blue strip shows up (indicating that the gauge is shifted to one side), the modulator is defective.

NOTE: Some vacuum modulators are of the diaphragm type and do not have a bellows.

Chapter 10 review questions

Select the *one* correct, best, or most probable answer to each question. Then check your answers against the correct answers given at the end of the book.

1. In transmissions used on the big engines, the clutches have:
   a. fewer plates
   b. the same number of plates
   c. more plates
   d. bigger springs
2. The type 350C transmission should have the oil changed every:
   a. year
   b. 12,000 miles [19,312 km]
   c. 24,000 miles [38,624 km]
   d. 100,000 miles [160,000 km]
3. Only two in-car adjustments are required on the type 350 transmission:
   a. selector-lever linkage and throttle linkage
   b. linkage and band
   c. band and throttle-valve linkage
   d. band and clutch linkage
4. Mechanic A says that when you check oil pressure on the 350C automatic transmission, you are actually checking line pressure. Mechanic B says you are checking 1-2 clutch and 2-3 clutch pressures. Who is right?
   a. A only
   b. B only
   c. both A and B
   d. neither A nor B
5. Several parts can be removed from the 350 transmission while it is still in the car, including the:
   a. vacuum modulator and direct clutch
   b. valve-body assembly and extension housing
   c. governor assembly and front band
   d. front and rear bands

6. Low line pressures could be due to low oil level, governor or valve-body trouble, or:
   a. pump trouble
   b. clutch slippage
   c. band slippage
   d. defective vacuum modulator
7. High line pressure in all ranges could be due to vacuum-modulator or governor trouble, stuck pressure regulator or boost valve, or a:
   a. stuck shift valve
   b. stuck detent valve
   c. faulty accumulator
   d. defective clutch
8. Slipping in reverse could be caused by low oil pressure, a stuck pressure regulator or boost valve, or a:
   a. defective governor
   b. slipping direct or low-and-reverse clutch
   c. sticking detent valve
   d. slipping band
9. Slipping in all ranges could be caused by low oil pressure, trouble in the valve body or 1-2 accumulator, or:
   a. incorrect manual-valve-linkage adjustment
   b. defective governor action
   c. stuck shift valves
   d. defective direct clutch
10. A rough 2-3 upshift could be caused by trouble in the valve body, a stuck 2-3 shift valve, faulty 2-3 accumulator action, or:
    a. a defective governor
    b. faulty vacuum-modulator action
    c. band not releasing smoothly
    d. band not applying smoothly

CHAPTER **11**

GENERAL MOTORS 200C AUTOMATIC-TRANSMISSION SERVICE

After studying this chapter, and with the proper instruction and equipment, you should be able to:

1. Explain how to check transmission fluid and add fluid if necessary.
2. Describe how to change the fluid.
3. Discuss the use of the trouble-diagnosis chart and explain how it can help locate causes of trouble.
4. Explain how to use the converter-clutch trouble-diagnosis chart.
5. Describe how to remove and install the transmission.
6. Discuss the procedure of disassembling and assembling the transmission.
7. Explain how to make transmission adjustments.
8. Perform the services listed above.

11-1 Introduction to 200C service This chapter describes the trouble-diagnosis checks, adjustments, removal, overhaul, and installation of the General Motors type 200C automatic transmission. Chapter 8 describes the construction and operation of this transmission.

General service procedures for all automatic transmissions are covered in Chap. 9. The procedures described in that chapter also apply to the type 200C transmission. Also, many of the service procedures on the General Motors type 350C automatic transmission, described in Chap. 10, also apply to the 200C transmission.

11-2 Checking and adding transmission fluid The instructions on the type 350C automatic transmission for checking and adding transmission fluid also apply to the type 200C automatic transmission. Refer to ✿ 10-3 for the procedure.

11-3 Towing instructions The towing instructions in ✿ 10-4 for the 350C transmission also apply to the 200C transmission.

11-4 Adjustments Two adjustments may be required on the 200C automatic transmission: linkage to the manual valve and linkage to the accelerator (for the detent downshift). The adjustments differ among the various car models. Refer to the shop manual for the

car you are working on to find the proper procedures and specifications. Figures 11-1 to 11-4 show the adjustments on some cars.

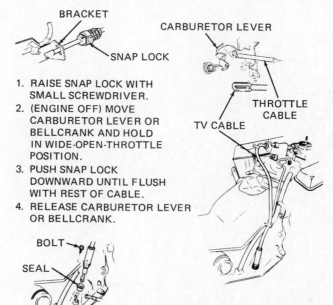

1. RAISE SNAP LOCK WITH SMALL SCREWDRIVER.
2. (ENGINE OFF) MOVE CARBURETOR LEVER OR BELLCRANK AND HOLD IN WIDE-OPEN-THROTTLE POSITION.
3. PUSH SNAP LOCK DOWNWARD UNTIL FLUSH WITH REST OF CABLE.
4. RELEASE CARBURETOR LEVER OR BELLCRANK.

Fig. 11-1 Throttle-valve (TV) cable adjustment. (*Oldsmobile Division of General Motors Corporation*)

143

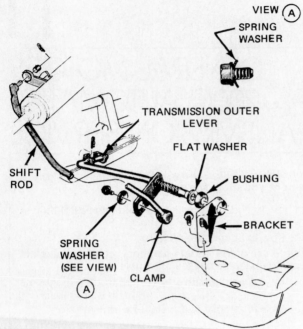

VIEW (A)
SPRING WASHER

TRANSMISSION OUTER LEVER

FLAT WASHER

BUSHING

SHIFT ROD

BRACKET

SPRING WASHER (SEE VIEW)

CLAMP

(A)

SHIFT-ROD ADJUSTMENT

1. WITH SHIFT-ROD CLAMP SCREW LOOSENED, SET TRANSMISSION OUTER LEVER IN NEUTRAL POSITION.
2. HOLD UPPER SHIFT LEVER AGAINST NEUTRAL POSITION STOP IN UPPER STEERING COLUMN. (DO NOT RAISE LEVER).
3. TIGHTEN SCREW IN CLAMP ON LOWER END OF SHIFT ROD TO SPECIFIED TORQUE.
4. CHECK OPERATION:
   A. WITH KEY IN "RUN" POSITION AND TRANSMISSION IN "REVERSE," BE SURE THAT KEY CANNOT BE REMOVED AND THAT STEERING WHEEL IS NOT LOCKED.
   B. WITH KEY IN "LOCK" POSITION AND SHIFT LEVER IN "PARK," BE SURE THAT KEY CAN BE REMOVED, THAT STEERING WHEEL IS LOCKED, AND THAT THE TRANSMISSION REMAINS IN PARK WHEN THE STEERING COLUMN IS LOCKED.
   C. WITH BRAKES FIRMLY APPLIED, CHECK TO MAKE SURE THAT THE STARTER WILL NOT WORK IN ANY SHIFT-LEVER POSITION EXCEPT NEUTRAL AND PARK.

Fig. 11-2 Column-shift linkage adjustment on some car lines. (*Oldsmobile Division of General Motors Corporation*)

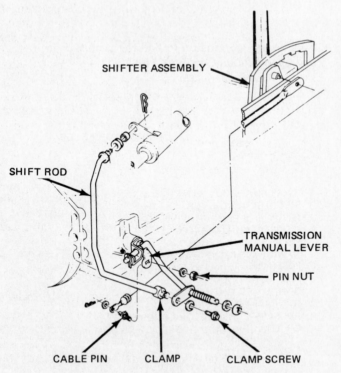

SHIFTER ASSEMBLY

SHIFT ROD

TRANSMISSION MANUAL LEVER

PIN NUT

CABLE PIN   CLAMP   CLAMP SCREW

SHIFT CABLE ADJUSTMENT

1. LOOSEN SHIFT-ROD CLAMP SCREW. LOOSEN PIN IN TRANSMISSION MANUAL LEVER.
2. PLACE SHIFT LEVER IN "PARK" POSITION. PLACE TRANSMISSION MANUAL LEVER IN "PARK" POSITION AND IGNITION KEY IN LOCK POSITION.
3. TIGHTEN CABLE PIN NUT TO SPECIFICATIONS.
4. ROTATE THE TRANSMISSION MANUAL LEVER FULLY AGAINST THE "PARK" STOP, THEN RELEASE THE LEVER.
5. PULL SHIFT ROD DOWN AGAINST LOCK STOP TO ELIMINATE LASH AND TIGHTEN CLAMP SCREW TO SPECIFICATIONS.
6. CHECK OPERATION:
   A. MOVE SHIFT HANDLE INTO EACH GEAR POSITION AND SEE THAT TRANSMISSION MANUAL LEVER IS ALSO IN DETENT POSITION.
   B. WITH KEY IN "RUN" POSITION AND TRANSMISSION IN "REVERSE," BE SURE THAT KEY CANNOT BE REMOVED AND THAT STEERING WHEEL IS NOT LOCKED.
   C. WITH KEY IN "LOCK" POSITION AND TRANSMISSION IN "PARK," BE SURE THAT KEY CAN BE REMOVED AND THAT STEERING WHEEL IS LOCKED.
   D. ENGINE MUST START IN PARK AND NEUTRAL.
   E. WITH BRAKES FIRMLY APPLIED, CHECK TO MAKE SURE THAT THE STARTER WILL NOT WORK IN ANY SHIFT-LEVER POSITION EXCEPT NEUTRAL AND PARK.

Fig. 11-3 Console-shift linkage adjustment on some car lines. (*Oldsmobile Division of General Motors Corporation*)

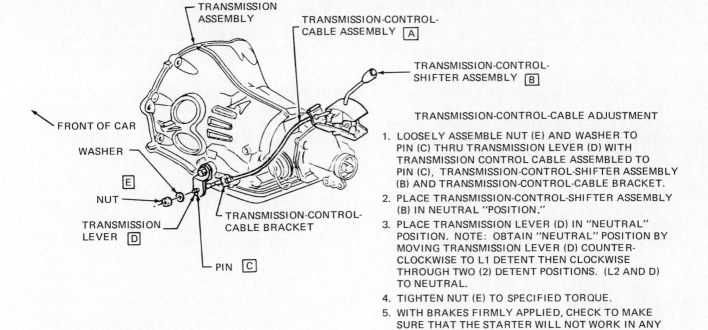

TRANSMISSION-CONTROL-CABLE ADJUSTMENT

1. LOOSELY ASSEMBLE NUT (E) AND WASHER TO PIN (C) THRU TRANSMISSION LEVER (D) WITH TRANSMISSION CONTROL CABLE ASSEMBLED TO PIN (C), TRANSMISSION-CONTROL-SHIFTER ASSEMBLY (B) AND TRANSMISSION-CONTROL-CABLE BRACKET.
2. PLACE TRANSMISSION-CONTROL-SHIFTER ASSEMBLY (B) IN NEUTRAL "POSITION."
3. PLACE TRANSMISSION LEVER (D) IN "NEUTRAL" POSITION. NOTE: OBTAIN "NEUTRAL" POSITION BY MOVING TRANSMISSION LEVER (D) COUNTER-CLOCKWISE TO L1 DETENT THEN CLOCKWISE THROUGH TWO (2) DETENT POSITIONS. (L2 AND D) TO NEUTRAL.
4. TIGHTEN NUT (E) TO SPECIFIED TORQUE.
5. WITH BRAKES FIRMLY APPLIED, CHECK TO MAKE SURE THAT THE STARTER WILL NOT WORK IN ANY SHIFT LEVER POSITION EXCEPT NEUTRAL AND PARK.

Fig. 11-4 Console-shift linkage adjustment on some car lines. (*Oldsmobile Division of General Motors Corporation*)

✸ 11-5 Transmission trouble-diagnosis procedure The basic theory of trouble diagnosis is covered in ✸ 9-6. The chart in ✸ 11-6 lists various transmission troubles, possible causes, and checks or corrections. This chart applies to both the earlier type 200 automatic transmission and to the later 200C (with converter clutch). The chart in Fig. 10-7 covers troubles in the

converter clutch. Figure 10-8 shows how to prepare test governors for the three General Motors transmissions using the converter clutch.

As part of the trouble-diagnosis procedure, connect a pressure gauge to the transmission hydraulic system. Then note the oil pressure under varying operating conditions. The purpose of the test procedure is to check the oil pressures at specific engine speeds with the selector lever in the various ranges. During the tests, you should also note the quality of the shifts—whether there is engine flare-up or the shifts are harsh (✸ 9-8). Figure 8-4 shows in chart form the transmission elements that are holding in the various ranges and gears. The test procedure recommended for the 200C automatic transmission is:

1. Check transmission-fluid level and conditions (✸ 10-3). Add fluid if necessary.
2. Check and adjust control linkages (manual-valve and detent).
3. Connect the oil-pressure gauge.
4. Connect the tachometer so engine rpm can be read.
5. Apply brakes.
6. With engine and transmission at normal operating temperature, run the engine at the speeds specified in the chart in Fig. 11-5. Move the selector lever to the positions shown in the chart and read the oil pressures. Note that the total running time for the four selector-lever drive positions—R, D, 2, and 1—should not exceed 2 minutes. A longer testing time could overheat and damage the transmission.
7. If further information is needed, test the car on the road or on a chassis dynamometer (✸ 9-8).
8. Refer to the trouble-diagnosis chart (✸ 11-6) to determine possible causes of troubles that show up

RANGE	kPa	psi
Park at 1000 rpm	345	50
	310–414	45–60
	414–551	60–80
	482–628	70–90
Reverse at 1000 rpm*	760–860	110–125
	793–1206	115–175
	1103–1447	160–210
Neutral at 1000 rpm	310–415	45–60
	415–551	60–80
	517–620	75–90
Drive at 1200 rpm	275–380	40–55
	415–551	60–80
	551–655	80–95
Intermediate and low at 1000 RPM*	517–622	90–105
	724–862	105–125
	896–1034	130–150
	827–965	120–140

*Note: Total running time for this combination not to exceed two minutes, and brakes must be applied at all times.

Fig. 11-5 Chart of specified pressures in the various selector-lever positions and engine speeds for certain models of cars. (*Oldsmobile Division of General Motors Corporation*)

during the tests. In most cases, the chart will help you locate causes so that you can make the necessary corrections. Sometimes it will be necessary to remove the transmission from the car for disassembly so that defective parts can be replaced. However, other trouble may be solved by replacement of parts without removing the transmission from the car.

9. If the converter clutch does not engage, engages improperly, or is engaged at all times, refer to the detailed diagnostic procedure in the chart in Fig. 10-7. The chart also explains how to check the thermal vacuum valve and the low vacuum switch. Note that the chart in Fig. 10-7 applies to the converter-clutch system in 350C, 250C, and 200C transmissions.

❁ **11-6 General Motors 200C trouble-diagnosis chart** The chart accompanying this section lists complaints, possible causes, and checks or corrections for the General Motors 200C automatic transmission.

General Motors 200C Trouble-Diagnosis Chart

COMPLAINT	POSSIBLE CAUSE	CHECK OR CORRECTION
1. No drive in drive range D1, D2, D3	a. Low oil level	Add oil, check for leaks.
	b. Manual linkage	Readjust.
	c. Low or no oil pressure, oil level OK	Check for: stuck pressure-regulator valve, plugged oil screen, defective pump.
	d. Control-valve assembly	Check for: disconnected manual valve, gaskets off location.
	e. D1 forward clutch not engaging or roller clutch not holding	Check for: plugged feed passage, defective clutch (forward or roller), stuck valve.
	f. D2 forward clutch or band not holding	Check for: plugged feed passage, defective clutch, defective servo, worn or defective band, stuck valve.
	g. D3 forward or direct clutch not engaging	Check for: plugged feed passage, defective clutch, stuck valve.
	h. Converter damaged	Replace converter.
2. High or low oil pressures	a. Throttle or manual-valve linkage defective or misadjusted	Service, adjust.
	b. TV exhaust-valve lifter rod bent or binding—causes high oil pressure	Straighten and loosen.
	c. Pressure-regulator valve binding or plug leaks	Free up, replace defective parts.
	d. Control-valve plunger binding	Check throttle valve or plunger, shift TV valve, intermediate-boost valve (if binding, incorrect pressures in 2 and 1 ranges), reverse-boost valve (if binding, incorrect pressures in R only).
	e. Reverse or intermediate boost orifice plugged in valve-body spacer plate	Remove, clean spacer plate.
	f. Pump trouble	Service, replace pump.
3. 1-2 and 2-3 shifts at full throttle only	a. Throttle-valve cable misadjusted	Readjust.
	b. TV exhaust-valve lifter rod binding or mispositioned	Straighten, position.
	c. Throttle valve or plunger sticking	Service control-valve assembly.
	d. Control-valve assembly gaskets leaking	Replace, tighten screws correctly.
4. First speeds only, no 1-2 upshift	a. Throttle-valve cable misadjusted	Readjust.
	b. Governor defective	Service, replace.
	c. Intermediate servo defective	Service, replace.
	d. 1-2 shift or throttle valve stuck	Service control-valve assembly.
	e. Control-valve assembly gaskets leaking	Replace.
	f. Orifices plugged in valve-body spacer	Clean, replace.

COMPLAINT	POSSIBLE CAUSE	CHECK OR CORRECTION
5. First and second speeds only—no 2-3 shift	*a.* Throttle-valve cable	Readjust.
	b. Governor defective	Service, replace.
	c. Intermediate servo defective	Service, replace.
	d. Valves stuck or spacer-plate orifices plugged	Service, clean, replace.
	e. Pump assembly defective	Service, replace.
	f. Direct clutch not holding	Clear feed circuit, service clutch.
6. Drives in neutral	*a.* Manual-valve link misadjusted	Readjust.
	b. Forward clutch not disengaging	Service, check exhaust ball for sticking.
	c. Cross-leakage in pump or case	Service pump or case.
7. No drive, or slips in reverse	*a.* TV or manual-valve linkage misadjusted	Readjust.
	b. Pump assembly defective	Service, replace.
	c. Valves binding in valve body	Service valve body.
	d. Valve-body spacer or gasket defective or misaligned	Service, replace.
	e. Direct or low-and-reverse clutch defective	Service, replace.
8. Slips on 1-2 shift	*a.* Low oil level	Check, add oil.
	b. TV cable not adjusted	Readjust.
	c. Servo piston and cover defective	Service, check seals, piston.
	d. Valves binding in valve body	Service valve body—throttle and TV valve, 1-2 accumulator valve.
	e. 1-2 accumulator defective	Service, replace.
	f. Forward-clutch feed in pump not aligned	Check, service pump.
	g. Forward clutch not holding	Check feed, clutch.
	h. Intermediate band not holding	Check servo, band.
9. Slips on 2-3 shift	*a.* Oil level low	Check, add oil.
	b. TV cable not adjusted	Readjust.
	c. Servo oil-seal ring damaged	Replace.
	d. TV valve binding	Service control-valve body.
	e. Valve-body spacer plate and gaskets misplaced or leaking	Clean, service.
	f. Pump-assembly problems	Gasket or seals leaking, channels cross-feeding.
	g. Direct clutch slow to act	Check piston, housing, seals, plates, check ball.
10. Rough 1-2 shift	*a.* TV cable misadjusted	Readjust.
	b. Servo-assembly trouble	Check oil-seal ring, band-apply pin.
	c. Valves sticking in valve body	Check TV plunger and valve, TV shift valve, 1-2 accumulator valve.
	d. Accumulator piston trouble	Check oil ring, piston, piston bore, spring.
11. Rough 2-3 shift	*a.* TV cable misadjusted	Readjust.
	b. Servo-assembly trouble	Check exhaust hole for plugging, check piston seals.
	c. Valves sticking in valve body	Check TV plunger and valve, shift TV valve for binding, direct-clutch check ball.
12. No engine braking—range 2	*a.* Servo trouble	Check small oil seal on piston.
	b. Band damaged	Service, replace.
	c. Valves or balls sticking in valve body	Check reverse check ball, shift TV check ball, intermediate-boost valve.

COMPLAINT	POSSIBLE CAUSE	CHECK OR CORRECTION
13. No engine braking—range 1	a. Low overrun-clutch valve in valve body binding	Check, service valve body.
	b. Low-and-reverse clutch trouble	Piston seals defective, clutch not holding, clutch-housing-to-case seal or plug damaged or not seating.
14. No part-throttle or detent downshift	a. Throttle-valve cable misadjusted	Readjust.
	b. Valve-body troubles	Check 2-3 TV and TV valve bushing passages, shift TV and throttle valve for binding, gasket and spacer-plate placement.
15. Shift points high or low	a. Throttle-valve cable misadjusted	Readjust.
	b. Governor seals broken or missing	Check shaft-to-cover seal ring, cover O ring.
	c. Pressure-regulator valve in pump	Check for sticking.
	d. Valve-body troubles	Check TV exhaust-valve lifter, 1-2 or 2-3 valves, TV plunger, throttle valve, shift TV valve and check ball 1, gaskets, and spacer-plate placement.
16. Won't hold in park	a. Manual linkage misadjusted	Readjust.
	b. Manual detent roller and spring	Check bolt that holds roller to control-valve assembly; check pin and roller for damage.
	c. Inside detent lever and pin	Check for wear or looseness.
	d. Parking-lock trouble	Check parking pawl, actuator rod or plunger, bracket.
17. Engine coolant in transmission	Leak has developed in oil cooler inside radiator. Transmission must be removed for complete rebuilding, cleaning, and replacement of seals, composition-faced clutch plates, nylon washers, and speedometer and governor gears. Flush converter. Repair or replace cooler and flush cooler lines.	
18. Transmission noise	Noise could come from low oil level, damaged pump gears, or damaged or worn internal transmission bearings or other parts. Sometimes, when the transmission is grounded to the car body (insulators missing), normal transmission noise will sound loud. A clunk at a very low speed might be caused by damaged governor springs.	

✸11-7 On-the-car repairs Several repairs, and removal and replacement of parts, can be done with the transmission on the car. These include:

Adjusting linkages (✸ 11-4)

Checking and adding fluid (✸ 11-3)

Shift-control-lever replacement

Throttle-valve-cable replacement

Governor replacement

Pressure-regulator-valve replacement

Valve-body replacement

Intermediate-servo replacement

Speedometer driven-gear replacement

Rear-extension-housing-oil-seal replacement

These services are covered in following sections. Some additional parts can be removed and replaced with the transmission on the car. These include the oil-filler pipe and seal ring, oil pan and gasket, inside detent lever and parking-brake actuator rod, oil screen, throttle lever and bracket, manual-detent roller and spring, manual shaft and seal, oil line connector, and parking pawl. Removal and replacement of these parts are covered later in the chapter, where the disassembly and assembly of the transmission are described.

✸11-8 Shift-control lever To remove and replace the shift-control lever (Fig. 11-6), raise the vehicle on a lift to disconnect the shift rod from the actuating lever. Lower the vehicle and remove the four cover-mounting screws and the cover. Disconnect the electrical connector to the combination starter-safety, backup-lamp, and seat-belt warning switch. Disconnect the quadrant lamp. Remove the switch. Then remove the four shift-control-assembly screws and the shift-control assembly.

To install, refer to Fig. 11-6 and reverse the above procedure.

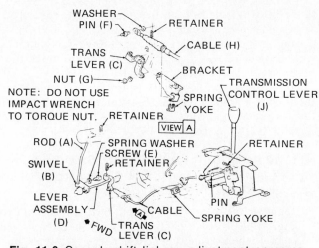

Fig. 11-6 Console-shift linkage adjustment on some cars equipped with the GM 200C automatic transmission. (*Chevrolet Motor Division of General Motors Corporation*)

❉ 11-9 Throttle-valve-cable replacement To remove and replace the throttle-valve cable (Fig. 11-7), remove the air cleaner. Then:

1. Push up on the bottom of the snap lock and release lock and cable.
2. Compress the locking tabs and disconnect the snap-lock assembly from the bracket.
3. Disconnect the cable from the carburetor lever.
4. Remove the clamp around the filler tube and the screw and washer attaching the cable to the transmission, and disconnect the cable.
5. To install the cable, first install a new seal into the transmission. Lubricate the seal with transmission fluid.

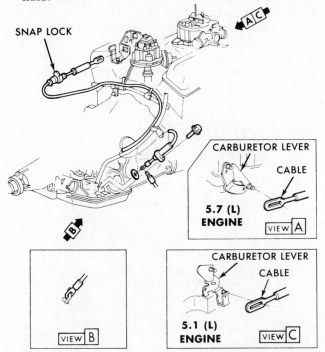

Fig. 11-7 Throttle-valve (TV) cable adjustment on some car lines. (*Chevrolet Motor Division of General Motors Corporation*)

6. Connect the transmission end of the cable and secure it to the case with bolt and washer.
7. Route the cable in front of the filler tube and clamp it to the filler tube at the filler-tube bracket.
8. Pass the cable through the bracket and engage the locking tube of the snap lock on the bracket. Connect the cable to the carburetor lever and install the air cleaner.
9. To adjust, disengage the snap lock and move the carburetor lever to wide open. Push the snap lock flush and return the carburetor lever to the closed position.

❉ 11-10 Governor replacement Disconnect the negative battery terminal. On cars with air conditioning, remove the heater-core-cover screws from the heater assembly. Disconnect the wire connector and move the cover out of the way (hose attached).

1. Raise the vehicle on a lift and remove the exhaust-pipe-to-torque-converter bolts. Let the torque converter and exhaust pipe hang down.
2. Put a safety stand under the transmission to support it. Then remove the rear transmission-support bolts.
3. Scribe marks on the drive shaft and companion flange so they can be reconnected in their original position. Then remove the drive shaft.
4. Let the transmission move down slightly for clearance to remove the governor.
5. Remove the governor retainer ring and cover. Discard the two seal rings.
6. Remove the governor assembly and the governor-to-case washer (Fig. 11-8). If the washer falls into the case, use a magnet to retrieve it. If you cannot get it, use a new washer on reassembly.
7. Reinstallation of the governor is the reverse of removal. Do not use a hammer to install the governor. It should slide in without requiring too much force.

❉ 11-11 Pressure-regulator-valve replacement Drain the transmission fluid from the oil pan and remove the pan and screen. Discard the gasket. Then

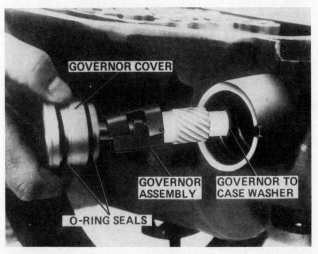

Fig. 11-8 Removing the governor. (*Chevrolet Motor Division of General Motors Corporation*)

149

Fig. 11-9 Removing the pressure-regulator valve. (*Chevrolet Motor Division of General Motors Corporation*)

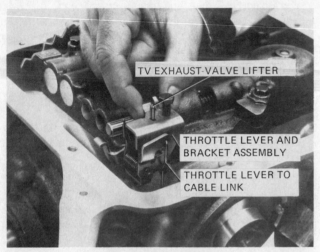

Fig. 11-10 Removing throttle-valve-lever-and-bracket assembly. (*Chevrolet Motor Division of General Motors Corporation*)

Fig. 11-11 Removing manual-detent-roller-and-spring assembly. (*Chevrolet Motor Division of General Motors Corporation*)

push down on the pressure-regulator-valve bore plug with a small screwdriver or special tool (Fig. 11-9). This compresses the spring so you can remove the retaining ring. Slowly release the spring force and remove the bore plug, valve, spring, and guide.

To install the parts, put the spring in first, followed by the guide and the valve, stem end out. Then install the bore plug, hole side out. Compress the spring and install the retainer ring.

✿ **11-12 Valve-body replacement** Drain the transmission fluid and remove the oil pan and screen. Discard the gasket. Disconnect the detent cable. Then:

1. Remove the throttle-lever-and-bracket assembly (Fig. 11-10). Do not bend the throttle-lever link.
2. Remove the manual-detent-roller-and-spring assembly (Fig. 11-11).
3. Support the valve assembly and remove the retaining bolts.
4. Hold the manual valve with your fingers and remove the valve assembly, spacer plate, and gaskets together to keep four check balls in the valve body and a fifth check ball in the spacer plate from dropping out (Fig. 11-12).
5. Lay the assembly, spacer-plate side up, on a clean surface. Overhaul of the valve body is covered in ✿ 11-28.
6. The intermediate-band anchor pin and reverse-clutch-cup plug may come out. If they do, be sure the anchor pin is positioned on the intermediate band after valve-body installation. Otherwise the transmission may be damaged.

✿ **11-13 Intermediate-servo replacement** Disconnect the grounded battery cable. Release the parking brake. Raise the vehicle on a lift.

To provide clearance, place a safety stand under the transmission and remove the rear transmission support. Let transmission drop down slightly and insert a 2-inch [51-mm] block to the right side of the transmission. This moves the transmission to the left. Use the

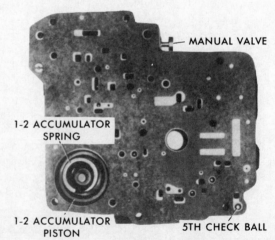

Fig. 11-12 Valve body, showing location of fifth check ball. (*Chevrolet Motor Division of General Motors Corporation*)

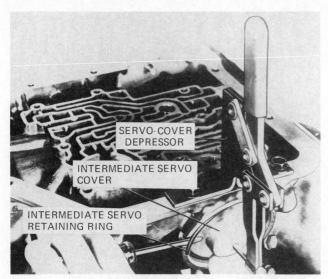

Fig. 11-13 Removing servo assembly. (*Chevrolet Motor Division of General Motors Corporation*)

◄ 5th CHECK BALL (TV EXHAUST)

Fig. 11-14 Removing the fifth check ball. (*Chevrolet Motor Division of General Motors Corporation*)

servo-cover depressor as shown in Fig. 11-13 to push the servo cover in so the retaining ring can be removed. Figure 11-13 shows the transmission upside down, as it would be during a major overhaul with the transmission in a holding fixture.

Remove the cover (discard the seal ring) and the servo piston and apply pin. Reinstall parts by reversing the removal procedure.

✿ 11-14 Speedometer driven gear The speedometer driven gear can be removed by disconnecting the speedometer cable and removing the retainer bolt, retainer, gear, and O-ring seal. Use a new O-ring seal on reinstallation.

✿ 11-15 Rear-extension-housing oil seal Scribe lines on the drive shaft and companion flange so they can be reconnected in their original position. Then remove the drive shaft. Pry the old seal out of the extension housing with a screwdriver (Fig. 10-18). Drive a new seal into place with the special seal driver. Then reinstall the drive shaft and adjust fluid level.

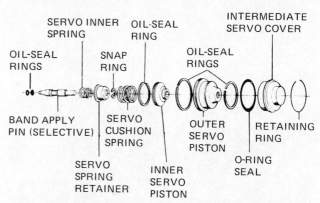

Fig. 11-15 Disassembled servo. (*Chevrolet Motor Division of General Motors Corporation*)

✿ 11-16 Transmission removal and installation The removal and installation procedure is very similar to that for the 350C transmission (✿ 10-14). On vehicles with air conditioning, remove the five heater-core-cover screws. Disconnect the wire connector and move the cover, with hoses attached, out of the way. Watch for shims at the rear support (Fig. 10-15). Save and reinstall any shims in the same position. After removing the transmission, install it in a transmission-holding fixture (Fig. 10-16).

✿ 11-17 Cleaning and inspecting transmission parts None of the moving parts require forcing to remove or install them. Bushings require special tools to remove and install them (✿ 11-29). Special checks of transmission parts to be made during disassembly are covered in ✿ 9-11. All old oil seal rings and old gaskets should be discarded.

✿ 11-18 Transmission disassembly How to remove the oil pan and valve body is covered in ✿ 11-12. From the case, remove the 1-2 accumulator spring and the fifth check ball (Fig. 11-14).

Remove the governor as explained in ✿ 11-10. Remove the intermediate servo as explained in ✿ 11-13. Figure 11-15 shows the servo components. At this point, check the band-apply pin length.

1. **Checking the band-apply pin** Install tools and dial indicator as shown in Fig. 11-16. Apply specified torque to the hex nut on the side of the gauge. Note dial-indicator needle travel. If it is not correct (see factory specifications) another pin will be required on reassembly.

2. **Removing front-unit parts** Before removal, check end play. Install the special sleeve on the output shaft. Then bolt the support fixture on the end of the case (Fig. 11-17). Turn transmission pump side up.

Remove one pump-to-case bolt and washer and install the end-play checking fixture as shown in Fig. 11-18. Force the output shaft upward by turning the adjusting screw (Fig. 11-17). When the white or scribed line on the sleeve begins to disappear, stop turning the wing screw. Set the dial indicator to zero.

151

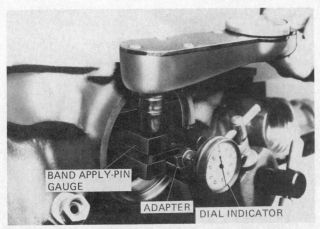

Fig. 11-16 Checking intermediate-band apply pin for proper length. (*Chevrolet Motor Division of General Motors Corporation*)

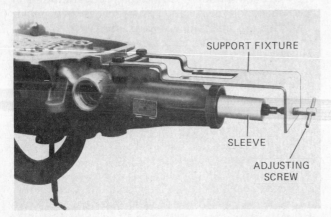

Fig. 11-17 Installing support fixtures on rear end of case. (*Chevrolet Motor Division of General Motors Corporation*)

Fig. 11-18 Checking front-unit end play. (*Chevrolet Motor Division of General Motors Corporation*)

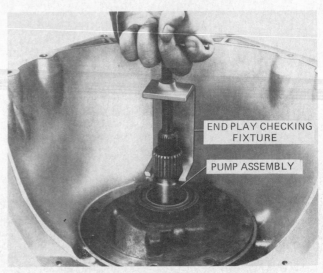

Fig. 11-19 Removing pump assembly. (*Chevrolet Motor Division of General Motors Corporation*)

Pull the turbine shaft up with the checking fixture attached to it (Fig. 11-19). Note dial-indicator needle travel. If it is excessive, the end play is too great. A thicker selective washer will be required on reassembly. This washer is located between the output shaft and the turbine shaft (Fig. 11-20).

a. Removing the pump—Remove the dial indicator but use the checking fixture to pull the pump out after removing the rest of the pump-to-case bolts and washers (Fig. 11-19). Lay the pump aside on a clean surface for disassembly.

Careful: The 200C transmission has an exhaust-line solenoid and a converter-clutch apply-valve assembly. (These are not in the earlier 200 transmission.) Do not damage the solenoid leads when removing the pump. See ✿ 11-27 and Figs. 11-40 and 11-41.

b. Forward and direct clutch—Lift up on the turbine shaft to remove the direct- and forward-clutch assemblies. Separate the two clutches and lay them aside on a clean surface for disassembly.

c. Remove the intermediate band and anchor pin from the case.

3. Front-unit parts   Remove the selective washer (Fig. 11-20). It might come off with the turbine shaft.

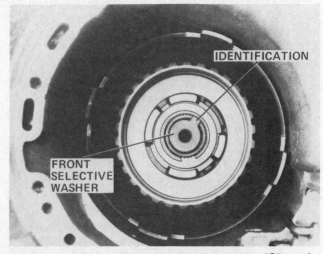

Fig. 11-20 Location of front selective washer. (*Chevrolet Motor Division of General Motors Corporation*)

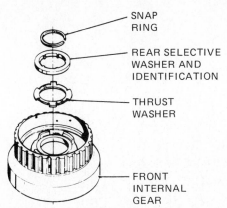

Fig. 11-21 Location of the thrust washer and selective washer in the front internal gear. (*Chevrolet Motor Division of General Motors Corporation*)

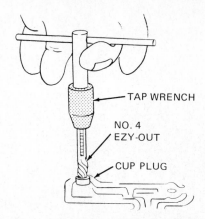

Fig. 11-23 Removing cup plug and seal. (*Chevrolet Motor Division of General Motors Corporation*)

Then check the rear-unit end play as follows. Loosen the adjusting screw on the output shaft (Fig. 11-17) and push the output shaft down. Then install a dial indicator with an extension resting on the end of the output shaft. Set the dial indicator to zero.

Turn the adjusting screw (Fig. 11-17) in to raise the output shaft until the white or scribed line on the sleeve begins to disappear. Then read the end play on the dial indicator. If it is not correct, a new selective washer will be required on reassembly. This selective washer is located between the front-internal-gear thrust washer and the snap ring (Fig. 11-21). Proceed as follows:

a. Remove snap ring, washers, and front internal gear.
b. Remove the front planetary-carrier assembly and thrust roller bearing (Fig. 11-22).
c. Remove the front sun gear and thrust-bearing assembly.
d. Remove the input drum and rear sun-gear assembly. Also, remove the thrust washer with four tangs.
e. Remove the low-and-reverse clutch housing. First, grind about ¾ inch [20 mm] off the end of a No. 4 Ezy-Out and use it to remove the housing-to-case cup plug (Fig. 11-23). Then remove the beveled snap

ring holding the clutch housing in the case. On reinstallation, this snap ring goes flat side against the clutch housing. Grasp the clutch housing with the special tool (Fig. 11-24) and move it up and down to remove the clutch housing. Then remove the large spacer ring that is under the housing.

4. **Rear gear parts** Grasp the output shaft and lift out the rest of the rear-unit parts. This includes the roller clutch, the thrust washer with four tangs, the rear carrier assembly, the clutch plates, and the roller thrust bearing. Then turn the transmission so the rear is up and remove the tools (Fig. 11-17).

5. **Manual shaft and parking pawl** Figure 11-25 shows the manual shaft and parking pawl. Use this illustration as a guide if you must replace any part. Use a No. 4 Ezy-Out to remove the steel cup plug and a No. 3 Ezy-Out to remove the parking pawl shaft.

✿ **11-19 Servicing subassemblies** Following sections describe the servicing of the subassemblies removed from the transmission during disassembly.

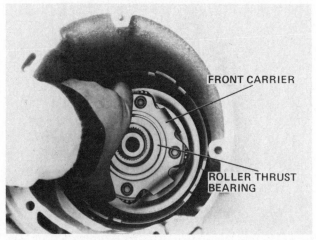

Fig. 11-22 Removing front carrier and thrust roller bearing. (*Chevrolet Motor Division of General Motors Corporation*)

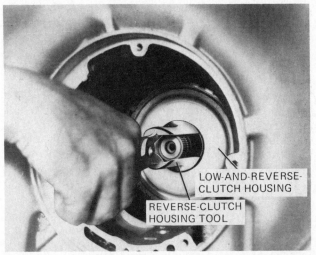

Fig. 11-24 Removing the low-and-reverse clutch housing. (*Chevrolet Motor Division of General Motors Corporation*)

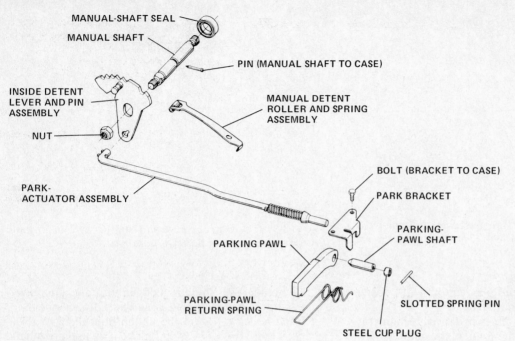

MANUAL-SHAFT SEAL

MANUAL SHAFT

PIN (MANUAL SHAFT TO CASE)

INSIDE DETENT
LEVER AND PIN
ASSEMBLY

MANUAL DETENT
ROLLER AND SPRING
ASSEMBLY

NUT

PARK-
ACTUATOR ASSEMBLY

BOLT (BRACKET TO CASE)

PARK BRACKET

PARKING-
PAWL SHAFT

PARKING PAWL

PARKING-PAWL
RETURN SPRING

SLOTTED SPRING PIN

STEEL CUP PLUG

Fig. 11-25 Manual-shaft and parking-shaft parts. (*Chevrolet Motor Division of General Motors Corporation*)

❂ 11-20 Replacing the speedometer drive gear

Two types of drive gears have been used, nylon and steel.

1. To replace the nylon gear, depress the drive-gear clip and slide the gear off, tapping it lightly with a plastic hammer. Put the drive-gear clip in the correct hole in the output shaft (the same hole it was in originally). Slide the drive gear into place, aligning the slot in the gear with the clip.

2. To replace the steel gear, use a gear puller (Fig. 11-26) to remove the old gear. Press the new gear on the shaft until the gear is the distance shown in Fig. 11-27 from the end of the shaft.

❂ 11-21 Roller-clutch-and-rear-carrier assembly

The roller clutch and the rear carrier are shown in Fig. 11-28. Put these parts together and install them, and the rear internal gear, as follows.

1. Install the rear internal gear on the output shaft, hub end first. Then install the thrust bearing with the inside diameter against the internal gear.

2. Install the thrust washer in the rear carrier (Fig. 11-28).

3. If a roller has come out of the roller-clutch cage, compress the energizing spring and insert the roller in the cage.

4. Install the roller clutch in the rear carrier. Push the roller-clutch race, spline side out, into the roller clutch, turning it counterclockwise as it enters.

5. Install the tanged thrust washer on the rear carrier. Retain it with petroleum jelly.

6. Install the rear-carrier assembly in the rear internal gear (Fig. 11-29).

7. Install the tools on the rear of the case (Fig. 11-17).

8. Turn the case vertical, pump end up.

9. Install the rear-unit parts in the case. Turn the in-

ternal gear so the parking-pawl lugs index with the parking-pawl tooth.

10. Look through the parking-pawl case slot and adjust the height of the internal-gear parking-pawl lugs to align flush with the parking-pawl tooth. The adjustment is made by turning the adjusting screw (Fig. 11-17).

11. Make sure the speedometer drive gear is visible through the speedometer-gear bore. If it is not visible, it has been installed incorrectly.

PULLER

SPEEDOMETER DRIVE GEAR

SPEEDOMETER GEAR PULLER

Fig. 11-26 Removing steel speedometer-drive gear from output shaft. (*Chevrolet Motor Division of General Motors Corporation*)

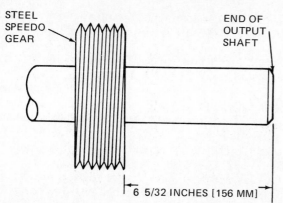

Fig. 11-27 Positioning the steel speedometer-drive gear. (*Chevrolet Motor Division of General Motors Corporation*)

⚙ 11-22 Low-and-reverse clutch The low-and-reverse clutch is shown disassembled in Fig. 11-30. To disassemble it, remove the snap ring and the spring retainer. After inspection of all parts, including the clutch plates (⚙ 9-11), reassemble the parts. Use new inner and outer piston seals. To install the piston in the housing, use a seal protector to cover the housing hub and protect the inner seal.

Use a small screwdriver or piston-installing tool (Fig. 11-31) to work the outer seal down into the clutch housing. Lubricate the seals with transmission fluid before installation. To install the clutch in the case, proceed as follows:

1. Soak the clutch plates in transmission fluid at least 30 minutes and install, starting with a flat steel plate. Then alternate composition and steel plates.
2. Install the large spacer ring in the case, followed by the clutch-housing assembly. Use the reverse-clutch-housing tool (Fig. 11-24) to hold the housing. Align the clutch-housing feed hole with the feed hole in the case.

3. If the low-and-reverse clutch housing does not seat past the case snap-ring groove, remove the tool. Use the rear sun gear and input drum as a tool. Install this assembly in the case. Rotate the sun gear back and forth, tapping lightly with the input drum to align the roller-clutch race and the low-and-reverse clutch hub splines.
4. Remove the input drum and the rear sun gear. Now try to install the snap ring. You may have to loosen the adjusting screw holding the output shaft in position (Fig. 11-17) to install the snap ring. You might then have to repeat the procedure in item 3. The flat side of the snap ring goes against the clutch housing.

⚙ 11-23 Front gear parts The rear sun gear and the input drum are part of the front gear arrangement (Fig. 11-32). If the sun gear has been removed from the input drum, install and secure it with the snap ring. Put the thrust washer with four tangs on the input drum, retaining it with petroleum jelly. The tangs fit into holes in the drum. Install the assembly in the case, over the low-and-reverse clutch housing.

1. Install the front sun gear with the identification mark (the drilled spot or groove) against the input-drum-to-rear-sun-gear snap ring.
2. Install the thrust bearing and race assembly with the needle bearings against the gear.
3. Install on the front carrier the roller thrust bearing with the smaller diameter against the carrier. Retain with petroleum jelly.
4. Install the front carrier with the roller thrust bearing over the front sun gear.
5. Install the thrust washer on the front internal gear and retain it with petroleum jelly. Then install the front internal gear with the thrust washer over the front carrier.
6. Install the proper thickness of rear selective washer

Fig. 11-28 Roller clutch and rear carrier, disassembled. (*Chevrolet Motor Division of General Motors Corporation*)

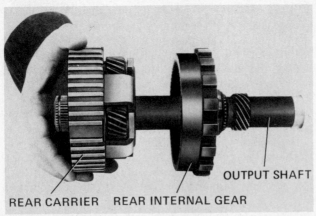

Fig. 11-29 Installing rear carrier into rear internal gear. (*Chevrolet Motor Division of General Motors Corporation*)

(Fig. 11-21) as determined during disassembly (✿ 11-18, item 2). Secure it with the snap ring.

✿ **11-24 Front-unit parts** The direct clutch is shown disassembled in Fig. 11-33. To disassemble it, remove the large snap ring and the clutch backing plate and clutch plates. Then use the clutch-spring compressor (Fig. 11-34) to compress the springs so the small snap ring can be removed. Remove the retainer and spring assembly, release-spring guide, and other internal parts (Fig. 11-33).

On reassembly, use new inner and outer piston seals. Use the inner-seal protector when installing the piston in the housing. Use the piston-installing tool (Fig. 11-31) to work the piston outer seal into place in the housing. Lubricate the seals with transmission fluid before installing the piston. Be careful when working the large

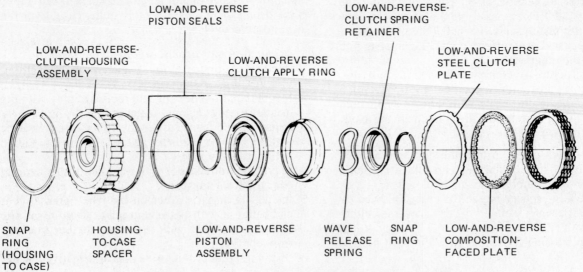

Fig. 11-30 Low-and-reverse-clutch housing, disassembled. (*Chevrolet Motor Division of General Motors Corporation*)

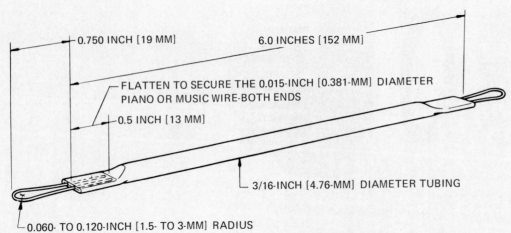

Fig. 11-31 How to make a piston-installing tool. (*Chevrolet Motor Division of General Motors Corporation*)

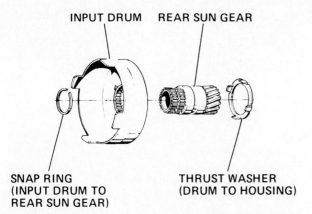

INPUT DRUM   REAR SUN GEAR

SNAP RING
(INPUT DRUM TO
REAR SUN GEAR)

THRUST WASHER
(DRUM TO HOUSING)

Fig. 11-32 Sun gear and input drum. (*Chevrolet Motor Division of General Motors Corporation*)

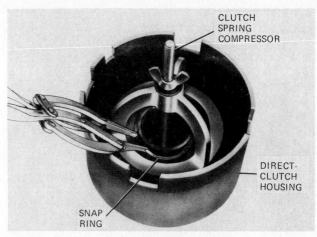

CLUTCH
SPRING
COMPRESSOR

DIRECT-
CLUTCH
HOUSING

SNAP
RING

Fig. 11-34 Using a clutch-spring compressor and snap-ring pliers to remove the snap ring and release the direct-clutch-spring tension. (*Chevrolet Motor Division of General Motors Corporation*)

piston seal past the large direct-clutch snap-ring groove. This groove is sharp-edged and could cut the seal.

Install the release-spring guide on the piston so the omitted rib is directly over the check-ball assembly in the face of the piston. Then install the retainer-and-spring assembly. Use the spring-compressor tool to compress the springs so the snap ring can be installed (Fig. 11-34).

Lubricate and install the clutch plates in the clutch housing, starting with a flat steel plate and alternating with composition plates. Install the backing plate, chamfered side out. Install the large snap ring to secure the plates. This assembly is now ready to install in the case. Set it aside temporarily.

✿ 11-25 Forward-clutch assembly The forward-clutch assembly is shown disassembled in Fig. 11-35. To disassemble it, put the assembly on the bench clutch down, with the turbine shaft through a hole in the bench.

1. Remove the large snap ring and the backing plate, the steel and composition plates, and the waved steel plate.

2. Install the forward-clutch-spring compressor as shown in Fig. 11-36 and remove the snap ring.
3. Take out the tool, clutch-retainer-and-spring assembly, apply ring, and piston assembly. Remove and discard the old piston-seal rings.
4. Discard the old turbine-shaft seal rings. Install new seal rings.
5. Install new seals on the piston with the lips facing away from the clutch-apply ring.
6. Lubricate the seals with transmission fluid and use the piston-installing tool (Fig. 11-31) on both the inner and outer seals to install the piston in the clutch housing. Be careful when inserting the piston in the housing that the outer ring is not cut by the large snap-ring groove.
7. Use the spring compressor (Fig. 11-36) to compress the spring so you can install the snap ring.
8. Soak the clutch plates in transmission fluid for at least 30 minutes and install them, starting with the

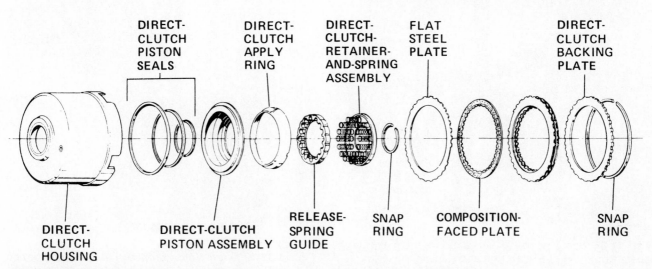

DIRECT-
CLUTCH
PISTON
SEALS

DIRECT-
CLUTCH
APPLY
RING

DIRECT-
CLUTCH-
RETAINER-
AND-SPRING
ASSEMBLY

FLAT
STEEL
PLATE

DIRECT-
CLUTCH
BACKING
PLATE

DIRECT-
CLUTCH
HOUSING

DIRECT-CLUTCH
PISTON ASSEMBLY

RELEASE-
SPRING
GUIDE

SNAP
RING

COMPOSITION-
FACED PLATE

SNAP
RING

Fig. 11-33 Disassembled view of the direct-clutch assembly. (*Chevrolet Motor Division of General Motors Corporation*)

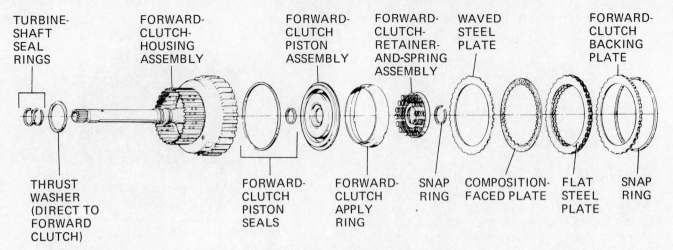

TURBINE-SHAFT SEAL RINGS FORWARD-CLUTCH-HOUSING ASSEMBLY FORWARD-CLUTCH PISTON ASSEMBLY FORWARD-CLUTCH-RETAINER-AND-SPRING ASSEMBLY WAVED STEEL PLATE FORWARD-CLUTCH BACKING PLATE

THRUST WASHER (DIRECT TO FORWARD CLUTCH) FORWARD-CLUTCH PISTON SEALS FORWARD-CLUTCH APPLY RING SNAP RING COMPOSITION-FACED PLATE FLAT STEEL PLATE SNAP RING

Fig. 11-35 Forward-clutch assembly, disassembled. (*Chevrolet Motor Division of General Motors Corporation*)

waved steel plate and alternating composition and steel plates.

9. Install the backing plate chamfered side up and secure it with the large snap ring.

10. Install the thrust washer on the front of the housing, securing it with petroleum jelly.

✿ **11-26 Installing forward and direct clutches** Put the direct-clutch assembly, clutch plates up, over a hole in the bench. Align the clutch-composition-plate teeth to make it easier to install the forward clutch.

1. Install the forward-clutch assembly in the direct-clutch assembly. Hold the direct-clutch housing and turn the forward clutch back and forth to mate the direct-clutch plates with the forward clutch.

2. When properly seated, the forward-clutch housing will be the distance above the direct-clutch housing shown in Fig. 11-37.

3. Lift the assembly, holding the two clutches together, and turn it upside down so the turbine shaft is up. Put the installing tool over the turbine shaft and install the assembly in the case. Rotate the assembly until it drops down into position. When correct, the distance between the direct-clutch housing and the pump face in the case should be as shown in Fig. 11-38.

✿ **11-27 Pump assembly** How to remove the pump during the transmission disassembly procedure was explained in ✿ 11-18. You were cautioned against damaging the solenoid leads on the 200C transmission (Fig. 11-39). The pumps for the 200 and the 200C transmission are alike except that the 200C pump has a solenoid and a converter-clutch apply valve assembly. The 200 pump is shown disassembled in Fig. 11-40. It does not have these two additional components.

Figure 11-41 shows how the solenoid is mounted on

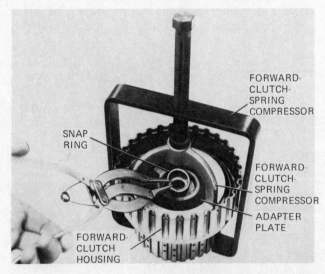

FORWARD-CLUTCH-SPRING COMPRESSOR

SNAP RING

FORWARD-CLUTCH-SPRING COMPRESSOR

ADAPTER PLATE

FORWARD-CLUTCH HOUSING

Fig. 11-36 Using a spring compressor and snap-ring pliers to remove the snap ring and release the forward-clutch-spring tension. (*Chevrolet Motor Division of General Motors Corporation*)

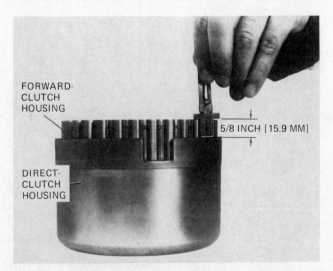

FORWARD-CLUTCH HOUSING

DIRECT-CLUTCH HOUSING

5/8 INCH [15.9 MM]

Fig. 11-37 Forward clutch properly seated in the direct clutch. (*Chevrolet Motor Division of General Motors Corporation*)

Fig. 11-38 Direct- and forward-clutch assemblies properly installed. (*Chevrolet Motor Division of General Motors Corporation*)

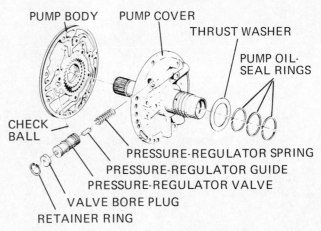

Fig. 11-40 Pump assembly, disassembled. (*Chevrolet Motor Division of General Motors Corporation*)

the pump. Figure 11-42 shows the converter-clutch actuator valve, apply valve, and apply-valve bushing. These are the components that work with the governor pressure valve to control the converter clutch (✿ 4-23 and 4-24).

1. To remove the pressure-regulator valve, remove the retainer ring, holding the valve-bore plug in with a small screwdriver to take up the spring force.

NOTE: The pump is body-side down with the drive shaft sticking in a hole in the bench. To keep the drive shaft from turning, put a bolt or screwdriver through a hole in the pump and the bench.

2. To remove the converter-clutch apply valve (Fig. 11-42), remove the retaining pin.

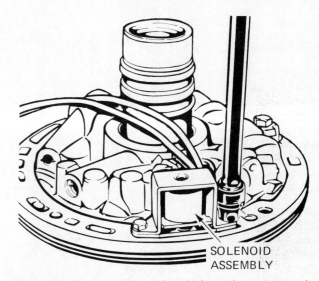

Fig. 11-41 Removing the solenoid from the pump on the 200C transmission. (*Buick Motor Division of General Motors Corporation*)

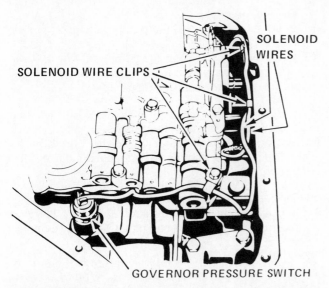

Fig. 11-39 Locations of the solenoid wires and governor pressure switch on the 200C transmission. (*Pontiac Motor Division of General Motors Corporation*)

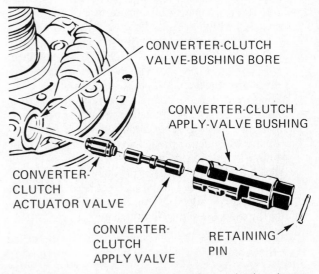

Fig. 11-42 Location of the apply valve and related parts on the pump in the 200C transmission. (*Pontiac Motor Division of General Motors Corporation*)

159

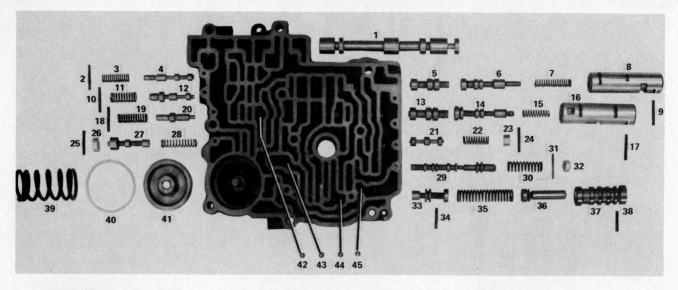

1. MANUAL VALVE
2. RETAINING PIN
3. INTERMEDIATE BOOST SPRING
4. INTERMEDIATE BOOST VALVE
5. 2-3 SHIFT VALVE
6. 2-3 THROTTLE VALVE
7. 2-3 THROTTLE VALVE SPRING
8. 2-3 THROTTLE VALVE BUSHING
9. RETAINING PIN
10. RETAINING PIN
11. LOW OVERRUN CLUTCH SPRING
12. LOW OVERRUN CLUTCH VALVE
13. 1-2 SHIFT VALVE
14. 1-2 THROTTLE VALVE
15. 1-2 THROTTLE VALVE SPRING

16. 1-2 THROTTLE VALVE BUSHING
17. RETAINING PIN
18. RETAINING PIN
19. DIRECT-CLUTCH EXHAUST SPRING
20. DIRECT-CLUTCH EXHAUST VALVE
21. REVERSE-BOOST VALVE
22. REVERSE-BOOST SPRING
23. REVERSE-BOOST BORE PLUG
24. RETAINING PIN
25. RETAINING PIN
26. 1-2 ACCUMULATOR BORE PLUG
27. 1-2 ACCUMULATOR VALVE
28. 1-2 ACCUMULATOR VALVE SPRING
29. SHIFT TV VALVE
30. SHIFT TV SPRING

31. RETAINING PIN
32. SHIFT TV BORE PLUG
33. THROTTLE VALVE
34. RETAINING PIN
35. THROTTLE-VALVE SPRING
36. THROTTLE-VALVE PLUNGER
37. THROTTLE-VALVE PLUNGER BUSHING
38. RETAINING PIN
39. 1-2 ACCUMULATOR SPRING
40. 1-2 ACCUMULATOR PISTON SEAL
41. 1-2 ACCUMULATOR PISTON
42. CHECK BALL #4
43. CHECK BALL #3
44. CHECK BALL #2
45. CHECK BALL #1

(NOTE: 5TH CHECK BALL
IS IN THE CASE)

Fig. 11-43 Valve body, disassembled. (*Pontiac Motor Division of General Motors Corporation*)

3. To remove the solenoid, remove the attaching bolts (Fig. 11-41).
4. Separate the pump cover from the pump body by removing the attaching bolts. Do not drop the check ball. Keep it separate from the five check balls in the case, which are smaller.
5. Remove gears from pump body.
6. Replace bushing or seal if needed. To install a new seal, coat the outside with nonhardening sealing compound and install it with a special tool.
7. When installing the gears, the driven gear identification marks go down against the gear pocket.
8. Install the drive gear with identification marks up.
9. Put the ball in the pocket in the pump body and retain it with petroleum jelly.
10. Attach the pump body to the pump cover.
11. After installing the pressure-regulator valve, the clutch-apply valve and components, and the solenoid, install the pump in the case.

NOTE: If the pump oil-seal rings need replacement, replace them before installing the pump. Lubricate them with transmission fluid.

12. Before installing the pump, put the intermediate band into the case. Make sure the intermediate-band anchor-pin lug is aligned with the anchor-pin hole in the case.
13. Put the thrust washer in place, retaining it with petroleum jelly.
14. Put a new gasket on the pump and retain it with petroleum jelly.
15. Use guide pins to align the pump and lower it into position. Remove guide pins and install all but one of the pump bolts, finger tight.

NOTE: If the turbine shaft cannot be rotated as the pump is being pulled into place, the forward or direct clutch has not been properly installed. All the clutch plates are not indexed. Remove the clutches and correct the condition.

16. Torque the bolts to specifications. Install the end-play checking fixture as shown in Fig. 11-18 to recheck the front-unit end play (✿ 11-18, item 2).

✿ **11-28 Control-valve-body assembly** Figure 11-43 shows the control-valve-body assembly disassembled. This illustration is your guide as you disassemble and reassemble the control-valve assembly. Use care to avoid mixing the valves and other parts. When you remove a valve train, place its component parts on the bench in the exact order they occupy in the assembly.

Careful: Valves, bushings, and other components in the valve body are not interchangeable!

Refer to ✿ 11-12 for the valve-body installation procedure.

✿ **11-29 Bushing replacement** Bushings that might require replacement are used in the:

Pump cover (two bushings)

Pump body

Direct clutch (two bushings)

Front internal gear

Rear sun gear

Low-and-reverse clutch housing

Rear and carrier bushing

Case bushing (at rear)

The procedure is the same for all. Use the specific service tool to press out the old bushing. Then use the correct tool to press the new bushing into place.

Chapter 11 review questions

Select the *one* correct, best, or most probable answer to each question. Then check your answers against the correct answers given at the end of the book.

1. Mechanic A says the main difference between the 200 and the 200C transmission is that the 200C has a converter clutch. Mechanic B says the 200C has a solenoid and apply valve on the pump that the 200 does not have. Who is right?
   a. A only
   b. B only
   c. both A and B
   d. neither A nor B
2. The two adjustments required on the 200C are the:
   a. band and clutch
   b. manual-valve and band linkages
   c. throttle-valve and band linkages
   d. manual- and throttle-valve linkages
3. Mechanic A says the trouble-diagnosis procedure for the 200C requires a road test. Mechanic B says the test can be made in the shop. Who is right?
   a. A only
   b. B only
   c. both A and B
   d. neither A nor B
4. The 200C oil pressure is checked at:
   a. 1000 and 1200 engine rpm
   b. 200 and 1000 rpm
   c. high engine speed and idling
   d. engine idle only
5. The two testing instruments required for the trouble-diagnosis procedure are:
   a. an oil-pressure gauge and a dwell meter
   b. a tachometer and a voltmeter
   c. an ammeter and a voltmeter
   d. an oil-pressure gauge and a tachometer

6. In drive range, first gear, the holding members are the:
   a. direct and forward clutches
   b. forward and roller clutches
   c. band and direct clutch
   d. band and forward clutch
7. In drive range, third gear, the holding members are the:
   a. direct and forward clutches
   b. forward and roller clutches
   c. band and direct clutch
   d. band and forward clutch
8. Mechanic A says that installing the test governor when analyzing torque-converter-clutch trouble is done to simulate high engine speed. Mechanic B says the purpose is to increase governor pressure. Who is right?
   a. A only
   b. B only
   c. both A and B
   d. neither A nor B
9. Mechanic A says the purpose of the converter clutch is to bypass the converter at highway speed. Mechanic B says the purpose is to prevent slippage through the converter. Who is right?
   a. A only
   b. B only
   c. both A and B
   d. neither A nor B
10. Among the parts that can be removed with the transmission in the vehicle are the:
    a. governor and front clutch
    b. valve body and governor
    c. forward and direct clutches
    d. oil pump and governor

CHRYSLER TORQUEFLITE TRANSMISSION SERVICE

After studying this chapter, and with proper instruction and equipment, you should be able to:

1. Explain how to use the diagnostic charts to find causes of trouble.
2. Describe the procedures of making diagnostic tests of the TorqueFlite including hydraulic and air-pressure, road, and stall tests.
3. Discuss manual-linkage and throttle-linkage adjustments.
4. Explain how to make band adjustments.
5. List the services that can be performed with the transmission in the car and explain how each is done.
6. Describe the procedure of removing and installing the transmission.
7. Discuss the procedure of disassembling and reassembling the transmission and the servicing procedures for the subassemblies.
8. Perform the services listed above.

12-1 Introduction to TorqueFlite service This chapter describes the trouble diagnosis, checks, adjustments, in-car service, removal, overhaul, and installation of the TorqueFlite transmission (Fig. 12-1). Chapter 5 discusses the construction and operation of this transmission.

There are three basic models of TorqueFlite—the A-727, the A-904, and the MA-904A. The later versions of the A-727 and the A-904 have the automatic-lockup torque converter (✿ 5-8). The basic servicing procedures for all of these TorqueFlite models are similar. Variations in the procedures for specific models are discussed as required.

For example, the A-904 has one large spring in its front clutch, while the A-727 has several small springs. There are also differences in the rear clutches, planetary-gear trains, and overrunning clutches. The pre-1977 models did not have the automatic-lockup torque converter.

This chapter does not discuss the TorqueFlite model A-404 transaxle, which is described in Chap. 15. Servicing of the transaxle TorqueFlite is covered in Chap. 16.

Plates 9 to 13 in the *Color Plates of Automatic-Transmission Hydraulic Circuits,* sixth edition, show the positions of the various valves and the hydraulic pressures during various TorqueFlite operating conditions.

✿ 12-2 Operating cautions For mountain driving, either with heavy loads or when pulling a trailer, use the 2 (second) or 1 (low) selector-lever position when on upgrades requiring a heavy throttle for ½ mile [0.8 km] or more. This reduces the possibility of transmission and converter overheating (see ✿ 9-5).

The TorqueFlite transmission will not permit starting the engine by pushing or towing.

When towing the vehicle, use a rear-end pickup or remove the drive shaft, as explained in ✿ 9-5. This protects the transmission from damage.

✿ 12-3 TorqueFlite trouble diagnosis Figures 12-2 to 12-6 are the trouble-diagnosis charts prepared by Chrysler for the A-727 and A-904 TorqueFlites. Figure 12-2 is the "abnormal-noise" chart. This chart tells you what to do to locate the trouble, and the remedies to eliminate the troubles. Figure 12-3 is the "vehicle will not move" chart, outlining the procedures to follow to locate and correct the trouble. Figure 12-4 explains where to look for leaks and what should be done to correct them.

Figure 12-5 is the general TorqueFlite diagnosis chart. To use it, you locate the condition causing the complaint at the top. Then run down from it to find the X's in the same column. When you find an X, move horizontally to the right to locate a possible cause. You

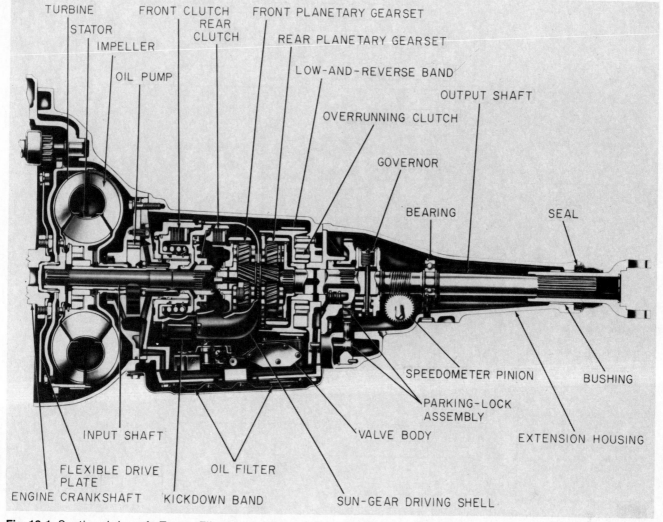

Fig. 12-1 Sectional view of a TorqueFlite automatic transmission. (*Chrysler Corporation*)

might find several X's in the vertical column and therefore several possible causes.

For example, suppose the condition is no kickdown or normal downshift. You find this in the top row and, running down, you find four X's. These four X's lead you to the right, where the possible causes are listed. These include (going from the bottom up) valve-body malfunction or leakage, incorrect throttle-linkage adjustment, governor malfunction, and kickdown-servo-band or linkage malfunction.

Figure 12-6 relates possible troubles with the lockup torque converter. If no lockup occurs, there are seven X's or possible causes to consider, from a faulty input shaft or seal ring to a faulty oil pump.

❖ 12-4 TorqueFlite diagnostic tests Diagnosis begins with checks of the fluid level and condition (❖ 12-5) and linkage adjustment (❖ 12-6 and 12-7). Then the car should be road-tested if necessary to determine whether or not the trouble has been eliminated. Road testing is discussed in ❖ 9-8. Certain transmission tests can also be made on the chassis dynamometer (❖ 1-15).

Figure 12-7 shows the shift speeds and governor pressures for various car models, engines, axle ratios, and tire sizes. Note that changes in tire sizes or axle ratios change the speed at which the shifts occur.

If adjusting the fluid level, linkages, and bands does not eliminate the trouble, pressure and stall tests should be performed to help pinpoint the cause.

❖ 12-5 Fluid level and condition Figure 12-8 shows the dipstick used in the TorqueFlites. Note that it calls for checking when the engine and transmission are hot and with the engine idling. The vehicle should be on level ground. Refer to ❖ 9-3 for details on checking fluid level, adding fluid, and analyzing the condition of the fluid. Specific procedures required on TorqueFlite transmissions are covered in ❖ 12-24.

❖ 12-6 Manual-linkage adjustment Manual-linkage adjustment can be checked by noting whether or

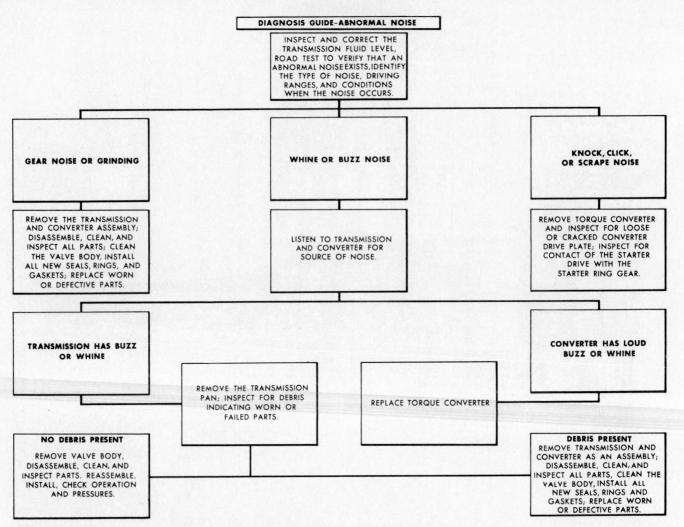

Fig. 12-2 Diagnosis guide to locate trouble if the complaint is abnormal noise. (*Chrysler Corporation*)

not the neutral safety switch operates normally. Move the selector lever slowly into P. If the starter operates, the P position is correct. Then move the selector lever slowly into N. If the starter operates correctly, the N position is correct. But if the starter does not operate in either position, adjustment is required. Figures 12-9, 12-10, and 12-11 show the adjustment procedures for different models.

1. **Column shift (Figs. 12-9 and 12-10)** The adjustable swivel block should be free to slide on the shift rod. Disassemble and clean parts if it sticks. Put the selector lever in P. With the adjustment-swivel lock screw loose, move the shift lever on the transmission all the way to the rear detent position (P). Tighten the lock screw to specifications. Recheck to make sure the starter will operate in N and P.

NOTE: If linkage rods must be disassembled from levers which have plastic grommets, new grommets should be installed. The old grommet must be cut away and a new one snapped into place with pliers. See ❀ 13-5 for the Ford procedure.

2. **Console shift (Fig. 12-11)** With the swivel lock screw loose, move the selector lever to P and move the shift lever on the transmission all the way to the rear (to P). Tighten the swivel lock screw to specifications and recheck to make sure the starter operates in N and P.

❀ **12-7 Throttle-linkage adjustment** The throttle-linkage adjustment must be correct to get proper transmission operation. It controls the speed at which shifts take place, the shift quality, and part-throttle downshifts. If the setting is short, early shifts and slippage between shifts will occur. If the setting is too long, shifts may be delayed and harsh, and part-throttle downshifts may be too sensitive. Here is the adjustment procedure (Figs. 12-12 to 12-14).

The engine must be at normal operating temperature and the choke should be disconnected to make sure the throttle is not on the fast-idle cam. Raise the vehicle on a lift.

Loosen the adjustment-swivel lock screw. Slide the swivel along the flat side of the throttle rod to make

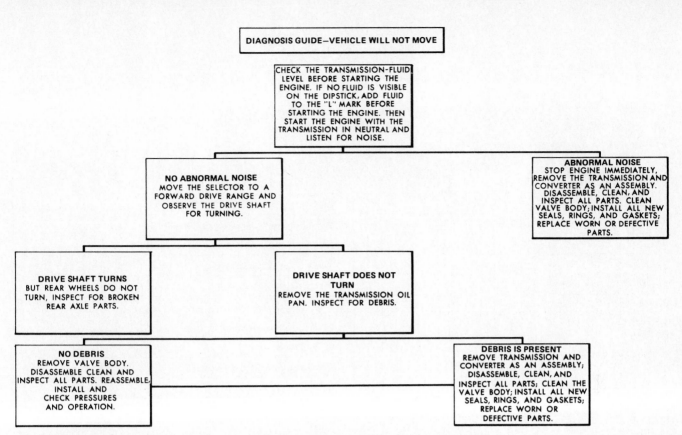

Fig. 12-3 Diagnosis guide to locate trouble if the vehicle will not move. (*Chrysler Corporation*)

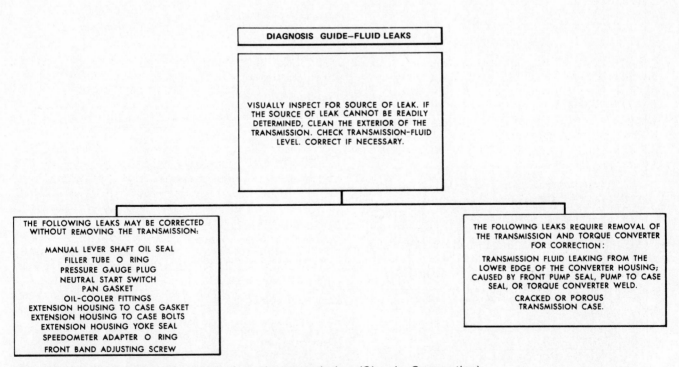

Fig. 12-4 Diagnosis guide if fluid leaks from the transmission. (*Chrysler Corporation*)

Fig. 12-5 Diagnosis chart relating various trouble conditions with possible causes. (*Chrysler Corporation*)

CONDITION columns (left to right):
1. Harsh engagement from neutral to D or R
2. Delayed engagement from neutral to D or R
3. Runaway upshift
4. No upshift
5. 3-2 kickdown runaway
6. No kickdown or normal downshift
7. Shifts erratic
8. Slips in forward drive positions
9. Slips in reverse only
10. Slips in all positions
11. No drive in any position
12. No drive in forward drive positions
13. No drive in reverse
14. Drives in neutral
15. Drags or locks
16. Grating, scraping, growling noise
17. Buzzing noise
18. Hard to fill, oil blows out filler tube
19. Transmission overheats
20. Harsh upshift
21. Delayed upshift
22. Slips in reverse or manual low

1	2	3	4	5	6	7	8	9	10	11	12	13	14	15	16	17	18	19	20	21	22	POSSIBLE CAUSE
X																			X			Faulty lockup clutch
															X							Overrunning-clutch inner race damaged
			X				X						X	X								Overruning clutch worn, broken, or seized
									X	X	X		X	X								Planetary gearsets broken or seized
													X									Rear clutch dragging
X	X						X			X	X	X										Worn or faulty rear clutch
													X					X				Insufficient clutch-plate clearance
																		X				Faulty cooling system
														X				X				Kickdown-band adjustment too tight
X																			X			Hydraulic pressure too high
																	X					Breather clogged
																	X					High fluid level
	X	X	X			X		X				X									X	Worn or faulty front clutch
	X	X	X	X	X																X	Kickdown-servo band or linkage malfunction
		X		X	X																X	Governor malfunction
X	X	X	X			X		X				X									X	Worn or broken reaction-shaft support seal rings
		X				X															X	Governor support seal rings broken or worn
															X							Output-shaft bearing and/or bushing damaged
							X			X												Overrunning clutch not holding
		X													X				X	X		Kickdown band out of adjustment
	X	X	X	X	X	X													X	X		Incorrect throttle-linkage adjustment
X																						Engine idle speed too low
X	X		X		X	X	X	X								X	X					Aerated fluid
X						X		X		X												Worn or broken input-shaft seal rings
X				X	X	X	X											X			X	Faulty oil pump
X	X			X	X				X	X							X					Oil filter clogged
X		X		X	X				X	X								X				Incorrect gearshift-control linkage adjustment
X	X	X	X	X	X	X	X	X	X	X					X			X				Low fluid level
X							X			X												Low-reverse servo, band, or linkage malfunction
X	X	X	X	X	X	X	X	X	X	X	X	X	X			X					X	Valve-body malfunction or leakage
							X			X			X	X								Low-reverse band out of adjustment
X	X	X	X	X	X	X	X	X	X	X	X								X	X		Hydraulic pressures too low
X																		X				Engine idle speed too high
													X									Stuck lockup valve
																		X				Stuck switch valve

166

No lockup	Will not unlock	Stays locked up to too low a speed in direct	Locks up or drags in low or second	Stalls or is sluggish in reverse	Loud chatter during lockup engagement	Vibration or shudder during lockup engagement (cold)	Vibrations after lockup engagement	Vibration when revved in neutral	Overheating: oil blowing out dipstick or pump seal	POSSIBLE CAUSE
X		X	X							Faulty oil pump
	X	X					X			Sticking governor valve
			X						X	Plugged cooler, lines or fittings
	X		X							Valve-body malfunction
X	X								X	Stuck switch valve
X	X	X								Stuck lockup valve
X	X	X	X							Stuck fail-safe valve
X					X	X				Faulty torque converter
								X		Out of balance
X										Failed locking clutch
					X					Leaking turbine hub seal
							X			Align exhaust system
						X	X			Tune engine
X										Faulty input shaft or seal ring
							X			Throttle linkage misadjusted

Fig. 12-6 Diagnosis chart for the lockup torque converter. (*Chrysler Corporation*)

sure the preload spring action is not restricted. If it is, disassemble and clean the parts.

Hold the transmission lever firmly forward against its internal stop and tighten the swivel lock screw to specifications. Lower the vehicle and reconnect the choke. Test linkage freedom by moving the throttle rod to the rear and releasing it to make sure it returns fully forward.

✿ 12-8 Front- (kickdown-) band adjustment Band adjustments are not recommended for normal passenger-car usage. Adjustments are recommended only if the vehicle is in severe service such as police, taxi, or trailer-towing service. The adjustment screw for the front band is on the left side of the transmission case (Fig. 12-15). To adjust, loosen the locknut and back it off several turns.

Test the adjusting screw to make sure it is free to turn. Then use the special band-adjusting tool to tighten the screw to specifications. Finally, back off the screw the specified numbers of turns. Typically, this is two turns for the A-904 and 2½ turns for the A-727. Hold the adjusting screw and tighten the locknut to specifications.

✿ 12-9 Low-reverse band (rear) Band adjustments are not recommended for normal passenger-car usage— only if the vehicle is in severe service. The adjusting screw for this band is inside and the oil pan must be removed to get to it (Fig. 12-16). First, raise the vehicle and drain the transmission fluid by loosening the oil pan. Then remove the oil pan.

Loosen the locknut and back off the nut about five turns. Use the special band-adjusting tool to tighten the band-adjusting screw to specifications. Then back off the screw the specified number of turns. This varies from two to seven turns, according to manufacturer's specifications.

✿ 12-10 Hydraulic-pressure tests Checking hydraulic pressures with the selector lever in the various positions will usually pinpoint causes of transmission troubles. Before making these checks, be sure the fluid level and condition are OK, and that the linkages have been properly adjusted.

Install an engine tachometer and raise the vehicle on a lift that allows the rear wheels to turn. Position the tachometer so it can be read under the vehicle. Disconnect the throttle rod and shift rod so they can be controlled from under the vehicle. Attach 100 psi [689

Car line Engine	HNFG 225		HNFG 318		ETSX 360-4		ETSX 360-2	
Axle ratio	2.76		2.45		3.21		2.45	
Standard tire	D78 × 14		FR78 × 15		GR78 × 15		GR78 × 15	
	mph	**km/h**	**mph**	**km/h**	**mph**	**km/h**	**mph**	**km/h**
Closed throttle 1-2	9–11	14–18	11–13	18–21	9–11	14–18	11–14	18–23
Closed throttle 2-3	13–15	21–24	15–18	24–29	12–15	19–24	16–19	26–31
Closed throttle lockup	27–36	43–58	32–43	51–69	—	—	26–30	42–48
Detent 1-2	23–30	37–48	27–35	43–56	22–33	34–53	28–36	45–58
Detent 2-3	46–66	74–106	55–78	89–126	49–68	79–109	56–80	90–129
WOT 1-2	36–46	43–74	42–54	68–87	39–48	63–77	43–55	69–89
WOT 2-3	63–73	101–117	74–87	119–140	65–75	105–121	76–88	122–142
Kickdown limit 3-2	58–68	93–109	69–81	111–130	60–70	97–113	70–83	113–134
Kickdown limit 3-1	28–38	45–61	33–44	53–71	31–40	50–64	33–45	53–72
Closed throttle 3-1	9–11	14–18	11–13	18–21	9–11	14–18	11–14	18–23
Part throttle 3-2 kickdown limit	30–41	48–66	35–49	56–79	33–44	53–71	36–50	58–80
Governor pressure 15 psi	18–19	29–31	22–23	35–37	17–18	27–29	22–23	35–37
Governor pressure 50 psi	45–50	72–80	54–60	87–97	48–53	77–85	55–61	89–98
Governor pressure 75 psi	64–70	103–113	76–83	122–134	66–71	106–114	78–84	126–135

Fig. 12-7 Chart showing governor pressure and car speeds at which shifts occur. Governor pressure should be from 0 to 1.5 psi at standstill, or downshift may not occur. Figures given are typical for other models. Changes in tire size or axle ratio will cause shift points to occur at corresponding higher or lower vehicle speeds. (*Chrysler Corporation*)

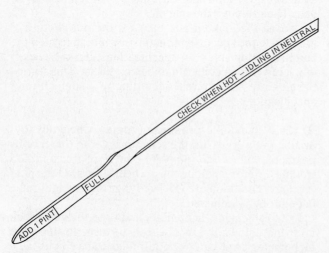

Fig. 12-8 Dipstick for TorqueFlite transmission. (*Chrysler Corporation*)

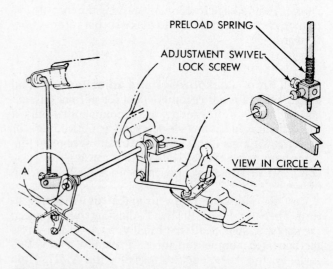

Fig. 12-9 Column-gearshift linkage on some car lines. (*Chrysler Corporation*)

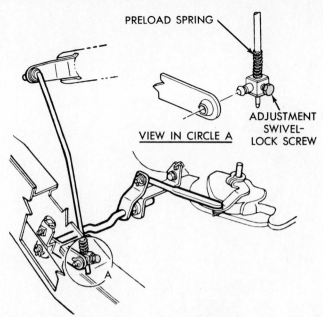

Fig. 12-10 Column-gearshift linkage on other car lines. (*Chrysler Corporation*)

kPa] gauges to the ports for the tests being run (Figs. 12-17 and 12-18). A 300 psi [2068 kPa] gauge is required to check reverse pressure.

Four tests are required, corresponding to the four operating positions of the selector lever—1, 2, D, and R.

✿ 12-11 Selector lever in 1 The first test checks pump output, pressure regulation, and the condition of rear-clutch and rear-servo hydraulic circuits. Attach gauges to the line and rear-servo ports (Fig. 12-17). Run the engine at 1000 rpm. Move the selector lever on the transmission all the way forward to the 1 position. Read pressures as the throttle lever on the transmission is moved from full forward to full rearward.

Line pressure should read 54 to 60 psi [372 to 413 kPa] with throttle forward and should gradually increase as the lever is moved to the rear to 90 to 96 psi

[621 to 662 kPa]. The rear-servo pressure should read within 3 psi [21 kPa] of line pressure.

✿ 12-12 Selector lever in 2 The second test checks pump output, pressure regulation, and the condition of the rear-clutch and lubrication hydraulic circuits.

Connect gauges to the line-pressure port and tee into the rear cooler-return line (Fig. 12-18). Operate the engine at 1000 rpm. Move the selector lever one detent rearward from the full-forward position. This is selector-lever position 2. Move the throttle lever on the transmission from full forward to full rearward. Pressures should be the same as for the test with the selector lever in position 1 (✿ 12-11).

✿ 12-13 Selector lever in D The third test is for pump output, pressure regulation, and the condition of the front and rear clutches and the lockup-clutch hydraulic circuits.

Attach gauges to the line and front-servo release ports. Operate the engine at 1600 rpm. Move the selector lever on the transmission two detents rearward from the full-forward position. This is selector position D. Move the throttle lever on the transmission from full-forward to full-rearward position. Line pressure should read 54 to 60 psi [372 to 413 kPa] with the throttle lever forward and should gradually increase as the lever is moved rearward.

The front-servo release pressure should read within 3 psi [21 kPa] of line pressure, up to the downshift point.

✿ 12-14 Selector lever in R Attach a 300 psi [2068 kPa] gauge to the rear-servo-apply port. Operate the engine at 1600 rpm with the selector lever on the transmission moved four detents to the rear from the full-forward position. This is R. The rear-servo pressure should read 160 to 270 psi [1103 to 1862 kPa]. This tests pump output, pressure regulation, and the condition of the front-clutch and rear-servo hydraulic circuits.

Move the selector lever on the transmission to D. Rear-servo pressure should drop to zero. This tests for leakage into the rear servo.

✿ 12-15 Analyzing hydraulic-pressure tests Here are the conclusions you can draw from the test results.

1. Proper pressures, minimum to maximum, mean the pump and pressure regulator are working properly.
2. Low pressure in D, 1, and 2 but correct pressure in R indicates rear-clutch-circuit leakage.
3. Low pressure in D and R but correct pressure in 1 means front-clutch-circuit leakage.
4. Low pressure in R and 1 but correct pressure in 2 indicates rear servo-circuit leakage.
5. Low line pressure in all positions indicates a defective pump, a clogged filter, or a stuck pressure-regulator valve.

✿ 12-16 Governor pressure If the transmission shifts at the wrong vehicle speeds even though the

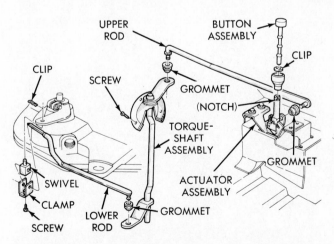

Fig. 12-11 Typical-gearshift linkage. (*Chrysler Corporation*)

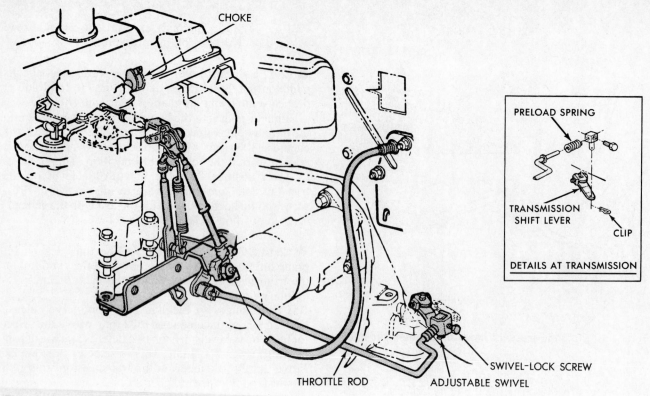

Fig. 12-12 Throttle-linkage adjustment on some cars with six-cylinder engines. (*Chrysler Corporation*)

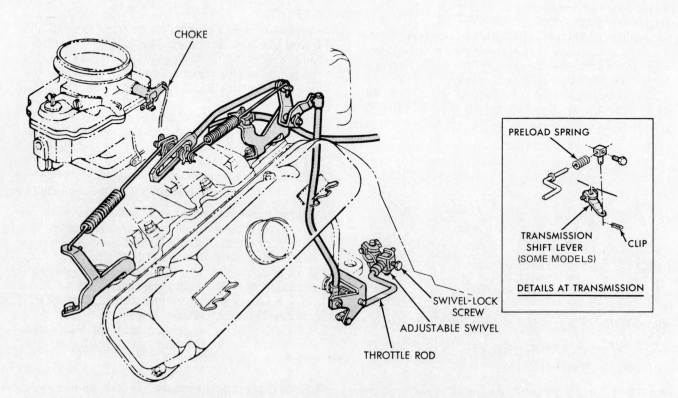

Fig. 12-13 Throttle-linkage adjustment on some cars with V-8 engines. (*Chrysler Corporation*)

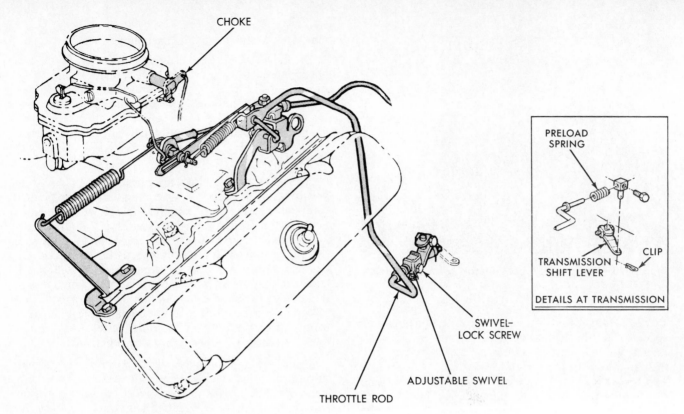

Fig. 12-14 Throttle-linkage adjustment on many cars with V-8 engines. (*Chrysler Corporation*)

throttle rod is correctly adjusted, suspect the governor. Connect a 100 psi [690 kPa] gauge at the governor pressure port (Fig. 12-18). Operate the transmission in D and compare pressures and speeds with those shown in the chart (Fig. 12-7). Incorrect pressures mean the governor valve or weights are sticking.

✿ 12-17 Throttle pressure There is no direct throttle-pressure test. Incorrect throttle pressure should be suspected if part-throttle upshift speeds are delayed or occur too early. Engine runaway on shifts can also be an indication of low throttle pressure. Adjusting throttle pressure requires removal of the valve body, as explained later. But the throttle pressure should not be adjusted until the throttle linkage has been checked and corrected if necessary.

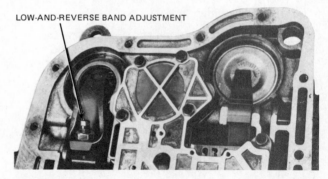

Fig. 12-16 Low-and-reverse-band adjustment. (*Chrysler Corporation*)

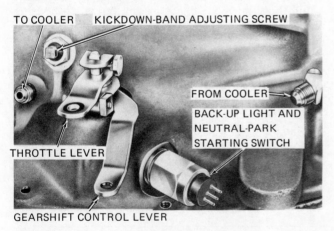

Fig. 12-15 External controls and adjustments. (*Chrysler Corporation*)

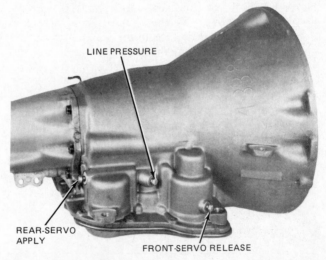

Fig. 12-17 Pressure-test locations on right side of case. (*Chrysler Corporation*)

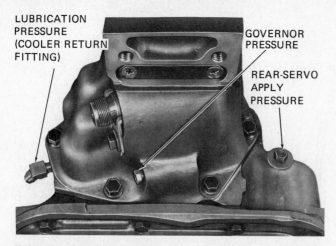

Fig. 12-18 Pressure-test locations on rear end of case. (*Chrysler Corporation*)

❀ 12-18 Stall test The stall test determines engine speed at full throttle in the D (drive) and R (reverse) ranges with the car held stationary by the brakes. The test is made with the selector lever in D or in R only. This test checks the torque-converter stator clutch and the holding ability of the transmission clutches. D checks the rear and overrunning clutches. R checks the front clutch and rear band. Check the transmission-fluid level and bring the engine up to operating temperature before beginning the stall test.

CAUTION: Both the parking and service brakes must be fully applied and both front wheels blocked while making this test. Do not allow anyone to stand in front or in back of the car during the test!

Do not hold the throttle wide open for more than 5 seconds at a time! If you have to make more than one stall check, run the engine at 1000 rpm in neutral for 20 seconds between runs to cool the transmission. If engine speed exceeds the maximum limits shown in the chart in Fig. 12-19, release the throttle at once. This condition indicates clutch slippage.

To make the stall test, connect a tachometer that can be read from the driver's seat, block the front wheels, apply both brakes, move the selector lever to D, and open the throttle wide. Read the maximum engine speed and compare it with the reading in the chart (Fig. 12-19) for the transmission being tested.

Careful: Make the test in 5 seconds or less!

1. **Stall speed above specifications** If engine speed increases by more than 200 rpm above the specification in the chart, clutch slippage is indicated. Release the throttle at once to avoid damage to the transmission. Clutch slippage requires further checking. Transmission hydraulic-control- and air-pressure checks should be made, as explained in ❀ 12-10 to 12-17 and 12-19.

2. **Stall speed below specifications** Low stall speed with a properly operating engine indicates torque-converter stator-clutch problems. A road test is necessary to determine what is wrong. If stall speeds are 250 to 350 rpm below specifications and the vehicle operates properly at highway speeds but has poor acceleration, the stator clutch is slipping. If the stall speed and acceleration are normal but abnormally high throttle opening is required to maintain highway speed, the stator clutch has seized (in models without lockup torque converter only). With either of these stator defects, the torque converter must be replaced.

3. **Noise** A whining or siren-like noise is normal during stall tests with many converters. However, loud, metallic noises indicate loose internal parts or interference due to a defective converter. To confirm that the noise originates in the converter, operate the engine at light throttle in D and N with the car on a lift and listen under the transmission.

❀ 12-19 Air-pressure tests Even though the hydraulic pressures check out correctly, the clutches or bands may not be working right. Clutches and servos can be checked by air pressure to see if they are working properly. Apply air pressure to various points as shown in Fig. 12-20. The compressed air must be clean and dry. Use a pressure of 30 psi [207 kPa].

1. **Front clutch** Apply air pressure to the front-clutch apply hole and listen for a dull thud, which indicates the front clutch is operating. Hold the air pressure for a few seconds to check for oil leaks.

2. **Rear clutch** Apply air pressure to the rear-clutch apply hole and listen for a dull thud, which indicates the rear clutch is operating. If you don't hear a dull thud, put your fingertips on the clutch housing and again apply air pressure. If the clutch engages, you can feel it.

ENGINE, CUBIC INCHES	TRANSMISSION TYPE	CONVERTER DIAMETER, INCHES	STALL RPM
225	A-904	10¾	1800–2100
318 FED./CAL.	A-904-LA	10¾	1700–2000
225	A-904-HD	10¾	1800–2100
318 FED./CAL.	A-727	10¾	1700–2000
360-4 H.D.	A-727	10¾	2150–2450
360-2	A-904-LA	10¾	1775–2075
360-2	A-727	10¾	1775–2075
360-4 FED./CAL.	A-904-LA	10¾	1700–2000

Fig. 12-19 Chart showing stall speeds for various models and applications of the TorqueFlite transmission. (*Chrysler Corporation*)

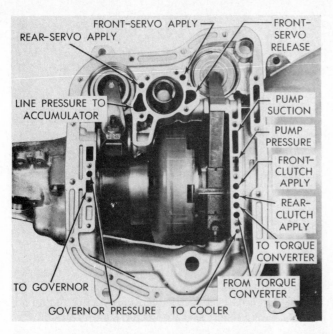

Fig. 12-20 Passage to which air pressure is applied to test the clutches and band. (*Chrysler Corporation*)

3. **Kickdown (front) servo** Apply air pressure into the front-servo apply hole. The front band will apply if the servo is working. Remove air pressure. Spring force should release the band.

4. **Low-and-reverse (rear) servo** Apply air pressure to the rear-servo hole. If the rear servo is operating, the rear band will apply. Remove air pressure. Spring force should release the band.

5. **Conclusions** If the clutches and servos operate properly but no upshifts or erratic upshifts occur, there is probably trouble in the valve body.

✿ 12-20 Line-pressure and throttle-pressure adjustments An incorrect throttle-pressure setting will

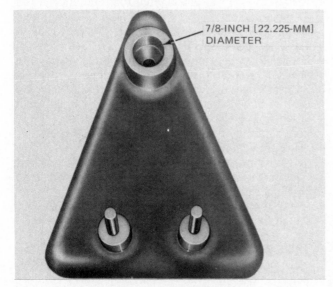

Fig. 12-21 Valve-body repair stand. (*Chrysler Corporation*)

Fig. 12-22 Line-pressure adjustment. (*Chrysler Corporation*)

cause incorrect line-pressure readings even though line-pressure adjustment is correct. Always check and correct the throttle-pressure before adjusting line pressure. Adjustments are made to the valve body with the body removed and mounted on a repair stand (Figs. 12-21 to 12-23).

The approximate line-pressure adjustment is shown in Fig. 12-22, measured from the valve body to the inner edge of the adjusting nut. The adjustment can be varied slightly one way or the other to get the proper line pressure. One complete turn of the adjusting screw changes closed-throttle line pressure almost 2 psi [14 kPa]. Turn the adjusting screw counterclockwise to increase line pressure.

To measure throttle pressure, insert the spacing tool, as shown in Fig. 12-23, between the throttle-lever cam and the kickdown valve. Push in on the tool to compress the kickdown valve against its spring so the throttle valve is completely bottomed inside the valve body. Then turn the throttle-lever stop screw with an Allen wrench until the head of the screw touches the throttle-lever tang. Be sure the adjustment is made with the spring fully compressed and the valve bottomed in the valve body.

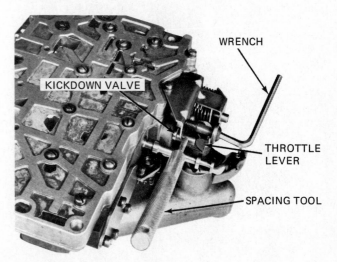

Fig. 12-23 Throttle-pressure adjustment. (*Chrysler Corporation*)

12-21 Fluid leaks If fluid leaks are suspected, note the color of the fluid that has leaked. Transmission fluid is red. If the leak is not red, it is probably engine-oil leakage. If the leakage is not around the oil pan (due to a loose pan or bad gasket), it could be from a defective or worn seal or bushing, loose front-pump-to case bolts, a loose kickdown-lever-shaft access plug, or defective welds in the torque converter itself. The transmission and torque converter can be removed from the engine, sealed, and tested for leaks by applying low pressure with an air pump (Figs. 12-24 and 12-25).

12-22 TorqueFlite in-car service Several services can be performed in the car without removing the complete transmission. These include lubrication (🌼 12-5) and replacing the backup and neutral starting switch, speedometer pinion, and some seals. These are discussed in following sections.

12-23 Cleaning and inspecting transmission parts During and after disassembly, clean and inspect all parts for wear or other damage. Refer to 🌼 9-11 for details. If the transmission shows signs of internal damage, foreign material may have been circulating in the fluid. This means completely cleaning all parts and lines that carry the fluid—clutches, servos, lines, valves and valve body, fluid cooler, and lines. A new or factory-reconditioned torque converter and a new filter will be required. The transmission should be filled with fresh fluid, as explained in 🌼 12-24.

12-24 Transmission fluid Checking the level and condition of the transmission fluid is covered in 🌼 9-3. Only Dexron automatic-transmission fluid (Mopar Dexron and Dexron II) should be used in the TorqueFlite transmission. Other fluids may damage the transmission.

NOTE: Periodic changing of the fluid and filter is not recommended for passenger cars in normal service.

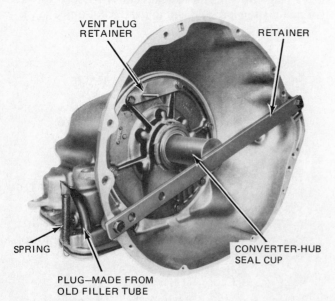

Fig. 12-24 Transmission prepared for pressure tests. (*Chrysler Corporation*)

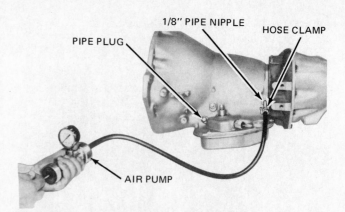

Fig. 12-25 Pressurizing the transmission. (*Chrysler Corporation*)

However, if the vehicle is in severe service, then the fluid and filter should be changed periodically. Also, the bands should be adjusted. Refer to the manufacturer's recommendations for frequency of the changes.

A typical procedure follows.

1. Raise the car on a lift. Put a drain container with a large opening under the transmission-oil pan.
2. Loosen pan bolts except at one corner. Tap the pan to break it loose, and slowly remove the bolts to allow the oil to drain.
3. Remove the access plate and drain plug from the front of the converter, and allow the converter to drain.

NOTE: If the oil has been contaminated by transmission failure, the torque converter must be either flushed out or replaced. Some TorqueFlite service instructions explain how to flush out the converter with solvent. Other TorqueFlite instructions recommend exchanging the old torque converter for a new or factory-reconditioned unit. Follow the manufacturer's recommendations for the model you are servicing.

4. Install a new filter and tighten the retaining screws to specifications.
5. Clean the oil pan and reinstall it with a new gasket. Tighten bolts to specifications.
6. Pour the specified amount of automatic-transmission fluid (Dexron II) through the filler tube.
7. Start and idle the engine for 2 minutes. Then, with the parking brake on or with the foot brake held down, move the selector lever through each position, returning it to the neutral position.
8. Add sufficient fluid to bring the fluid level to the ADD ONE PINT mark. After the transmission has reached operating temperature, recheck. The level should be between FULL and ADD ONE PINT.
9. Install the dipstick and cap so that the cap is fully seated to keep dirt from entering the transmission.

12-25 Backup and neutral starting switch The location of the neutral starting switch is shown in Fig. 12-26. To check the switch, remove the wiring connector and check for continuity with a 12-volt test light between the center pin of the switch and the transmis-

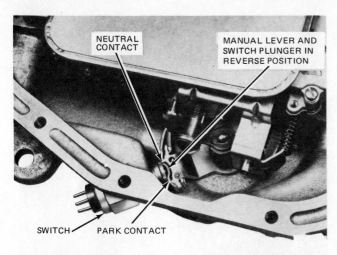

Fig. 12-26 Installation of the start and backup switch. (*Chrysler Corporation*)

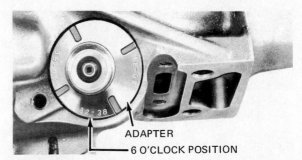

Fig. 12-28 Speedometer pinion and adapter installed but with retainer removed. (*Chrysler Corporation*)

sion case. The light should go on only when the transmission is in P or N. If the switch tests bad, check the gearshift-linkage adjustment before replacing the switch.

To replace the switch, unscrew it and allow fluid to drain into a container. Move the selector lever to P and then N. Check to see that the switch-operating-lever fingers are centered in the switch opening in the case. Install a new switch, tightening it to specifications. Retest the switch and add fluid to replace any that was lost.

The backup-light-switch circuit extends through the two outside terminals of the switch. To test the switch, remove the wiring connector from the switch and check for continuity between the two outside pins. Continuity should exist only with the transmission in R. No continuity should exist from either pin to the transmission case. If the switch is defective, replace it, as explained above.

⚙ **12-26 Speedometer pinion** Speedometer-pinion parts are shown in Fig. 12-27. To replace the pinion, remove the bolt and retainer and carefully work the adapter and pinion out of the extension housing. If you find transmission fluid in the cable housing, replace the pinion-and-seal assembly. The seal and retainer ring must be pushed in with a tool until the tool bottoms.

Careful: All parts must be clean!

Count the number of teeth on the pinion. Then rotate the adapter so that the number on the adapter which

corresponds with the number of teeth will be at the bottom, or 6 o'clock. For example, if the number of teeth were 34, then you would locate the 32-38 at 6 o'clock, as shown in Fig. 12-28.

Install the retainer and bolt, with the retaining tangs in the adapter-positioning slots. Tap the adapter firmly into the extension housing, and tighten the retainer bolt to specifications.

⚙ **12-27 Extension-housing yoke seal** To replace this seal, mark the parts for reassembly in the same relative positions. Then disconnect the driveshaft at the rear universal joint. Pull the shaft yoke out of the extension housing. Do not nick or scratch finished surfaces!

Use special tools to remove the old seal and install a new one. Reattach the drive shaft after pushing the shaft yoke carefully into the extension housing and onto the main-shaft splines.

⚙ **12-28 Extension-housing bushing and output-shaft** Some car models require unloading of both torsion bars and removal of the left torsion bar to provide clearance. Also, one side of the torsion-bar cross member must be dropped. Then proceed as follows.

1. Mark parts for reassembly and disconnect the drive shaft at the rear universal joint. Pull the drive shaft out of the extension housing.
2. Remove the extension-housing yoke seal and the speedometer-pinion-and-adaptor assembly (Fig. 12-27). Drain about two quarts [1.9 L] of fluid.
3. Remove the extension-housing-to-cross-member bolts. Raise the transmission slightly with a jack. Remove the center-cross-member-and-support assembly. Then remove the bolts attaching the extension housing to the transmission.
4. On the console-shift models, remove the bolts attaching the gearshift-shaft lower bracket to the extension housing. Swing the bracket out of the way. When removing or installing the housing, move the selector lever to 1 (first) so the parking-lock control rod is rearward. There, it can be disengaged or engaged with the parking-lock sprag.
5. Remove the screws, the plate, and the gasket from the bottom of the extension-housing mounting pad. Spread the large snap ring from the output-shaft bearing (Fig. 12-29). Then carefully tap the exten-

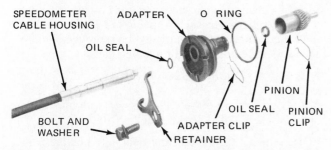

Fig. 12-27 Speedometer drive, disassembled. (*Chrysler Corporation*)

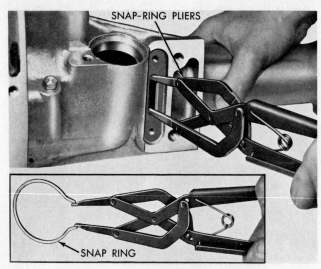

Fig. 12-29 Using snap-ring pliers to remove the snap ring from the output-shaft bearing. (*Chrysler Corporation*)

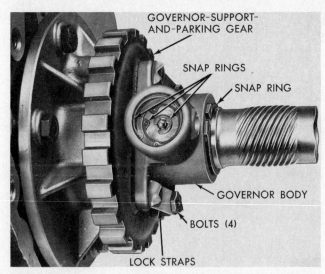

Fig. 12-31 Governor shaft-and-weight snap rings. (*Chrysler Corporation*)

sion housing off the output-shaft bearing. Carefully pull the housing rearward to move the parking-lock-control-rod knob past the parking sprag. Remove the housing.

6. To replace the bearing, use snap-ring pliers to remove the rear snap ring (Fig. 12-30). Install a new bearing and secure it with a new snap ring. The A-727 has a snap ring in front of the bearing. The A-904 does not.

7. To replace the bushing, press out the old bushing with the proper tool. Slide a new bushing on the installing end of the tool. Align the oil hole in the bushing with the oil slot in the housing. Drive the bushing into place. Install a new oil seal.

8. Use a new extension-housing gasket when installing the housing. Position the large-bearing snap ring in the housing, holding it with snap-ring pliers. Slide the extension housing onto the output shaft, guiding the parking-lock-control-rod knob past the parking sprag. When the housing is in position, release the large snap ring so it seats in the bearing outer groove.

9. Install the housing bolts. Install the gasket, plate, and two screws on the bottom of the housing mounting pad.

10. Install the center-cross-member-and-rear-mount

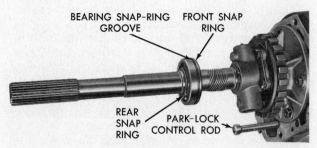

Fig. 12-30 Use snap-ring pliers to expand the rear snap ring so the bearing can be removed. (*Chrysler Corporation*)

assembly with retaining bolts. Lower the transmission so the housing can be attached.

11. On the console shift, align the gearshift-shaft lower bracket with the extension housing and secure with the retaining bolts.

12. Install the speedometer pinion and adaptor.

13. Install and attach the drive shaft.

14. Add fluid as necessary.

❋ 12-29 Governor service To remove the governor, the extension-housing-and-output shaft bearing must be removed. Governor parts are held together by snap rings, as shown in Figs. 12-31 and 12-32. Remove the small snap rings to slide the valve and valve shaft from the governor body. Remove the other snap rings to complete disassembly. The governor body is held to the governor-support-and-parking gear by bolts. Both can be slid off the output shaft by removing the shaft snap ring.

Thoroughly clean and inspect all governor parts. The major trouble in governors is sticking valves or weights. You can remove rough surfaces with a crocus cloth. Clean parts before reassembly.

When reinstalling the governor body and support on the output shaft, make sure the valve-shaft hole in the output shaft aligns with the hole in the governor body. Assemble all parts in their original places, following Fig. 12-32, and secure with snap rings.

❋ 12-30 Parking-lock components The parking-lock components are shown in Fig. 12-33. To replace them, the extension housing must be removed. Follow Fig. 12-33 in removing and replacing parts.

❋ 12-31 Valve-body assembly and accumulator piston To remove the valve body, raise the vehicle on a lift and remove the transmission oil pan, as explained in ❋ 12-24. Then proceed as follows:

1. Disconnect the throttle and gearshift linkage from the transmission levers. Loosen the clamp bolts and remove the levers.

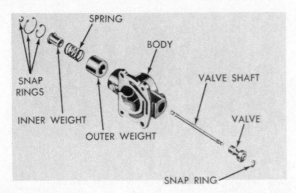

Fig. 12-32 Disassembled governor. (*Chrysler Corporation*)

2. Remove the E clip. Disconnect and remove the backup-light and neutral starting switch.
3. With the drain pan to catch oil, remove the 10 valve-body-to-transmission-case bolts, and lower the valve body from the case. Disconnect the parking-lock rod from the lever while lowering the valve body.
4. Withdraw the accumulator piston from the transmission case. Inspect the piston for scoring and the rings for wear or damage. Replace as necessary.
5. If the valve-body manual-lever-shaft seal requires replacement, drive it out of the case with a punch. Use a special tool to install a new seal.

NOTE: Service the valve body and internal parts as explained in ✿ 12-36.

To reinstall the valve body:

1. If the parking-lock rod was removed, insert it through the opening in the rear of the case with the knob positioned against the plug and sprag (see Fig. 12-33). Force the knob past the sprag, rotating the drive shaft if necessary.
2. Install the accumulator piston and position the accumulator spring in the valve body.
3. Put the manual lever in L and lift the valve body into approximate position. Connect the parking-lock rod to the manual lever and secure with an E clip. Then install the valve-body bolts finger tight.
4. Install the neutral starting switch. Put the manual lever in N. Shift the valve body as necessary to

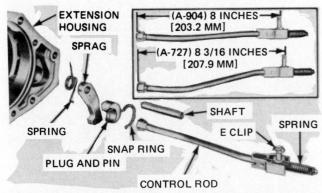

Fig. 12-33 Parking-lock components. (*Chrysler Corporation*)

center the neutral finger over the neutral-switch plunger. Then snug bolts down evenly and tighten to specifications.
5. Install the gearshift lever and tighten the clamp bolt. Make sure it is free to move in all lever positions. If binding exists, loosen the valve-body bolts and re-align the body.
6. Make sure the throttle-shaft seal is in place. Install the flat washer and throttle lever, and tighten the clamp bolt. Reconnect the linkage and check linkage adjustments.
7. Install the oil pan, using a new gasket. Add transmission fluid, as explained in ✿ 12-24, to bring fluid up to the proper level.

✿ **12-32 TorqueFlite out-of-car service** The following sections describe transmission removal, disassembly, servicing of subassemblies, reassembly, and installation.

CAUTION: The transmission and torque converter assembly is heavy. Use a transmission jack (Fig. 12-34) to support and remove it. The transmission and converter must be removed and installed as an assembly. Otherwise, the converter drive plate, pump bushing, and oil seal will be damaged. The drive plate cannot support a load. None of the transmission weight must be allowed to rest on the plate at any time.

Transmission removal. Disconnect the ground cable from the battery to avoid the danger of accidentally shorting the battery. This could happen if a hot terminal or wire were accidentally grounded.

NOTE: On some cars, the exhaust system must be dropped to provide enough clearance.

1. If there are engine-to-transmission struts, remove them.
2. Disconnect the cooler lines from the transmission.
3. Remove the starting motor and cooler-line bracket.
4. Raise the vehicle on a lift.
5. Remove the converter access cover. Remove the plug from the converter and drain oil.
6. Loosen the oil-pan bolts and the oil pan to allow fluid to drain.

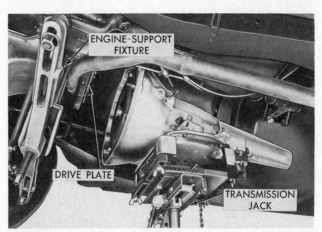

Fig. 12-34 Engine-support fixture and transmission jack for removing transmission from an engine. (*Chrysler Corporation*)

7. Reinstall the pan.
8. Mark converter and drive plate so they can be reinstalled in the same relationship. The crankshaft-flange-bolt circle and the four tapped holes in the front face of the converter have one hole offset so these parts must be installed in the original position. This maintains proper balance of the engine and converter.
9. Rotate the engine clockwise with a socket wrench on the vibration-damper bolt to get at the bolts attaching the torque converter to the drive plate. Remove the bolts.
10. Mark parts for reassembly and disconnect the drive shaft at the rear universal joint. Carefully pull the shaft out of the extension housing.
11. Disconnect the wire connector from the backup and neutral-starting switch.
12. Disconnect the gearshift rod and torque-shaft assembly.

NOTE: If linkage rods must be disconnected from the levers that have plastic grommets, the grommets must be replaced with new ones. Pry and cut out the old grommets. Snap new grommets in with pliers.

13. Disconnect the throttle rod from the lever at the left side of the transmission. Remove the linkage bell crank from the transmission if there is one.
14. Remove the oil-filler tube and the speedometer cable.
15. Install the engine-support fixture (Figs. 12-34 and 12-35) and position the transmission jack to support the transmission (Fig. 12-34). Raise the transmission slightly to relieve the load on the supports.
16. Remove the bolts attaching the transmission mount to the cross member and the cross member to the frame. Then remove the cross member.

NOTE: Some cars have a torsion-bar-anchor cross member, which remains in place. This requires a careful downward tilt of the front of the transmission as the transmission is removed. If there is a vibration-damping weight bolted to the rear of the extension housing, it must be removed to provide clearance.

17. Attach a small C clamp to the edge of the converter housing to hold the converter in place during transmission removal.
18. Remove all bell-housing bolts. Carefully work the transmission-and-converter assembly off the engine-block dowels and disengage the converter hub from the crankshaft.
19. Lower the transmission and remove the converter assembly by taking off the C clamp. Then slide the converter off.
20. Mount the transmission in a repair stand for service.

✿ 12-33 Starting-motor ring-gear replacement   If a ring gear on a lockup torque converter is damaged, the complete converter must be replaced. The ring gear on the torque converter without the lockup is replace-

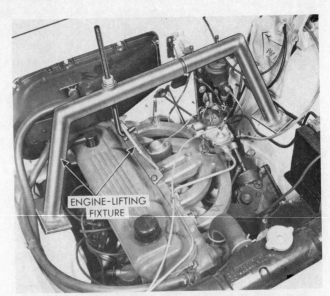

Fig. 12-35 Engine-lifting fixture for an in-line six-cylinder engine. (*Chrysler Corporation*)

able, as follows. The starting-motor ring gear is welded in four places to the outer diameter of the torque converter. If the ring gear must be replaced, these welds must be cut with a hacksaw or grinding wheel. Then the ring gear must be driven off with a brass drift punch and hammer.

Careful: Do not allow the converter to rest on its hub during this operation. This could damage the converter.

The new ring gear must be heated so that it expands slightly. Then it can be driven onto the torque converter. Finally, the ring gear must be welded in place, with the new welds in the same places as the old welds.

✿ 12-34 Transmission disassembly   Before disassembling the transmission, plug all openings and thoroughly clean the exterior, preferably with a steam cleaner, if available. Do not wipe parts with a shop towel. Instead, wash them in solvent and air-dry them. Do not scratch or nick finished surfaces. This could cause valve hanging or fluid leakage and faulty transmission operation. Proceed with disassembly as follows:

Fig. 12-36 Measuring drive-train end play with a dial indicator. (*Chrysler Corporation*)

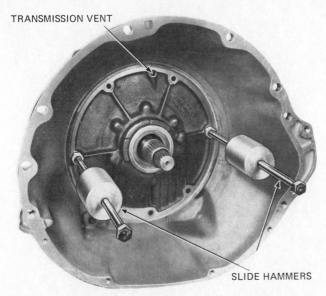

TRANSMISSION VENT

SLIDE HAMMERS

Fig. 12-37 Removing the pump and reaction-shaft support with slide hammers. (*Chrysler Corporation*)

1. First, measure the drive-train end play by mounting a dial indicator, as shown in Fig. 12-36. Move the input shaft in and out to check end play. Specifications are 0.030 to 0.089 inch [0.762 to 2.260 mm] for the A-904 transmission and 0.037 to 0.084 inch [0.939 to 2.134 mm] for the A-727 transmission. Write down the indicator reading for reference when reassembling the transmission.
2. Remove the oil pan and gasket.
3. Remove the valve body and accumulator piston and spring, as explained in ✿ 12-31.
4. To remove the extension housing, pull the parking-lock rod forward out of the case. Rotate the output shaft if necessary to align the parking gear and sprag to permit the knob on the end of the control rod to pass the sprag. Then proceed as follows:
   a. Remove the speedometer and adapter assembly.
   b. Remove extension-housing-to-transmission bolts.
   c. Remove the screws, plate, and gasket from the bottom of the extension-housing mounting pad. Spread the large snap ring on the output-shaft bearing with snap-ring pliers, as shown in Fig. 12-29. With the snap ring spread as much as possible, carefully tap the extension housing off the output shaft and bearing.
5. How to remove the governor and support is explained in ✿ 12-29. Now remove this assembly from the transmission.
6. The pump and reaction-shaft support come out next. First, tighten the front-band adjusting screw until the band is tight on the front-clutch retainer. This adjustment will hold the retainer when the pump is removed and thereby prevent damage to the clutch.
   Next, remove the oil-pump-housing retaining bolts. Then use two slide hammers, as shown in Fig. 12-37, to remove the pump.
7. Loosen the front-band adjuster, remove the band strut, and slide the band from the case.

8. Slide the front-clutch assembly from the case.
9. Pull on the input shaft to slide the input-shaft-and-rear-clutch assembly out of the case.

Careful: Do not lose the thrust washer located between the rear end of the input shaft and the forward end of the output shaft.

10. Support the output shaft and driving shell with your hands, and slide the assembly forward and out through the case.

Careful: Be very careful to avoid damaging the ground surfaces on the output shaft.

11. Remove the low-and-reverse drum. Then loosen the rear-band adjuster, and remove the band strut and link and the band. On the double-wrap band (A-904LA), loosen the band-adjusting screw to remove the band and then the low-and-reverse drum.
12. Note the position of the overrunning-clutch rollers and springs before disassembly so that you can reassemble everything in the original position. The rollers *must* be reinstalled exactly as they were originally. Slide out the clutch hub and remove the rollers and springs. If the overrunning-clutch cam or spring retainer is damaged, refer to ✿ 12-42 for the service procedure.
13. Next, remove the kickdown servo. The servo spring must be compressed by an engine-valve-spring compressor, as shown in Fig. 12-38, so that the snap ring can be removed. This allows removal of the rod guide, springs, and piston rod. Do not damage these parts during removal. Now, the piston can be withdrawn from the transmission case.

COMPRESSOR

Fig. 12-38 Compressing the kickdown-servo spring with an engine valve-spring compressor. (*Chrysler Corporation*)

179

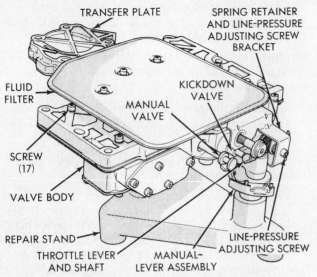

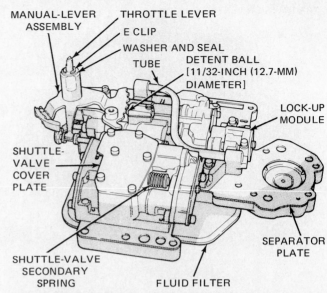

Fig. 12-39 Valve-body assembly mounted on repair stand. (*Chrysler Corporation*)

Fig. 12-40 Valve-body controls. (*Chrysler Corporation*)

14. Remove the low-and-reverse servo by compressing the piston spring with the engine-valve-spring compressor so that the snap ring can be removed. Then remove the spring retainer, spring, and servo piston and plug from the transmission case.

✿**12-35 Reconditioning subassemblies** Keep everything clean while working on transmissions. Use care

to avoid scratching or nicking finished surfaces. Crocus cloth can be used, within reason, to remove burrs and rough spots. When using it on valves, avoid rounding off sharp edges. These sharp edges on valves are essential for proper valve operation. They help prevent dirt particles from getting between the valve and body and possibly causing the valve to stick.

Use new seal rings when rebuilding an automatic

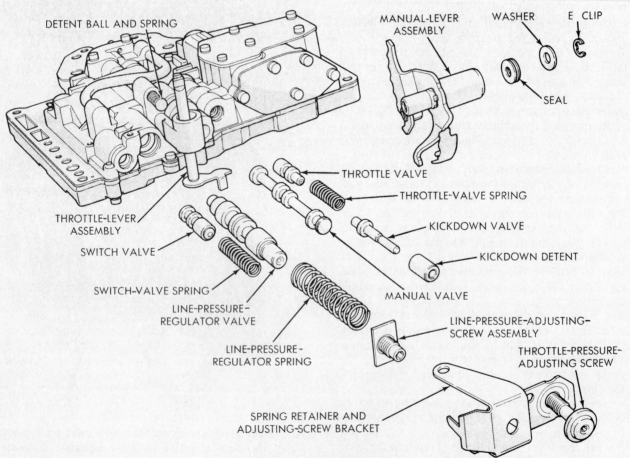

Fig. 12-41 Pressure regulators and manual control. (*Chrysler Corporation*)

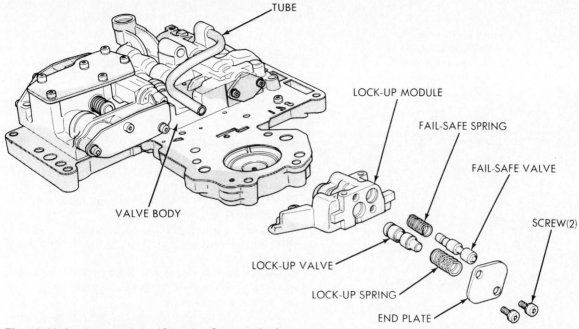

Fig. 12-42 Lockup module. (*Chrysler Corporation*)

transmission. Also, check all bushings and replace those that are worn or galled. Use the special bushing removal-and-replacement tools as required.

The following sections describe the service procedure for the various subassemblies removed from the TorqueFlite transmission.

✿ **12-36 Valve body** Use the valve-body repair stand shown in Fig. 12-39 to work on the valve body. Never clamp any part of the valve body in a vise. Doing so can distort the body and cause leakage, valve hanging, and other transmission trouble. Slide valves and plugs in and out of the valve body with care to avoid damage. Proceed as follows. (Figure 12-40 shows the valve body for the transmission with the lockup converter.)

1. With the valve body on the repair stand, as shown in Fig. 12-39, remove three screws and the filter.
2. Hold the spring retainer against the spring force and remove the three screws so that the spring retainer can be removed.
3. As the spring retainer is removed, the line and throttle-pressure adjusting screws and the line-pressure and switch-valve-regulator springs will come off (Fig. 12-41). Slide the switch valve and the line-pressure valve out of their bores.

NOTE: Tag all springs so you will not mix them up.

4. Remove the screws from the lockup module (Fig. 12-42). Earlier models without the lockup have a stiffener plate, which should be removed. Disassemble the lockup module. Tag the springs for identification.
5. Remove the transfer-plate screws and lift off the plate and separator plate (Fig. 12-43). Separate the parts.
6. Remove the seven balls from the valve body (Fig. 12-44).

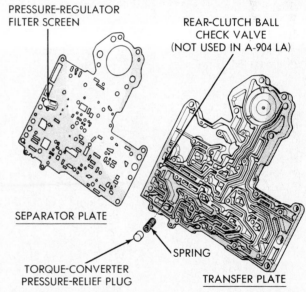

Fig. 12-43 Transfer and separator plates. (*Chrysler Corporation*)

7. Remove the shuttle-valve and governor plugs (Fig. 12-45).
8. Remove the shift valves and pressure-regulator-valve plugs (Fig. 12-46).
9. Clean and inspect parts. Soak parts in a suitable solvent for several minutes. Air-dry. Make sure all passages are clean. Check operating levers and shafts for wear, looseness, or distortion. Check mating surfaces for burrs, nicks, and scratches. Check valve springs for distortion and collapsed coils. Make sure all valves and plugs fit their bores freely. They should fall freely in the bores.
10. Reassembly is essentially the reverse of disassembly. Refer to Figs. 12-39 to 12-46. Tighten all screws to specifications.

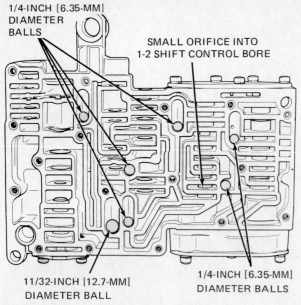

Fig. 12-44 Steel ball locations. (*Chrysler Corporation*)

Adjust throttle and line pressures, as shown in Figs. 12-22 and 12-23.

✿12-37 Other subassemblies In previous sections, the removal and replacement of the accumulator piston, extension-housing oil seal, parking-lock sprag, and governor-and-support assembly were described. Other subassemblies discussed in following sections are the oil pump, clutches, and planetary-gear train.

✿12-38 Oil pump and reaction-shaft support These are different for the A-904 and A-727 transmissions. Figures 12-47 and 12-48 show disassembled views of the two oil pumps. Both are disassembled as follows:

1. Remove bolts from the rear side of the reaction-shaft support and lift the support off the pump.
2. Remove the rubber seal ring from the pump-body flange.
3. Drive out the oil seal with a blunt punch.
4. Inspect parts as follows:
   a. Check the interlocking seal rings on the reaction-shaft support for wear or broken locks. Replace if necessary.
   b. On the A-904, check the thickness of the thrust washer, which should be 0.043 to 0.045 inch [0.99 to 1.13 mm]. Replace if worn.
   c. Check the machined surfaces for nicks and burrs, and check pump rotors for scores or pitting. Clean the rotors and pump body, and install the rotors to check clearances. Put a straightedge across the surface of the pump face and measure with a feeler gauge to the rotors. Clearance limits are 0.0015 to 0.003 inch [0.0381 to 0.0762 mm]. Rotor-tip clearance between rotor teeth should be 0.005 to 0.010

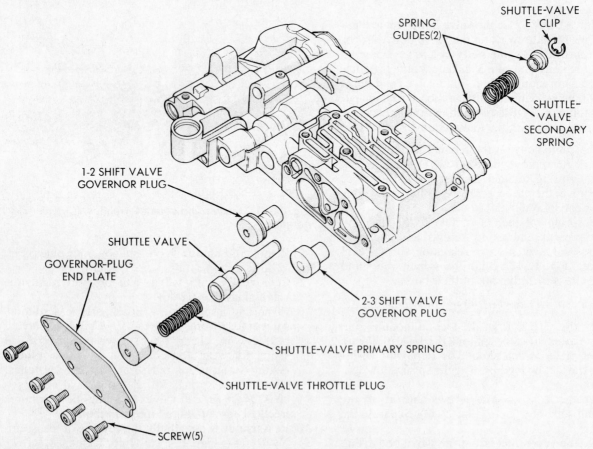

Fig. 12-45 Shuttle-valve and governor plugs. (*Chrysler Corporation*)

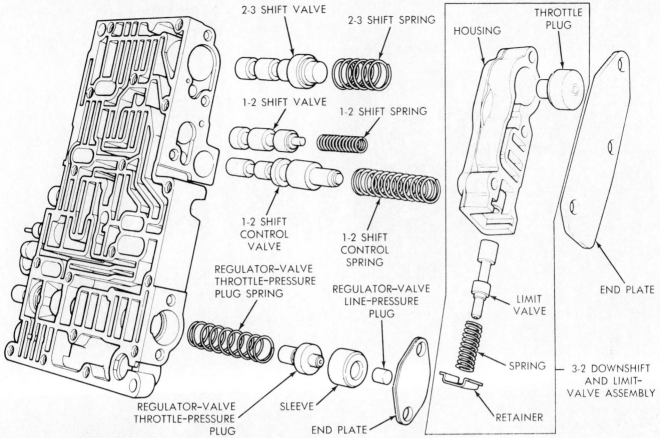

Fig. 12-46 Shift valves and pressure-regulator valve plugs. Parts enclosed in solid line are used on V-8 engines. (*Chrysler Corporation*)

inch [0.127 to 0.254 mm]. Clearance between the outer rotor and its bore should be 0.004 to 0.008 inch [0.102 to 0.203 mm]. Replace the rotors and body if worn.

5. Replace the oil-pump-body bushing, if worn, as shown in Fig. 12-49 for the A-904 and as shown in Fig. 12-50 for the A-727. Be careful not to cock the removal or installation tools because this could damage the bore in the oil-pump body. Stake the new bushing in place, as shown in Fig. 12-51. A gentle tap at each stake is all that is necessary. Use a narrow-bladed knife to remove high points around the stake that could interfere with the shaft.

6. Replace the reaction-shaft bushing if necessary (Figs. 12-52 and 12-53).

7. Assemble the A-904 pump, as shown in Fig. 12-54. Note that two pilot studs are used to align the pump body and that the rotors are installed in the support and aligned with the aligning tool. Rotate the aligning tool as the pump is brought down into place. Tighten pump-body bolts to specifications. Install a new pump-body oil seal, lip facing in, using a special installing tool.

8. Assemble the A-727 pump by putting the rotors in the housing and installing the reaction-shaft support with the vent baffle over the vent opening. Tighten

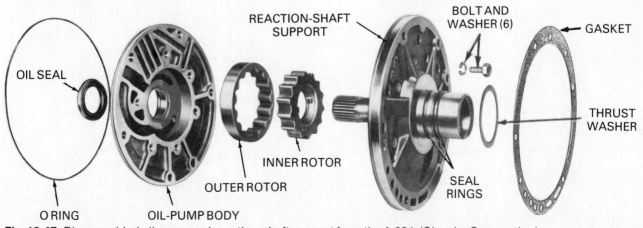

Fig. 12-47 Disassembled oil pump and reaction-shaft support from the A-904. (*Chrysler Corporation*)

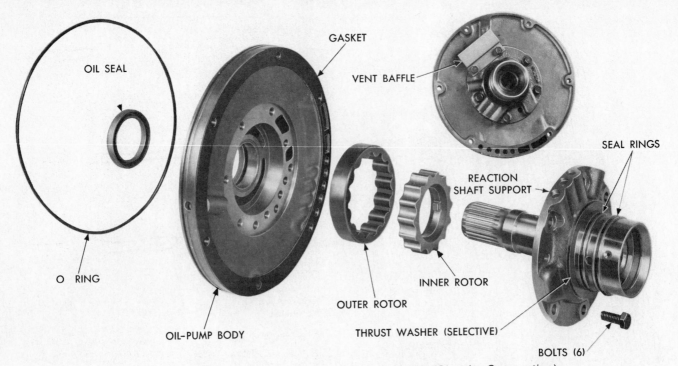

Fig. 12-48 Disassembled oil pump and reaction-shaft support from the A-727. (*Chrysler Corporation*)

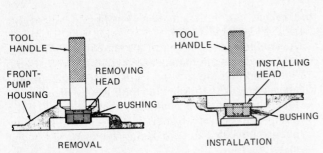

Fig. 12-49 Removal and installation of the pump-body bushing on the A-904. (*Chrysler Corporation*)

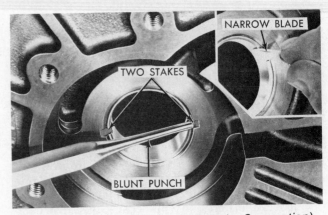

Fig. 12-51 Staking the bushing. (*Chrysler Corporation*)

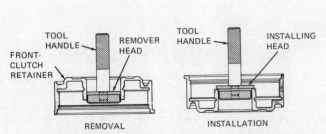

Fig. 12-50 Removal and installation of the pump-body bushing on the A-727. (*Chrysler Corporation*)

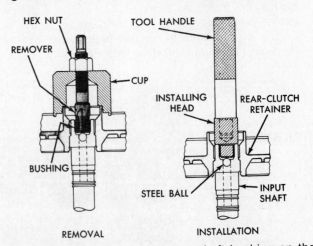

Fig. 12-52 Replacing the reaction-shaft bushing on the A-904. Note that the removal tool has threads which can be threaded into the bushing so that it can be pulled out. (*Chrysler Corporation*)

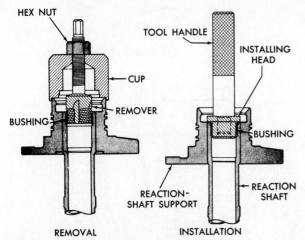

Fig. 12-53 Replacing the reaction-shaft bushing on the A-727. Note that the removal tool has threads which can be threaded into the bushing so that it can be pulled out. (*Chrysler Corporation*)

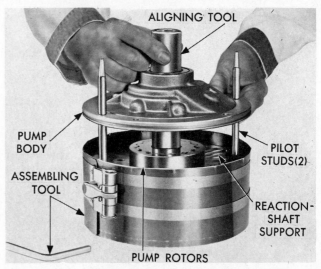

Fig. 12-54 Assembling the pump and reaction-shaft support for the A-904. (*Chrysler Corporation*)

the retaining bolts to specifications. Install a new pump-body oil seal, lip facing in, with a special installing tool.

✿ 12-39 Front clutch The front clutches are different for the A-904 and A-727 transmissions (Figs. 12-55 and 12-56). However, the disassembly procedures are similar, except for the piston-spring compressor that must be used. First, remove the large selective snap ring and remove the pressure plate and clutch plates. Next, use the spring compressor to remove the small snap ring, spring retainer, and spring or springs. Finally, bump the piston retainer on a wood block to remove the piston. Remove the seal rings from the piston.

Inspect the clutch plates and facings. Facings that are charred, pitted, glazed, or flaking require plate replacement. Steel plates that are burned, scored, or damaged should be replaced. Steel-plate-lug grooves in the clutch retainer must be smooth so that plates can

move freely. The band surface on the outside of the piston retainer must be smooth, and the ball check must be free. Surfaces in the retainer on which the piston seals slide must be smooth. Use new seals on reassembly.

If the retainer bushing needs replacement, follow the procedure shown in Fig. 12-57. Special tools are needed to remove and replace the bushing. The procedure is similar for the A-904 and the A-727.

On reassembly, make sure the lips on the seal rings face down, or into the piston retainer, when installed on the piston. Lubricate all parts with transmission fluid for easier installation. Use the spring compressor to compress the spring or springs so that the snap ring can be installed. Make sure the spring or springs are in the same location as on the original assembly. Note that on the A-727, there may be from six to fifteen springs.

After reassembly, check the clearance between the pressure plate and the selective snap ring, as shown in

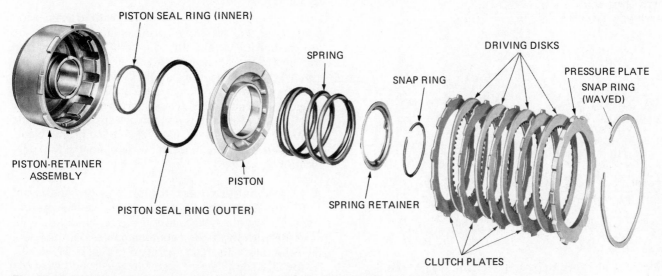

Fig. 12-55 Disassembled view of the A-904 front clutch. (*Chrysler Corporation*)

185

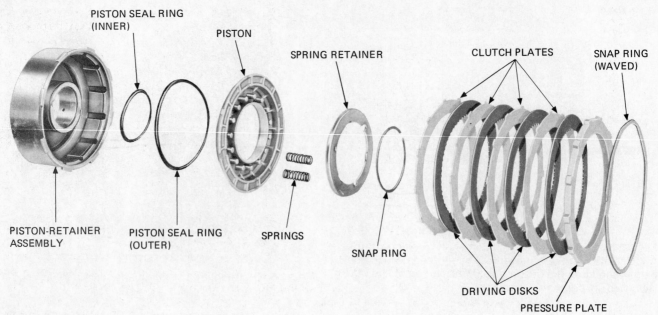

Fig. 12-56 Disassembled view of the A-727 front clutch. Only two of the fifteen springs required are shown. (*Chrysler Corporation*)

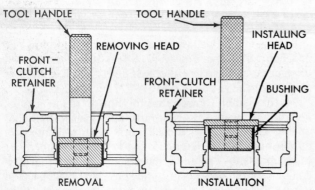

Fig. 12-57 Replacing the bushing in the A-904 front-clutch-piston retainer. (*Chrysler Corporation*)

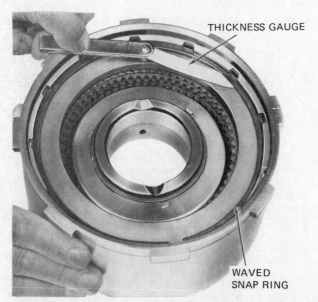

Fig. 12-58 Measuring front-clutch plate clearance. (*Chrysler Corporation*)

Fig. 12-58. The clearance varies with different engines. Check the shop manual for the proper specifications. There are snap rings of different thicknesses which can be installed to produce the correct clearance.

☸ 12-40 Rear clutch Figures 12-59 and 12-60 show the rear clutches for the A-904 and A-727 transmissions in disassembled views. To disassemble the rear clutch, remove the large selective snap ring and lift the pressure plate, clutch plates, and inner pressure plate out of the retainer. Then carefully pry one end of the wave spring out of its groove in the clutch retainer and remove the wave spring, spacer ring, and clutch-piston spring. Tap the retainer on a wood block to remove the piston. Take off the piston seals, noting the direction of the seal lips.

Inspect parts as for the front clutch, as explained in ☸ 12-39.

On reassembly, lubricate parts with transmission fluid. To install the wave spring, start one end in the groove and then work the ring into the groove from that end to the other end. Check the clearance on the clutch by having an assistant press down firmly on the outer pressure plate. Then measure with a thickness gauge between the selective snap ring and the pressure plate. If the clearance is not correct, install a snap ring of the correct thickness that will give the proper clearance. Check the manufacturer's shop manual for specifications.

☸ 12-41 Planetary-gear trains Figures 12-61 and 12-62 are disassembled views of the planetary-gear trains used in the A-904 and A-727 transmissions. With both, disassembly is essentially a process of removing the snap rings and sliding parts off the shaft. However, before disassembly, measure the clearance between the shoulder on the output shaft and the rear-ring-gear-

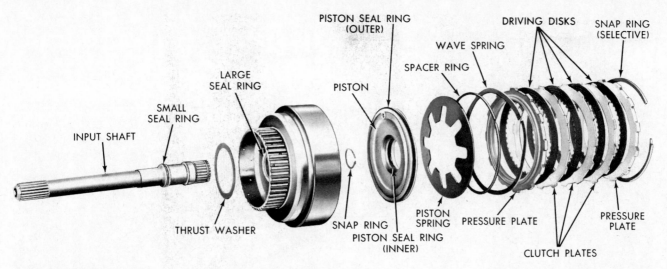

Fig. 12-59 Disassembled A-904 rear clutch. (*Chrysler Corporation*)

support hub, as shown in Fig. 12-63. This clearance is the gearset end play. If it is excessive, another selective snap ring of the proper thickness must be used to correct the end play.

Figures 12-61 and 12-62 show the proper relationship of parts during reassembly. Make sure all parts are in good condition and not burred, nicked, or worn.

❋12-42 Overrunning-clutch cam If the cam is worn, a condition which is rare, it can be replaced. On the A-904, it is held in place with rivets, which must be driven out. The new cam is then installed with retaining bolts. On the A-727, there is a setscrew that must be removed, followed by four bolts that hold the output-shaft support in place. The cam is then driven out of the transmission case by a punch inserted through the bolt holes.

❋12-43 Servos Figures 12-64 and 12-65 show the kickdown and low-and-reverse servos in disassembled views. Disassembly is accomplished by removing the snap rings.

❋12-44 TorqueFlite transmission reassembly Keep everything clean. Use transmission fluid to lubricate parts during reassembly of the transmission. Do not use force to install mating parts. Everything should slide into place easily. If force is required, something is not properly installed. Remove parts to find the trouble.

Reassemble the transmission as follows:

1. Insert the clutch hub and install the overrunning-clutch rollers and springs exactly as shown in Fig. 12-66.

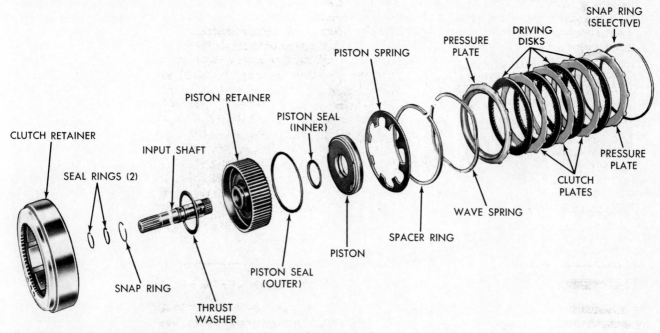

Fig. 12-60 Disassembled A-727 rear clutch. (*Chrysler Corporation*)

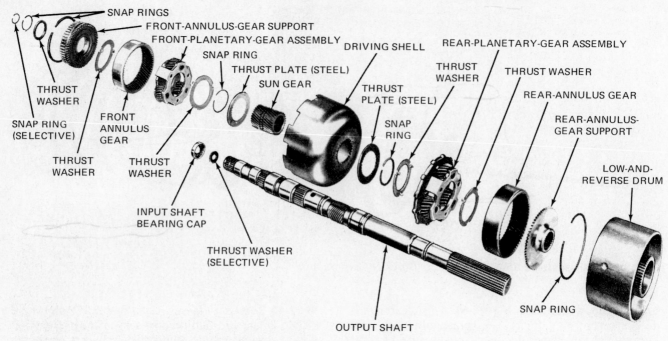

Fig. 12-61 Disassembled A-904 planetary-gear train. (*Chrysler Corporation*)

2. Install the low-and-reverse-servo piston in the case with a twisting motion and put the spring, retainer, and snap ring over the piston (Fig. 12-65). Use a compressor tool, as shown in Fig. 12-38, to compress the spring so that the snap ring can be installed.

3. Put the rear band in the case and install a short strut (if used). Connect a long link and anchor to the band, as shown in Fig. 12-67. Screw in the band adjuster just enough to hold the strut in place. Install the low-and-reverse drum. Make sure the link is installed as shown in Fig. 12-66.

4. Install the kickdown-servo piston with a twisting motion, followed by other parts, as shown in Fig. 12-64. Use a spring compressor, as shown in Fig.

12-38, to compress the spring so that the snap ring can be installed.

5. Install the planetary-gear assembly with the sun gear and driving shell, as follows: Support the assembly in the case and insert the output shaft through the rear support. Carefully work the assembly into the case, engaging the rear planetary lugs into the low-and-reverse-drum slots.

Careful: Do not damage finished surfaces on the output shaft!

6. The front and rear clutches, front band, oil pump, and reaction-shaft support are more easily installed with the transmission in an upright position. Proceed as follows:

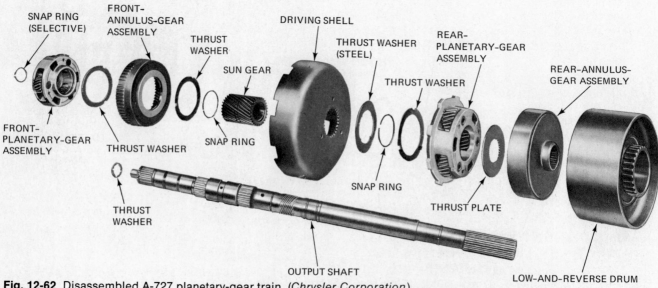

Fig. 12-62 Disassembled A-727 planetary-gear train. (*Chrysler Corporation*)

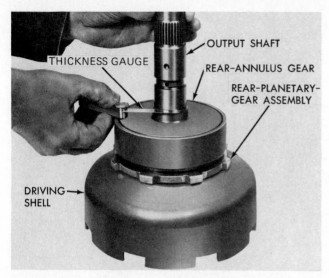

Fig. 12-63 Measuring the end play of the planetary-gear assembly. (*Chrysler Corporation*)

a. On the A-904 transmission, if the end play, shown being checked in Fig. 12-63, is not correct, a selective thrust washer of a different thickness will be required. Stick this thrust washer to the end of the output shaft with a coat of grease. On the A-727, apply a coat of grease on the input-to-output-shaft thrust washer (Fig. 12-62) to hold it on the end of the output shaft.

b. Align the front-clutch-plate inner splines and put the assembly in position on the rear clutch. Make sure the front-clutch-plate splines are fully engaged on the rear-clutch splines.

c. Align the rear-clutch-plate inner splines, grasp the input shaft, and lower the two clutch assemblies into the transmission case.

d. Carefully work the clutch assemblies in a circular action to engage the rear-clutch splines over the splines of the front annulus gear. Make sure the front-clutch drive lugs are fully engaged in the slots in the driving shell.

7. The front band and associated parts are shown in Fig. 12-68. Slide the band over the front-clutch assembly and install the parts, tightening the adjusting screw tight enough to hold the strut and anchor in place.

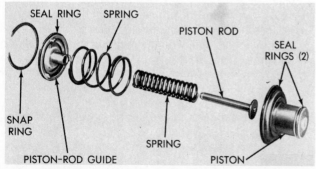

Fig. 12-64 Disassembled view of the kickdown servo. (*Chrysler Corporation*)

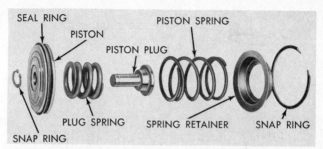

Fig. 12-65 Disassembled view of the low-and-reverse servo. (*Chrysler Corporation*)

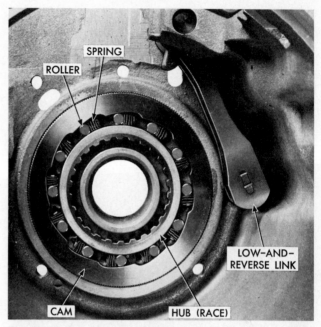

Fig. 12-66 Overrunning clutch and low-and-reverse-band link in place. (*Chrysler Corporation*)

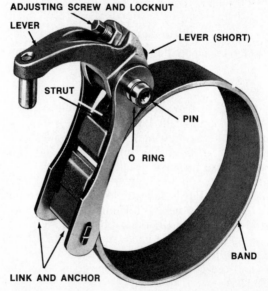

Fig. 12-67 Low-and-reverse brake band and linkage. (*Chrysler Corporation*)

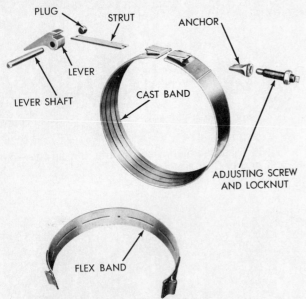

Fig. 12-68 Kickdown band and linkage with flex band for the A-727. (*Chrysler Corporation*)

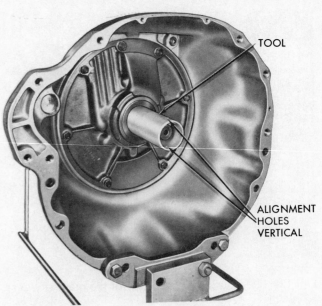

Fig. 12-70 Using a special tool to align the oil-pump rotors. (*Chrysler Corporation*)

Fig. 12-69 Installing the oil pump in the transmission case. (*Chrysler Corporation*)

8. Install the oil pump by first installing two pilot studs, as shown in Fig. 12-69. Put a new gasket over the studs. Put a new seal ring in the groove on the outer flange of the pump housing. Coat the seal with grease for easy installation. Install the pump in the case, tapping it down with a wood mallet if necessary. Put a deflector over the vent opening and install the pump-attaching bolts. Remove the studs and install the other bolts. Rotate the input and output shafts to make sure nothing is binding. Then tighten the bolts to specifications.

9. Install the governor and support, as explained in ✿ 12-29.

10. Install the extension-housing-and-output-shaft bearing, as explained in ✿ 12-28.

11. Install the valve-body assembly, as explained in ✿ 12-31.

12. Install the torque converter as follows:

   a. Use a special tool to turn the pump rotors, as shown in Fig. 12-70, until the holes in the tool are vertical. Remove the tool.

   b. Make sure converter-impeller-shaft slots are vertical so that they will engage with the oil-pump inner-rotor lugs. Then slide the converter into place.

   c. Check for full engagement by placing a straightedge on the face of the case and measuring from the straightedge to one of the front-cover mounting lugs. The distance should be at least ½ inch [12.7 mm].

   d. Attach a small C clamp to hold the converter in place.

✿ **12-45 Transmission installation** To install the transmission in the car, proceed as follows:

1. Check the converter drive plate for distortion or cracks. If it is replaced, torque the bolts attaching it to the crankshaft to specifications.

2. Coat the converter-hub hole in the crankshaft with wheel-bearing grease. Put the transmission assembly on a jack and raise it up under the car, aligned for installation. Rotate the converter so that the mark on it will align with the mark on the drive plate.

3. Carefully work the transmission assembly forward so that the converter hub enters the crankshaft hole.

4. Install and tighten the converter-housing bolts to specifications.

5. Install and tighten the two lower drive-plate-to-converter bolts to specifications.

6. Install the starting motor and connect the battery ground cable. Use the remote-control switch to rotate the converter and install and tighten the other two drive-plate-to-converter bolts to specifications.

7. Install the cross member. Tighten the exhaust system if it was loosened for transmission removal.

8. Check and adjust pressure, linkages, and bands, as described previously. Add fluid as necessary.

Chapter 12 review questions

Select the *one* correct, best, or most probable answer to each question. Then check your answers against the correct answers given at the end of the book.

1. Three diagnostic tests are to be made on the TorqueFlite: oil pressure, stall, and:
   a. hydraulic control
   b. shift point
   c. air pressure
   d. speed

2. To check the performance of the torque-converter stator clutch and transmission clutches, give the transmission the:
   a. pressure test
   b. air-pressure test
   c. shift-point test
   d. stall test

3. Stall speed more than 200 rpm above specifications indicates:
   a. poorly tuned engine
   b. clutch slippage
   c. band slippage
   d. throttle linkage out of adjustment

4. Stall speed 250 to 300 rpm below specifications indicates:
   a. torque-converter overrunning-clutch slippage
   b. transmission-clutch slippage
   c. band slippage
   d. manual valve stuck

5. Rear-servo apply pressure, as compared with line pressure in D, should be:
   a. higher
   b. about the same
   c. lower
   d. half as much

6. The hydraulic-control-pressure checks to be made include line pressure and kickdown-servo-release pressure, lubrication pressure, and:
   a. clutch-release pressure
   b. rear-servo-apply pressure
   c. front-servo pressure
   d. rear-servo-release pressure

7. The two hydraulic-control-pressure adjustments that can be made are to the line pressure and the:
   a. servo-apply pressure
   b. throttle pressure
   c. clutch-apply pressure
   d. converter pressure

8. Air-pressure tests check the actions of the:
   a. servos and clutches
   b. valves and pump
   c. servos and linkages
   d. converter and valves

9. In addition to the pressure adjustments, other adjustments to be made include throttle-rod linkage and:
   a. gearshift linkage and clutches
   b. throttle opening and gearshift linkage
   c. converter lockup
   d. gearshift linkage and band

10. If clutches and servos work properly but shifting is incorrect, the trouble is probably in the:
    a. pump
    b. valve body
    c. linkages
    d. converter

11. Mechanic A says the purpose of the converter lockup is to bypass the converter at highway speed. Mechanic B says when the converter locks up, there is no speed loss through the converter. Who is right?
    a. A only
    b. B only
    c. both A and B
    d. neither A nor B

12. To attach the converter to the drive plate during transmission installation, turn the drive plate with:
    a. a wrench
    b. the starting motor
    c. the output shaft
    d. the crankshaft

13. For normal service the recommendation on oil changes is:
    a. do not change oil
    b. change every 12,000 miles [19,312 km]
    c. change every 24,000 miles [38,624 km]
    d. change every year

14. The purpose of the stall test is to check the:
    a. bands
    b. oil pressures
    c. air pressures
    d. clutches

15. Mechanic A says the most complex and critical part of the transmission is the valve body. Mechanic B says the most complex and critical part of the trans-

mission is the compound planetary gearset. Who is right?

a. A only
b. B only
c. both A and B
d. neither A nor B

16. The purpose of the air-pressure tests is to find out whether or not the:
    a. pressures are sufficient
    b. clutches and servos work
    c. linkages are adjusted
    d. converter lockup works

17. To remove the governor, you first have to remove the:
    a. valve body
    b. extension housing
    c. converter
    d. transmission

18. Drive end play is checked by moving the:
    a. output shaft in and out
    b. torque converter back and forth
    c. input shaft in and out
    d. transmission forward and backward

19. A major difference between the front clutches for the A-904 and A-727 transmissions is that the A-727 has a series of small springs while the A-904 has:
    a. one large spring
    b. a flat tension spring
    c. no springs
    d. 24 to 36 coil springs

20. The bands can be adjusted:
    a. on the car only
    b. off the car only
    c. either on or off the car
    d. only with the transmission disassembled

FORD AUTOMATIC-
TRANSMISSION SERVICE

After studying this chapter, and with proper instruction and equipment, you should be able to:

1. Explain how to use the diagnostic charts to find causes of trouble.
2. Describe the procedures of making diagnostic tests of the transmission including road, stall, fluid-pressure, and air-pressure tests.
3. List and describe the services that can be performed with the transmission in the car.
4. Discuss the procedure of removing and installing the transmission in the car.
5. Explain how to disassemble the transmission, inspect and clean parts, and put the transmission back together again.
6. Perform the above services in the shop.

13-1 Introduction to Ford automatic transmission service The first part of this chapter describes the maintenance and trouble diagnosis of late-model Ford automatic transmissions (except for the overdrive transmission). The later part of the chapter describes the removal, overhaul, and installation of a C6 automatic transmission. This transmission is similar to the C4 and to the FMX. But the FMX has a different planetary gearset, which is similar to the gearset used in the overdrive transmission (Chap. 14). Refer to the shop manual that covers the transmission you are servicing. Servicing the Ford automatic overdrive transmission is covered in Chap. 14.

13-2 Maintenance and diagnostic checks In any diagnostic work, check the fluid level and condition, and linkage adjustments. Then make stall and road checks. Many of the checks and tests on Ford automatic transmissions are covered generally in Chap. 9.

13-3 Checking transmission fluid The fluid level should be checked every time the engine oil is changed (✹ 9-2 and 9-3). At the same time, check the condition of the fluid. If fluid must be added, be sure to use the fluid specified for the Ford transmission you are checking. Different Ford models use different fluids. Figure 13-1 shows the dipsticks from different Ford automatic transmissions and the fluid specified for each. If the fluid is contaminated, the hydraulic system, fluid

cooler, and lines must be thoroughly cleaned. Also, the transmission must be overhauled and the torque converter replaced. A new or factory serviced torque converter will be required. The contaminated torque converter cannot be flushed out in the field.

In normal operation, neither the fluid nor the filter requires periodic draining and replacement. However, if the vehicle is in severe service such as hauling a trailer or in police, taxicab, or door-to-door delivery operation, the fluid and filter should be replaced at frequent intervals. The bands should be adjusted even more frequently for severe service (✹ 13-7).

In severe service, transmissions used with four- and six-cylinder engines should have the fluid drained and refilled every 20 months or every 20,000 miles [32,000 km]. Transmissions used with V-8 engines should have this service every 22.5 months or 22,500 miles [36,000 km].

If the fluid is low, there is probably a leak. Check oil lines and fittings, the oil-pan gasket, converter and converter seals, speedometer-cable connections, and the engine coolant. If you find transmission fluid in the coolant, the cooler in the radiator is probably leaking. You can check it by disconnecting the line to the cooler, plugging one connection, and applying air pressure to the other. Remove the radiator cap to relieve any pressure buildup that might occur if the cooler is leaking. The converter can be checked for leakage, as explained in ✹ 13-32.

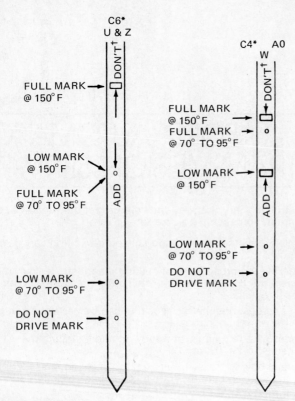

FULL MARK @ 150°F

LOW MARK @ 150°F

FULL MARK @ 70° TO 95°F

LOW MARK @ 70° TO 95°F

DO NOT DRIVE MARK

C6*
U & Z

DON'T†

ADD

FULL MARK @ 150°F

FULL MARK @ 70° TO 95°F

LOW MARK @ 150°F

LOW MARK @ 70° TO 95°F

DO NOT DRIVE MARK

C4*   A0
W

DON'T†

ADD

*USE CJ TYPE FLUID (ESP-M2C138-CJ OR DEXRON II SERIES D).
†DON'T ADD IF BETWEEN ARROWS.

Fig. 13-1 Dipsticks for different Ford transmissions. (*Ford Motor Company*)

✿ 13-4 Linkage adjustments There are two linkages, the kickdown linkage and the manual-valve linkage. Typical adjustments follow.

1. **Manual-valve-linkage adjustments** Two types of linkages are used for column shift, rod (Fig. 13-2), and cable (Fig. 13-3). Console shifts are rod (Fig. 13-4). To adjust the column-shift linkage:

a. Put selector lever in D. Hang a weight on the selector lever to make sure it remains against the D stop during the adjustment.
b. Loosen the shift-rod adjusting nut (Fig. 13-2).
c. On the cable shift, remove the nut at A (Fig. 13-3) and detach the cable from the transmission manual-lever stud.
d. Shift the transmission manual lever into D, which is the second detent position from full counterclockwise.
e. On the shift cable, put the cable end on the transmission-lever stud, aligning the flats on the stud with flats on the cable. Start the adjustment nut.
f. Make sure the selector lever has not moved from the D stop. Tighten the nut to specifications.
g. Check the transmission in all selector-lever detent positions.

To adjust console or floor shift linkage (Fig. 13-4):

a. Position the selector lever in D and hold in that position while shift adjustment is made.
b. Raise the vehicle and loosen the manual-lever shift-rod retaining nut. Move the transmission manual

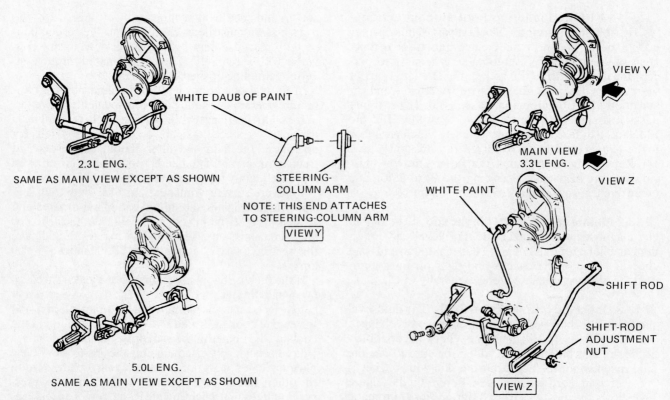

WHITE DAUB OF PAINT

2.3L ENG.
SAME AS MAIN VIEW EXCEPT AS SHOWN

STEERING-COLUMN ARM

NOTE: THIS END ATTACHES TO STEERING-COLUMN ARM

VIEW Y

5.0L ENG.
SAME AS MAIN VIEW EXCEPT AS SHOWN

VIEW Y

MAIN VIEW 3.3L ENG.

VIEW Z

WHITE PAINT

SHIFT ROD

SHIFT-ROD ADJUSTMENT NUT

VIEW Z

Fig. 13-2 Column-shift manual-linkage adjustments on different cars with rod-type linkage. (*Ford Motor Company*)

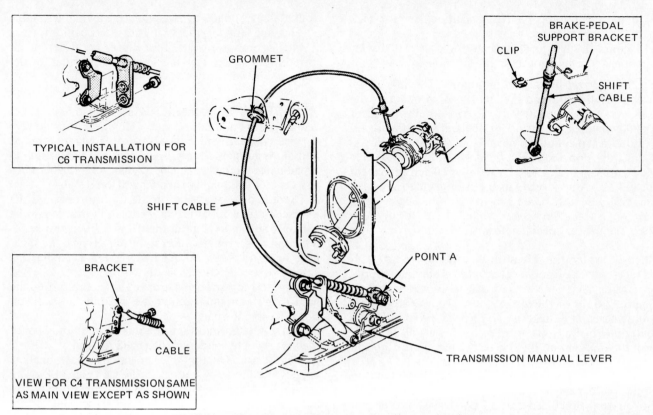

Fig. 13-3 Column-shift manual-linkage adjustments on several cars with cable-type linkage. (*Ford Motor Company*)

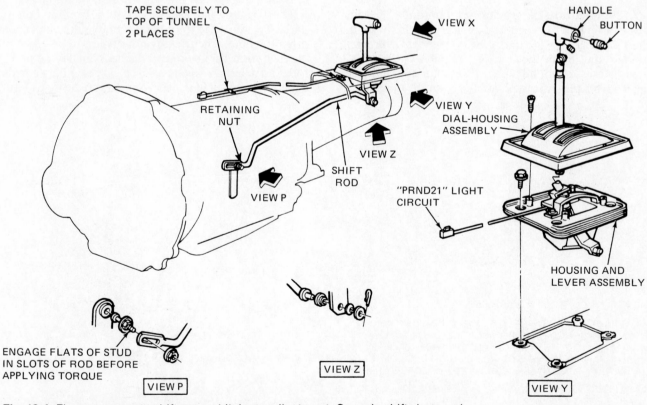

Fig. 13-4 Floor- or console-shift manual-linkage adjustment. Console shifts have rod-type linkage. (*Ford Motor Company*)

lever to D, the second detent position from the back of the transmission.

c. Tighten the nut to specifications. Check the operation of the selector lever in each position.

2. **Kickdown-linkage adjustment** The linkage between the throttle and the transmission operates the transmission downshift valve to cause a downshift or kickdown when the throttle is opened wide. Figure 13-5 shows a typical adjustment. Hold the downshift rod against the stop and adjust the downshift screw to provide the proper clearance, as shown. Then reconnect the linkage. A variety of linkage arrangements are used on Ford vehicles, related to the type of engine, carburetor, and transmission installed. Refer to the Ford shop manual for specific information.

✿ 13-5 Replacing the shift-linkage grommets If linkages must be disconnected at the grommets, the old grommets must be removed and new ones installed. Figure 13-6 shows the use of a Ford grommet-replacing tool to assist in doing this. To remove the rod, put the lower jaw of the tool between the lever and the rod, with the stop pin on the end of the rod. Force the rod out.

Remove the grommet from the lever by cutting the large shoulder off with a knife. Adjust the stop pin on the tool to within 0.5 inch [12.7 mm], as shown in Fig. 13-6. Coat the new grommet with lubricant and put it on the stop pin. Force it into the lever hole. Turn the grommet several times to make sure it is seated. Then squeeze the rod into the grommet until the stop washer seats against the grommet.

✿ 13-6 Neutral-start-switch adjustment The neutral-start switch is used on console models. It should be closed only with the selector lever in P or N to permit starting in those positions only. If the switch does not do its job and the starting motor does not work in P or N, the switch probably is out of adjustment.

Adjustment is made by loosening the switch-attaching screws and moving the switch to the position where normal operation results. Procedures are different for different cars. Refer to the Ford shop manual for details.

✿ 13-7 Band adjustment Band adjustment is not required for normal service. However, for severe service (taxicab and police service, towing a trailer, and stop-and-go delivery operation) the bands should be adjusted at 7500, 15,000, 30,000, and 45,000 miles [12,000, 24,000, 48,000, and 72,000 km].

There is only one band on the C6 transmission: the intermediate band. It is adjusted as follows: Raise the car on a lift and clean all dirt from around the band-adjustment-screw area. Remove and discard the locknut. Install a new locknut and tighten the adjusting screw to specifications, as shown in Fig. 13-7. Then back off the adjusting screw exactly 1½ turns. Hold the adjusting screw and tighten the locknut to specifications. Lower the car.

On the C4 transmission, there is an additional band: the low-reverse band, which is used instead of a clutch as on the C6. On the C4, the intermediate band is adjusted in the same way as in the C6 (Fig. 13-8). The low-reverse band on the C4 is adjusted in a similar way, but with a special tool. Figure 13-9 shows the special tools required for different cars using the C4. With a new locknut installed in place of the old one, tighten the adjusting screw until the tool handle clicks. It is preset to click at 10 lb-ft [13.6 N-m] torque. Then back off the adjusting screw three turns. Tighten the locknut to the proper torque, holding the screw stationary.

The FMX transmission has both a front and a rear band. To adjust the front band, remove the oil pan and drain the fluid. Then remove the fluid screen and clip. Next, loosen the locknut, pull back on the actuating rod, and insert a ¼-inch [6.35-mm] spacer between the adjusting screw and the servo-piston stem (Fig. 13-10).

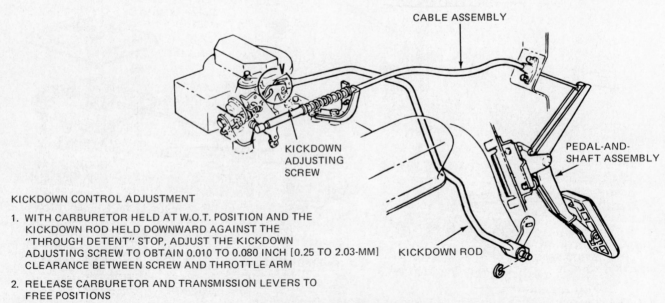

CABLE ASSEMBLY

KICKDOWN ADJUSTING SCREW

PEDAL-AND-SHAFT ASSEMBLY

KICKDOWN ROD

KICKDOWN CONTROL ADJUSTMENT

1. WITH CARBURETOR HELD AT W.O.T. POSITION AND THE KICKDOWN ROD HELD DOWNWARD AGAINST THE "THROUGH DETENT" STOP, ADJUST THE KICKDOWN ADJUSTING SCREW TO OBTAIN 0.010 TO 0.080 INCH [0.25 TO 2.03-MM] CLEARANCE BETWEEN SCREW AND THROTTLE ARM

2. RELEASE CARBURETOR AND TRANSMISSION LEVERS TO FREE POSITIONS

Fig. 13-5 Throttle and downshift linkage. (*Ford Motor Company*)

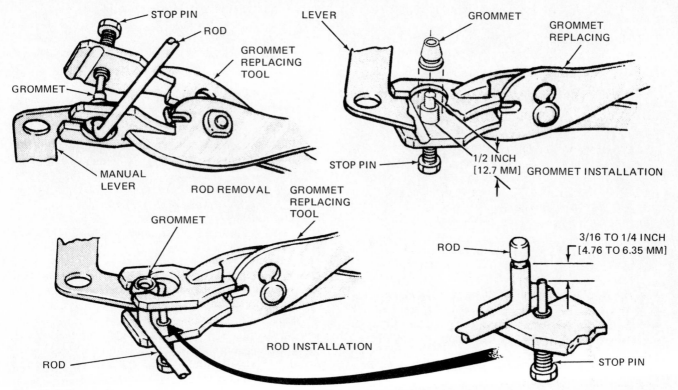

Fig. 13-6 Removing and installing a shift-linkage grommet. (*Ford Motor Company*)

Tighten the adjusting screw to 10 pound-inches [1.13 N-m] torque. Remove the spacer and tighten the screw an additional ¾ turn. Hold the adjusting screw stationary and tighten the locknut. Reinstall the fluid screen and clip. Then install the oil pan using a new gasket. Refill the transmission fluid, as described in ✿ 13-3.

To adjust the rear band on the FMX, remove all dirt from the adjusting-screw threads and oil the threads.

Then loosen the locknut. Use the special torque wrench, as shown in Fig. 13-11, and tighten the screw until the tool clicks. Then, back off the screw 1½ turns. Hold the screw and tighten the locknut to specifications.

Careful: The transmission can be damaged if the adjusting screw is not backed off exactly 1½ turns!

✿ **13-8 Transmission-trouble diagnosis** The fundamentals of trouble diagnosis are discussed in ✿ 9-6 and 9-7. Following pages have trouble-diagnosis charts for Ford transmissions. However, road, stall, and pressure checks of the transmission are covered first. These checks help to pinpoint trouble causes by showing you

Fig. 13-7 Adjusting the intermediate band on the C6. (*Ford Motor Company*)

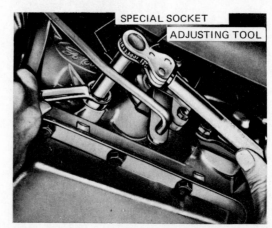

Fig. 13-8 Adjusting the intermediate band on the C4. (*Ford Motor Company*)

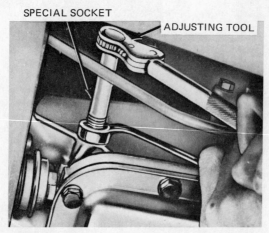

Fig. 13-9 Adjusting the low-reverse band on the C4. (*Ford Motor Company*)

exactly what the trouble is, such as rough shifts, no drive in some gear positions, slipping, and noisy transmission. The charts then list possible causes of each trouble and the corrections to be made.

❀ **13-9 Shift-point checks (road or shop test)** The engine should be in good condition, linkages properly adjusted, and the transmission fluid at the proper height. The shift points can be checked on the road or in the shop if a dynamometer or a vacuum pump is available.

Without the dynamometer, engine loading can be simulated by applying a varying vacuum to the vacuum diaphragm of the transmission. By varying the vacuum to simulate closed-to-open throttle and varying engine speed, you can make all the checks needed in the shop. Ford furnishes diagnostic guides, such as that shown in Fig. 13-12. They have spaces in which to record test data on shift points, stall speed, and pressure tests. Note that there are spaces in which to record the spec-

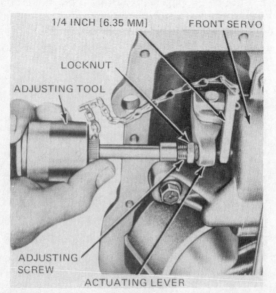

Fig. 13-10 Adjusting the front band on the FMX. (*Ford Motor Company*)

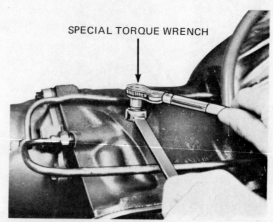

Fig. 13-11 Adjusting the rear band on the FMX. (*Ford Motor Company*)

ifications found in the shop manual. The actual test results are then compared with the specifications. This comparison usually identifies any trouble in the transmission.

1. To make the shift-point test in the shop, jack up the car and place it on safety stands, or raise the car on a lift, so that the drive wheels can rotate freely. Then disconnect the vacuum line from the transmission and plug the line. Connect the vacuum hose from a vacuum pump to the transmission. Start the engine and move the selector lever through all positions. Adjust the vacuum to 18 inches of mercury [457 mmHg] to simulate a light throttle. Move the lever to D and gradually increase engine speed. Note the upshift points. You can tell these because the speedometer needle will surge momentarily and you can feel the slight bump indicating the upshift. Slow the engine to idle.

Careful: Do not exceed 60 mph [97 km/h] on the speedometer!

2. Now simulate a high engine load or wide-open throttle by reducing the vacuum to 0 to 2 inches of mercury [0 to 51 mmHg]. Accelerate the engine slowly and note the 1-2 upshift point. The 2-3 shift point was checked in the previous test. Do not exceed 60 mph [97km/h]!

3. Next, test the kickdown system. To avoid running an unloaded engine at wide-open throttle, leave the vacuum adjusted to 0 to 2 inches of mercury [0 to 51 mmHg]. Manually hold the downshift linkage in the wide-open-throttle position. Accelerate the engine just enough to cause a 1-2 upshift. Do not exeed 60 mph [97 km/h]. Note that the wide-open-throttle 1-2 shift should be at a higher car speed than the high-engine-load test. This indicates that the kickdown system is working.

4. To make the shift-point tests on the highway, operate the car at various speeds and in the different selector-lever positions. Note whether all shifts take place properly.

Careful: Drive carefully and obey all traffic laws. Use a chassis dynamometer (❀ 1-15), if available, to test the transmission.

Automatic Transmission Diagnosis Guide
Ford Customer Service Division

General: This form must be completely filled in throughout the steps required to diagnose the condition covering transmission malfunction complaints (e.g., erratic shifting, slippage during shifts, failure to shift, harsh and delayed shifts, noise, etc.). It is not necessary to complete this form on complaints involving external leaks.

Transmission Model _____   Transmission Date Code/or Serial No. _____

R.O. No. _____   Axle Ratio _____   Tire Size _____

DIAGNOSIS PROCEDURE

Following steps will provide complete data necessary to perform an accurate diagnosis of transmission difficulties.

1. Check transmission fluid level   ☐ Room Temp.   ☐ Operating Temp.   ☐ OK   ☐ Overfilled   ☐ Low

2. Engine (CID) and Calibration Number _____

   Idle RPM in Drive   _____   _____   _____

                       Specification   As Received   Set To

   Check EGR System (if so equipped)

     Valve Operation   ☐ OK   ☐ Other (Explain) _____

     Restriction   ☐ OK   ☐ Other (Explain) _____

3. Check downshift and manual linkage   ☐ OK   ☐ Other (Explain) _____

4. Drive the car in each range, and through all shifts, including forced downshifts, observing any irregularities of transmission performance.

Throttle Opening	Range	Shift	Shift Points (MPH) Record Actual	Shift Points (MPH) Record Spec.
Minimum	D	1-2		
(Above	D	2-3		
12"	D	3-1		
Vacuum)	1	2-1		

Throttle Opening	Range	Shift	Shift Points (MPH) Record Actual	Shift Points (MPH) Record Spec.
To Detent (Torque Demand)	D	1-2		
	D	2-3		
		3-2		
Thru Detent (WOT)	D	1-2		
	D	2-3		
	D	3-2		
	D	2-1 or 3-1		

5. Control Pressure Test _____ AND _____ Stall Speed Data

   Transmission fluid must be normal operating temperatures. DO NOT hold throttle open over five seconds during tests.

   CAUTION: Release throttle immediately if slippage is indicated.

   After each stall test move selector lever to neutral with engine running at 1000 RPM to cool the transmission.

Engine RPM	Manifold Vacuum In-Hg	Throttle	Range	PSI Record Actual	PSI Record Spec.
Idle	Above 12	Closed	P		
			N		
			D		
			2		
			1		
			R		
As Required	10 ①	As Required	D, 2, 1		
As Required	Below 3	Wide Open	D		
			2		
			1		
			R		

Above Specified Engine RPM
1. Transmission slippage
2. Clutches or bands not holding

Below Specified Engine RPM
1. Poor engine performance, such as need for tune-up
2. Converter one way clutch slipping or improperly installed

Specified Engine RPM	Record Actual Engine RPM

① On units equipped with a dual area diaphragm, the front port of diaphragm must be vented to atmosphere (hose disconnected and plugged) during this check only.

After the tests, you should know the following items:

- CONTROL PRESSURE – Does the transmission have the CORRECT CONTROL PRESSURE?   ☐ Yes   ☐ No
- CONTROL VALVES – Beyond the manual valve are all the CONTROL VALVES FUNCTIONING?   ☐ Yes   ☐ No
- HYDRAULIC CIRCUITS – If the first two items check out good, then check the transmission's internal hydraulic circuits that are beyond the VALVE BODY. These circuits must be checked during transmission disassembly.

6. TORQUE CONVERTER AND OIL COOLER (where applicable)

   - Was torque converter flushed with a mechanical cleaner?   ☐ Yes   ☐ No
   - Was oil cooler flushed with a mechanical cleaner?   ☐ Yes   ☐ No

7. The problem was diagnosed to be: _____

8. If it was necessary to disassemble the transmission, record the actual problem found: _____

Fig. 13-12 Automatic-transmission diagnosis guide. (*Ford Motor Company*)

✿13-10 Control-pressure and stall checks These checks consist of measuring the control pressure in the hydraulic system under varying operating conditions and also checking for transmission slippage. Ford recommends the tests in the shop with a dynamometer or with the transmission tester, shown in Fig. 13-13. The transmission tester has a pressure gauge, vacuum gauge, and tachometer. The use of the transmission tester to make the checks is described below.

First, connect the tachometer to the ignition system so that you can measure engine speed. One clip goes on the distributor terminal on the ignition coil. The black-covered clip is attached to ground.

Next, connect the vacuum hose from the tester (Fig. 13-13) into the manifold vacuum line by means of a T fitting, as shown in Fig. 13-14. The third connection is the pressure hose to the control-pressure port. The connection to the C6 transmission is shown in Fig. 13-15. The control-pressure port has different locations in other transmissions; check the shop manual. With the three connections made, proceed as follows:

1. The engine must be in good condition with an intake-manifold vacuum reading of about 18 inches of mercury [457 mmHg] at idle.
2. The "Stall-Speed Data" chart in the diagnosis guide of Fig. 13-12 has spaces in which to record the specified engine rpm at which the stall test should be made, and in which to write the stall-test results.
3. The "Control-Pressure Test" chart in Fig. 13-12 has spaces in which to record the specified pressures under different conditions, and places to write the results of the pressure tests.
4. Prior to the test, find the specifications for the transmission and car being tested in the shop manual. These should be recorded on the diagnosis guide.
5. Now, with the transmission at normal operating temperature, apply the foot brake and make the stall check by shifting into the indicated selector-lever position, and increasing the engine speed to the specified value.

CAUTION: Do not allow anyone to stand in front or in back of the car during the test. If the brakes should slip, the car might lunge forward or backward.

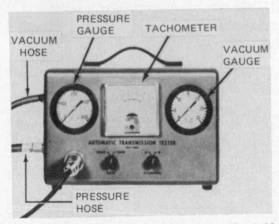

Fig. 13-13 Automatic-transmission tester. (*Ford Motor Company*)

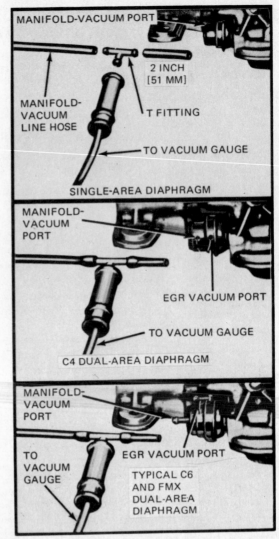

Fig. 13-14 Vacuum-line connections to check control pressure on various Ford automatic transmissions. (*Ford Motor Company*)

Careful: Do not hold the throttle open more than 5 seconds! After each stall check, return the selector lever to N and idle the engine for one minute or longer to cool the transmission. Release the throttle instantly if the transmission slips. This is indicated if the engine speed goes above specifications.

6. Excessive engine rpm in any selector position indicates transmission slippage.
7. Now make the pressure check by adjusting the engine idle to the specified speed. With a closed throttle, move the selector lever to all positions and note the pressures (Fig. 13-12).
8. Next, with a vacuum pump connected to the transmission instead of the intake manifold of the engine, make the pressure tests at a vacuum of 10 inches and 1 inch of mercury [254 and 25 mmHg]. These tests can be made with the engine idling and only the vacuum changed. It is not necessary to open the throttle wide to get the vacuums needed for the test. Note that the pressures should be checked in D, D2, and D1 with a vacuum of 10 inches of mercury [254

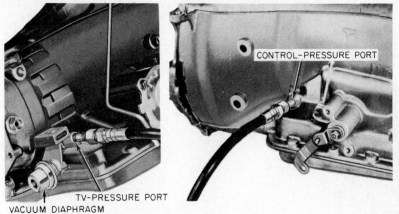

CONTROL-PRESSURE PORT

TV-PRESSURE PORT

VACUUM DIAPHRAGM

Fig. 13-15 Vacuum-diaphragm and control- and TV-pressure connecting points on C6 transmission. (*Ford Motor Company*)

mmHg] and in D, D2, D1, and R with a vacuum of 1 inch of mercury [25 mmHg].

9. The chart in Fig. 13-16 shows possible trouble causes if the pressures are not within the specifications. Improper pressures often require transmission removal and disassembly to find and fix the cause.

✿ 13-11 Air-pressure checks Further information on the condition of the clutches and servo can be ob-
tained by using air pressure to operate them. This procedure requires removal of the oil pan and control-valve body and draining of the fluid, as explained in ✿ 13-17. Then air pressure can be directed into the appropriate fluid passage to see if the band will apply or the clutch will engage. Figure 13-17 shows the transmission case for the C6 with the various fluid passages identified. Figure 13-18 shows the FMX transmission case with all fluid passages identified.

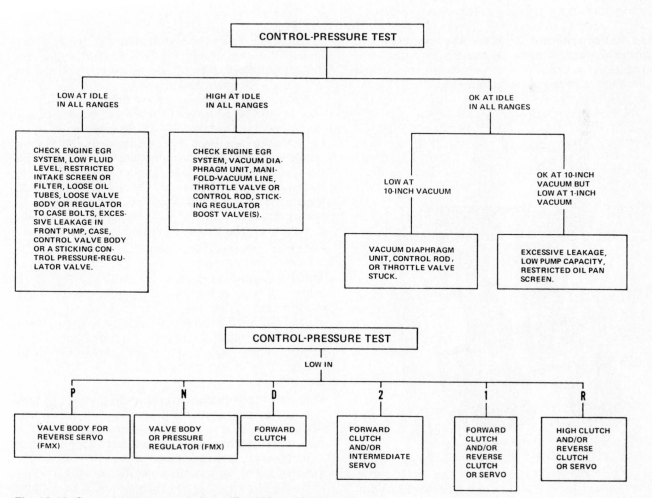

Fig. 13-16 Control-pressure test chart. (*Ford Motor Company*)

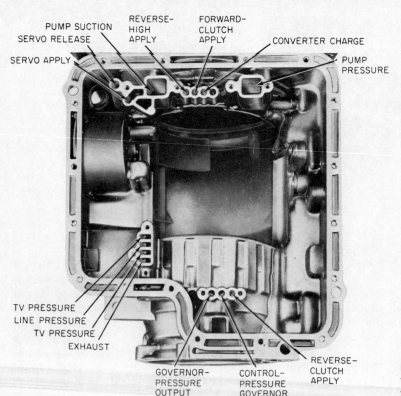

PUMP SUCTION
SERVO RELEASE
REVERSE-HIGH APPLY
FORWARD-CLUTCH APPLY
CONVERTER CHARGE
SERVO APPLY
PUMP PRESSURE

TV PRESSURE
LINE PRESSURE
TV PRESSURE
EXHAUST

GOVERNOR-PRESSURE OUTPUT
CONTROL-PRESSURE GOVERNOR
REVERSE-CLUTCH APPLY

Fig. 13-17 Points at which to apply air pressure to check the clutch and band action in the C6 transmission. (*Ford Motor Company*)

❋ **13-12 Vacuum-unit checks and adjustments**
The vacuum unit actuates the primary throttle valve, as explained in Chap. 6. The unit should be checked for leakage and for bellows failure. To check for leak-

age, connect the unit to a vacuum pump. Apply a vacuum of 18 inches of mercury [457 mmHg] to the vacuum unit and see if it will hold this vacuum. If it does not, the diaphragm is leaking.

Earlier Ford shop manuals carried an additional check (which is no longer included) as follows. Check the bellows, as shown in Fig. 13-19. Insert a rod in the unit. Make a reference mark on the rod with the rod bottomed in the hole. Then press down on the unit with the rod resting on a scale, as shown in Fig. 13-19. Increase force to 12 pounds [53.4 N]. If the mark remains visible, the bellows is OK. If the mark disappears before 4 pounds [17.8 N] is exerted, the bellows is defective. The complete vacuum unit must be replaced.

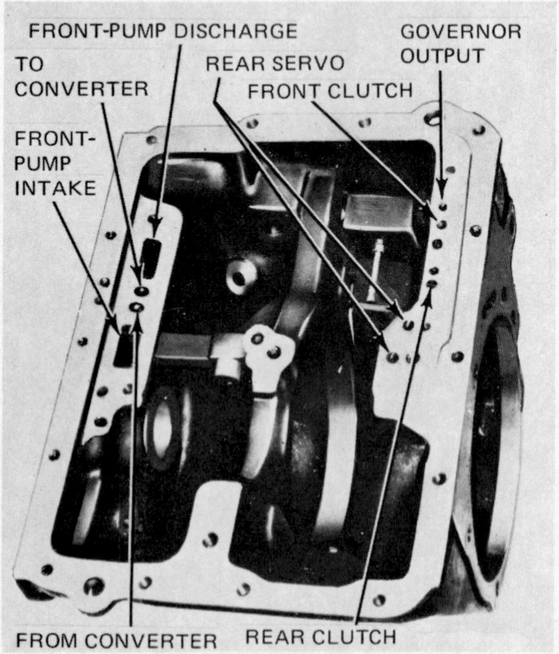

❋ **13-13 Diagnosis guides** The guide in Fig. 13-12 explains how to make various diagnostic tests of the transmission to determine what might be wrong and what is causing the trouble. In addition to these, there are diagnosis guides for specific transmissions. These guides can help you pinpoint every cause of trouble in a certain type of transmission.

❋ **13-14 Transmission trouble-diagnosis chart**
The accompanying trouble-diagnosis chart lists transmission complaints or troubles that could cause the driver to bring the car in for service. Each complaint has a list of possible causes of the trouble and checks or corrections. There is a separate noise diagnosis chart (Fig. 13-20) which shows the checking procedure to follow to locate and correct the problem.

FRONT-PUMP DISCHARGE
TO CONVERTER
REAR SERVO
GOVERNOR OUTPUT
FRONT CLUTCH
FRONT-PUMP INTAKE

FROM CONVERTER
REAR CLUTCH

Fig. 13-18 Points at which to apply air pressure to check the clutch and band action on the FMX transmission. (*Ford Motor Company*)

Automatic-Transmission Trouble-Diagnosis Chart

COMPLAINT	POSSIBLE CAUSE	CHECK OR CORRECTION
1. Slow initial engagement.	a. Improper fluid level	Add fluid as required.
	b. Damaged or improperly adjusted linkage	Repair or adjust linkage.
	c. Contaminated fluid	Replace fluid.
	d. Improper clutch engagement and band application, or oil control pressure	Perform control-pressure test.
2. Rough initial engagement in either forward or reverse.	a. Improper fluid level	Perform fluid-level check.
	b. High engine idle	Adjust idle to specifications.
	c. Looseness in the driveshaft, U joints	Repair as required.
	d. Incorrect linkage adjustment	Repair or adjust linkage.
	e. Improper clutch engagement, band application, or oil control pressure	Perform control-pressure test.
	f. Sticking or dirty valve body	Clean, repair, or replace valve body.
3. No drive in any gear.	a. Improper fluid level	Perform fluid-level check.
	b. Damaged or improperly adjusted linkage	Repair or adjust linkage.
	c. Improper clutch engagement or band application, or oil control pressure	Perform control-pressure test.
	d. Internal leakage	Check and repair as required.
	e. Valve body loose	Tighten to specification.
	f. Damaged or worn clutches	Perform air-pressure test.
	g. Sticking or dirty valve body	Clean, repair, or replace valve body.
4. No drive forward—reverse OK.	a. Improper fluid level	Perform fluid-level check.
	b. Damaged or improperly adjusted linkage	Repair or adjust linkage.
	c. Improper clutch engagement, band application, or oil-pressure control system	Perform control-pressure test.
	d. Damaged or worn forward clutch or governor	Perform air-pressure test.
	e. Valve body loose	Tighten to specification.
	f. Dirty or sticking valve body	Clean, repair, or replace valve body.
5. No drive, slips or chatters in reverse—forward OK.	a. Improper fluid level	Perform fluid level check.
	b. Damaged or improperly adjusted linkage	Repair or adjust linkage.
	c. Looseness in the drive shaft, U joints, or engine mounts	Repair as required.
	d. Reverse band out of adjustment or damaged reverse clutch	For C3, C4, or FMX—adjust reverse band. For C6—repair reverse clutch as required.
	e. Improper oil-pressure control	Perform control-pressure test.
	f. Damaged or worn reverse clutch or servo	Perform air-pressure test.
	g. Valve body loose	Tighten to specification.
	h. Dirty or sticking valve body	Clean, repair, or replace valve body.
6. No drive, slips or chatters in first gear in D. All other gears normal.	a. Damaged or worn one-way clutch	Repair or replace one-way clutch.
7. No drive, slips or chatters in second gear.	a. Improper fluid level	Perform fluid-level check.
	b. Damaged or improperly adjusted linkage	Repair or adjust linkage.
	c. Intermediate band out of adjustment	Adjust intermediate band.
	d. Improper band application, clutch engagement, or oil-pressure control	Perform control-pressure test.
	e. Damaged or worn servo and/or internal leaks	Perform air-pressure test.
	f. Dirty or sticking valve body	Clean, repair, or replace valve body.

COMPLAINT	POSSIBLE CAUSE	CHECK OR CORRECTION
	g. Polished, glazed intermediate band or drum	Replace or repair as required.
8. Starts in high—in D, drag or lockup at 1-2 shift point or in 2 or 1.	*a.* Improper fluid level	Perform fluid-level check.
	b. Damaged or improperly adjusted linkage	Repair or adjust linkage.
	c. Damaged or worn governor	Repair or replace governor; clean screen.
9. Starts in 2 or 3.	*a.* Improper fluid level	Perform fluid-level check.
	b. Damaged or improperly adjusted linkage	Repair or adjust linkage.
	c. Improper band application and/or clutch engagement, or oil-pressure control system	Perform control-pressure test.
	d. Damaged or worn governor	Perform governor check. Replace or repair governor; clean screen.
	e. Valve body loose	Tighten to specification.
	f. Dirty or sticking valve body	Clean, repair, or replace valve body.
	g. Cross-leaks between valve body and case mating surface	Replace valve body and/or case as required.
10. Shift points incorrect.	*a.* Improper fluid level	Perform fluid-level check.
	b. Improper vacuum-hose routing or leaks	Correct hose routing.
	c. Improper operation of EGR system	Repair or replace as required.
	d. Linkage out of adjustment	Repair or adjust linkage.
	e. Improper speedometer gear installed	Replace gear.
	f. Improper clutch engagement, band application, or oil-pressure control system	Perform shift test and control-pressure test.
	g. Damaged or worn governor	Repair or replace governor—clean screen.
	h. Dirty or sticking valve body	Clean, repair, or replace valve body.
11. No upshift at any speed in D.	*a.* Improper fluid level	Perform fluid-level check.
	b. Vacuum leak to diaphragm unit	Repair vacuum line or hose.
	c. Linkage out of adjustment	Repair or adjust linkage.
	d. Improper band application, clutch engagement, or oil-pressure control system	Perform control-pressure test.
	e. Damaged or worn governor	Repair or replace governor; clean screen.
	f. Dirty or sticking valve body	Clean, repair, or replace valve body.
12. Shifts from 1 to 3 in D.	*a.* Improper fluid level	Perform fluid-level check.
	b. Intermediate band out of adjustment	Adjust band.
	c. Damaged intermediate servo and/or internal leaks	Perform air-pressure test. Repair front servo and/or internal leaks.
	d. Polished, glazed band or drum	Repair or replace band or drum.
	e. Improper band application, clutch engagement, or oil-pressure control system	Perform control-pressure test.
	f. Dirty or sticking valve body	Clean, repair, or replace valve body.
13. Engine overspeeds on 2-3 shift.	*a.* Improper fluid level	Perform fluid-level check.
	b. Linkage out of adjustment	Repair or adjust linkage.
	c. Improper band application, clutch engagement, or oil-pressure control system	Perform control-pressure test.
	d. Damaged or worn high clutch and/or intermediate-servo	Perform air-pressure test. Repair as required.
	e. Dirty or sticking valve body	Clean, repair, or replace valve body.

COMPLAINT	POSSIBLE CAUSE	CHECK OR CORRECTION
14. Mushy 1-2 shift.	a. Improper fluid level	Perform fluid-level check.
	b. Incorrect engine performance	Tune, adjust engine idle as required.
	c. Improper linkage adjustment	Repair or adjust linkage.
	d. Intermediate band out of adjustment	Adjust intermediate band.
	e. Improper band application, clutch engagement, or oil-pressure control system	Perform control-pressure test.
	f. Damaged high clutch and/or intermediate servo or band	Perform air-pressure test. Repair as required.
	g. Polished, glazed band or drum	Repair or replace as required.
	h. Dirty or sticking valve body	Clean, repair, or replace valve body.
15. Rough 1-2 shift.	a. Improper fluid level	Perform fluid-level check.
	b. Incorrect engine idle or performance	Tune and adjust engine idle.
	c. Intermediate band out of adjustment	Adjust intermediate band.
	d. Improper band application, clutch engagement, or oil-pressure control system	Perform control-pressure test.
	e. Damaged intermediate servo	Air-pressure-check intermediate servo.
	f. Dirty or sticking valve body	Clean, repair, or replace valve body.
16. Rough 2-3 shift.	a. Improper fluid level	Perform fluid-level check.
	b. Incorrect engine performance	Tune and adjust engine idle.
	c. Improper band application, clutch engagement, or oil-control pressure system	Perform control-pressure test.
	d. Damaged or worn intermediate-servo release and high-clutch-piston check ball	Air-pressure-test the intermediate-servo apply and release and the high-clutch-piston check ball. Repair as required.
	e. Dirty or sticking valve body	Clean, repair, or replace valve body.
17. Rough 3-1 shift at closed throttle in D.	a. Improper fluid level	Perform fluid-level check.
	b. Incorrect engine idle or performance	Tune and adjust engine idle.
	c. Improper linkage adjustment	Repair or adjust linkage.
	d. Improper clutch engagement, band application, or oil-pressure control system	Perform control-pressure test.
	e. Improper governor operation	Perform governor test. Repair as required.
	f. Dirty or sticking valve body	Clean, repair, or replace valve body (or, on FMX, regulator assembly).
18. No engine braking in manual second gear.	a. Improper fluid level	Perform fluid-level check.
	b. Linkage out of adjustment	Repair or adjust linkage.
	c. Intermediate band out of adjustment	Adjust intermediate band.
	d. Improper band application, clutch engagement, or oil-pressure control system	Perform control-pressure test.
	e. Intermediate servo leaking	Perform air-pressure test of intermediate servo for leakage. Repair as required.
	f. Polished or glazed band or drum	Repair or replace as required.
19. Transmission noisy—valve resonance.	a. Improper fluid level	Perform fluid-level check.
	b. Linkage out of adjustment	Repair or adjust linkage.
	c. Improper band application, clutch engagement, or oil-pressure control system	Perform control-pressure test.
	d. Cooler lines grounding	Free up cooler lines.
	e. Dirty, sticking valve body	Clean, repair, or replace valve body.
	f. Internal leakage or pump cavitation	Repair as required.

COMPLAINT	POSSIBLE CAUSE	CHECK OR CORRECTION
20. Transmission overheats.	a. Improper fluid level	Perform fluid-level check.
	b. Incorrect engine idle or performance	Tune, or adjust engine idle.
	c. Improper clutch engagement, band application, or oil-pressure control system	Perform control-pressure test.
	d. Restriction in cooler or lines	Repair restriction.
	e. Seized one-way clutch	Replace one-way clutch.
	f. Dirty or sticking valve body	Clean, repair, or replace valve body.
21. Transmission fluid leaks.	a. Improper fluid level	Perform fluid-level check.
	b. Leakage at gasket, seals, etc.	Remove all traces of lubricant on exposed surfaces of transmission. Check the vent for free breathing. Operate transmission at normal temperatures and inspect for leakage. Repair as required.
	c. Vacuum diaphragm unit leaking	Replace diaphragm.

On-the-car repairs

✿ 13-15 Governor and extension-housing removal and replacement

Several transmission parts can be removed without removing the complete transmission from the car. These include the governor and extension housing, the servo, and the valve body. To remove the governor and extension housing, first raise the car on a lift. Then disconnect the parking-brake cable from the equalizer. On the Lincoln Continental, remove the equalizer. Proceed as follows:

1. Mark the U joint and flange for proper reassembly. Then disconnect the drive shaft from the rear-axle flange and remove the drive shaft from the transmission.
2. Disconnect the speedometer cable from the extension housing.
3. Remove the engine-rear-support-to-extension-housing attaching bolts. On the Lincoln, remove the reinforcement plate from under the transmission oil pan.
4. Place a jack under the transmission and raise it enough to remove the weight from the engine rear support.
5. Remove the bolt that attaches the engine rear support to the cross member and remove the support.
6. Put a drain pan under the transmission.
7. Lower the transmission and remove the extension-housing attaching bolts. Slide the housing off the output shaft and allow the fluid to drain.
8. Remove the governor-attaching bolts and slide the governor off the output shaft (Fig. 13-21).
9. Clean the governor parts with solvent and air-dry them. Check valves and bores for scores. Minor scores can be removed with crocus cloth. Heavier scores or other damage requires replacement of the governor assembly. Valves should slide through the bores of their own weight when dry. Fluid passages should be free. Mating surfaces should be smooth and flat.
10. On the C4, the oil screen should be removed from the collector body, cleaned, and air-dried.
11. Check the extension-housing bushing and rear seal. If either needs replacement, use the special tools required to remove the old part and install the new one.
12. To reinstall the governor and extension housing, first attach the governor to the distributor flange. Secure the governor with the attaching bolts, torqued to specifications.
13. Next, make sure the mating surfaces of the extension housing and transmission case are clean. Use

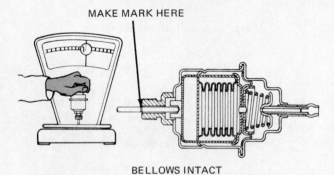

MAKE MARK HERE

BELLOWS INTACT

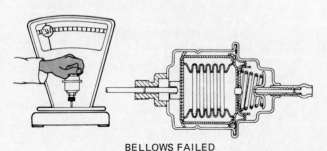

BELLOWS FAILED

Fig. 13-19 Checking vacuum-unit bellows. *(Ford Motor Company)*

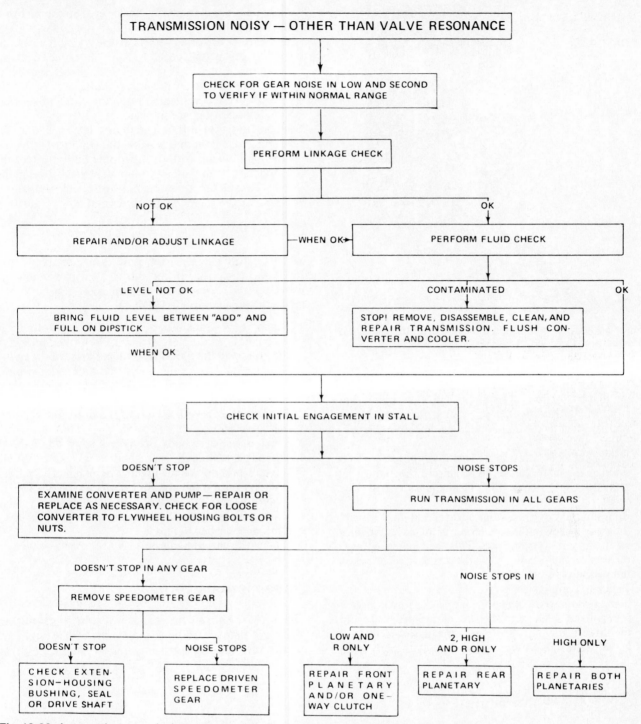

Fig. 13-20 Automatic-transmission noise trouble-diagnosis chart. (*Ford Motor Company*)

a new gasket and install the housing, securing it with the attaching bolts.

14. Raise the transmission high enough to position the engine rear support on the cross member. Secure the support with the attaching bolt and nut, torqued to specifications.

15. Lower the transmission and remove the jack. Install and torque the engine rear-support-to-extension-housing attaching bolts. On the Lincoln, install the reinforcing plate with attaching bolts.

16. Attach the speedometer cable and connect the parking-brake cable to the equalizer. On the Lincoln, connect the parking-brake equalizer. Adjust the parking brakes.

17. Install the drive shaft. Fill the transmission to the correct level with the specified transmission fluid.

✿ **13-16 Servo removal and replacement** Remove the servo by raising the car on a lift. Then:

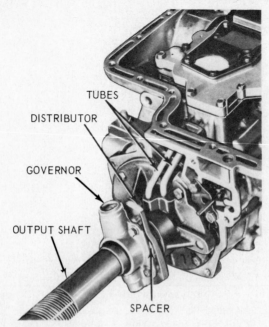

Fig. 13-21 Governor in position in the C6 transmission. (*Ford Motor Company*)

1. Remove the engine rear-support-to-extension-housing attaching bolts. Raise the transmission high enough to remove the weight from the engine rear support and remove the bolt that secures the engine rear support to the cross member. Remove the support.
2. Lower the transmission and remove the jack. Put a drain pan under the servo. Remove the bolts that attach the servo cover to the transmission case (Fig. 13-22).
3. Remove the cover, piston, spring, and gasket from the case, screwing the band-adjusting screw inward as the piston is removed. This will place enough tension on the band to keep the struts in place when the piston is removed.
4. To change the piston seals, apply air pressure to the port in the servo cover to remove the piston. Then remove the seals from the piston and the seal from the cover. Dip new seals in transmission fluid and install them. Coat two new gaskets with petroleum

jelly and install them on the cover. Dip the piston in transmission fluid and install it in the cover.
5. Put the servo spring on the piston rod and insert the piston rod in the case. Secure the cover with the attaching bolts, taking care to back off the band-adjusting screw as the cover bolts are tightened. Make sure the vent-tube retaining clip and service-identification tag are in place.
6. Raise the transmission high enough to install the engine rear support. Secure the support to the extension housing with the attaching bolts. Lower the transmission to install the support to the cross member. Torque the attaching bolt to specifications.
7. Remove the jack. Adjust the band as explained in ✿ 13-7.
8. Lower the car and add fluid as necessary.

✿ **13-17 Valve-body removal and replacement**
Raise the car on a lift and put a drain pan under the transmission. Loosen the transmission-pan bolts and allow the fluid to drain. Then:

1. Remove the transmission-pan attaching bolts from both sides to complete draining. Remove the rest of the bolts and the pan. Remove and discard the nylon shipping plug from the filler-tube hole. This plug is used to retain transmission fluid during shipping and should be removed when the pan is off.
2. Remove the valve-body attaching bolts and take off the valve body.
3. Service the valve body, as described in ✿ 13-25.
4. When reinstalling the valve body, make sure that the selector and downshift levers are engaged. Then install and torque the attaching bolts to specifications.
5. Clean the pan and gasket surfaces thoroughly. Use a new pan gasket and attach the pan with the bolts tightened to specifications.
6. Lower the car and add fluid as necessary.

Transmission overhaul

✿ **13-18 Transmission removal and installation**
There are many variations in engine-compartment arrangements and transmission supports in different car

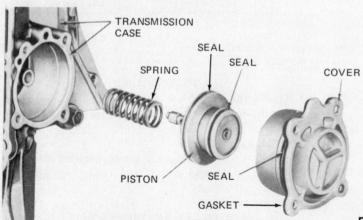

Fig. 13-22 Disassembled servo. (*Ford Motor Company*)

models. Refer to the shop manual for the car you are working on before proceeding with transmission removal. A typical removal procedure follows:

1. Raise the vehicle on a lift and drain the fluid from the transmission and converter.
2. Mark the U joint and flange for proper reinstallation. Then disconnect the drive shaft from the rear axle and slide the shaft rearward to remove it from the transmission. Install a seal-installation tool in the extension housing to prevent fluid leakage.
3. Disconnect the ground cable from the battery and remove the starting motor.
4. Remove the four converter-to-flywheel attaching nuts. Do not pry on the converter to turn it. Instead, use a wrench on the crankshaft pulley to turn the crankshaft so that you can gain access to the nuts.
5. Remove the rear-mount-to-cross-member attaching bolt.
6. Remove the two cross-member-to-frame attaching bolts.
7. Remove the two engine-rear-support-to-extension-housing attaching bolts.
8. Disconnect the downshift rod and the manual-linkage rod from the transmission levers. On console models, disconnect the column-lock rod from the transmission.
9. Remove the two bolts securing the bell crank bracket to the converter housing.
10. Raise the transmission with a transmission jack (Fig. 12-35) to provide clearance to remove the cross member. Remove the rear mount from the cross member and then remove the cross member.
11. Lower the transmission enough to gain access to the oil-cooler lines. Disconnect the lines from the transmission.
12. Disconnect the vacuum line from the transmission vacuum unit and detach the line from the retaining clip on the transmission.
13. Disconnect the speedometer cable from the extension housing.
14. Remove the bolt that secures the transmission oil-filler tube to the cylinder block. Lift the filler tube and dipstick from the transmission.
15. Use a chain to secure the transmission to the jack.
16. Remove the converter-housing-to-engine attaching bolts.
17. Carefully move the transmission away from the engine and lower it.

Careful: Don't let the converter slip off!

18. Remove the converter over a drain pan to catch the oil. Mount the transmission in a repair fixture, as shown in Fig. 13-23.
19. Reinstallation of the transmission is essentially the reverse of removal. When installing the oil filler tube, use a new O ring. Do not use a wrench on the converter-attaching nuts to turn the converter when you are attaching the converter to the flywheel. Instead, use a wrench on the crankshaft-pulley attaching nut. Adjust linkages and refill the transmission with the specified fluid when instal-

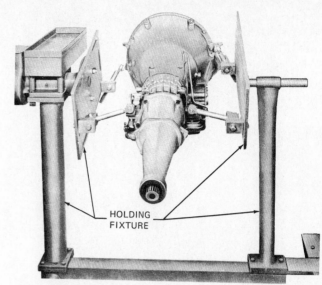

Fig. 13-23 Fixture to hold the transmission for repair. (*Ford Motor Company*)

lation is complete. Start the engine and check the transmission operation in all selector-lever positions.

✿ 13-19 Transmission-service precautions Cleanliness is important in transmission work. The tiniest piece of dirt or lint from a cleaning rag can cause a valve to hang up and prevent normal transmission operation. Under some conditions, this could damage the transmission. The outside of the transmission must be clean before it is opened. The workbench, tools, your hands, and all parts must be kept clean at all times. Other precautions for transmission service are discussed in ✿ 9-10.

✿ 13-20 Cleaning and inspecting transmission parts During and after disassembly, clean and inspect all parts for wear or other damage. Refer to ✿ 9-11 for details. If the transmission shows signs of internal damage to clutches or other parts, or if there is evidence that foreign material has been circulating in the oil, then all parts, including the oil cooler and lines, must be flushed out. A new oil filter should be installed on reassembly. A new converter will also be required. A contaminated converter cannot be flushed out satisfactorily in the field.

✿ 13-21 Transmission disassembly First, mount the transmission in a holding fixture, as shown in Fig. 13-23. The transmission gear train is shown in disassembled view in Fig. 13-24. Proceed with disassembly as follows:

1. Remove the 17 oil-pan attaching screws and the oil pan.
2. Remove the eight valve-body attaching screws and the valve body.
3. Attach a dial indicator to the front pump, as shown in Fig. 13-25. Install a special tool in the extension

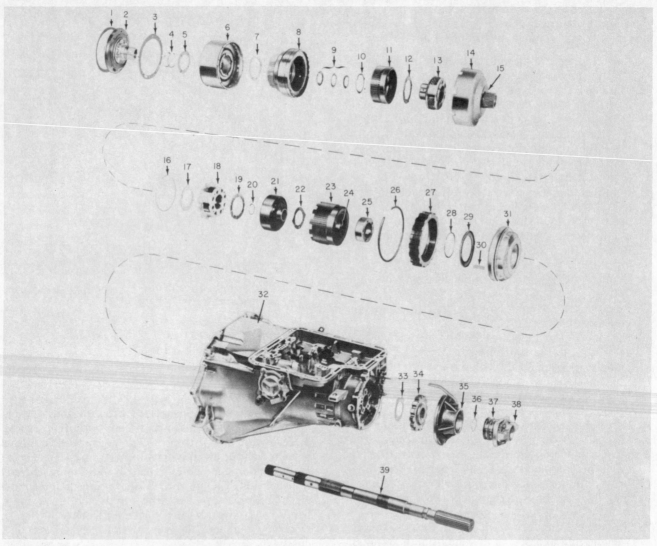

1. FRONT-PUMP SEAL RING
2. FRONT PUMP
3. GASKET
4. SEAL
5. NO. 1 THRUST WASHER (SELECTIVE)
6. REVERSE-HIGH-CLUTCH ASSEMBLY
7. NO. 2 THRUST WASHER
8. FORWARD-CLUTCH AS-SEMBLY
9. NO. 3 THRUST WASHER
10. NO. 4 THRUST WASHER
11. FORWARD-CLUTCH-HUB ASSEMBLY
12. NO. 5 THRUST WASHER
13. FORWARD-PLANET AS-SEMBLY
14. INPUT-SHELL-AND-SUN-GEAR ASSEMBLY
15. NO. 6 THRUST WASHER
16. SNAP RING
17. NO. 7 THRUST WASHER
18. REVERSE-PLANET AS-SEMBLY
19. NO. 8 THRUST WASHER
20. REVERSE RING GEAR
AND HUB-RETAINING RING
21. REVERSE RING GEAR AND HUB
22. NO. 9 THRUST WASHER
23. LOW-REVERSE CLUTCH HUB
24. ONE-WAY CLUTCH
25. ONE-WAY CLUTCH INNER RACE
26. SNAP RING
27. LOW-REVERSE CLUTCH
28. SNAP RING
29. LOW-REVERSE PISTON-
RETURN-SPRING RE-TAINER
30. RETURN SPRING
31. LOW-REVERSE PISTON
32. CASE
33. NO. 10 THRUST WASHER
34. PARKING GEAR
35. GOVERNOR-DISTRIB-UTOR SLEEVE
36. SNAP RING
37. GOVERNOR DISTRIBUTOR
38. GOVERNOR
39. OUTPUT SHAFT

Fig. 13-24 Disassembled view of the transmission drive train. (*Ford Motor Company*)

housing to center the shaft. Then pry the gear train to the rear of the case, as shown, and press the input shaft inward until it bottoms. Set the dial indicator to zero.

4. Pry the gear train forward and note the amount of gear-train end play as registered on the dial indicator. Record the end play. You will need this figure on reassembly.

5. Remove the vacuum unit, rod, and primary throttle valve. Slip the input shaft out of the pump.

6. Remove the front-pump attaching bolts. Pry the gear train forward, as shown in Fig. 13-26, to remove the pump.

7. Loosen the band adjustment screw and remove the two struts. Rotate the band 90° counterclockwise to align the ends with the slot in the case. Then slide the band off the clutch drum.

8. Remove the forward part of the gear train as an assembly, as shown in Fig. 13-27.

9. Remove the large snap ring that holds the reverse

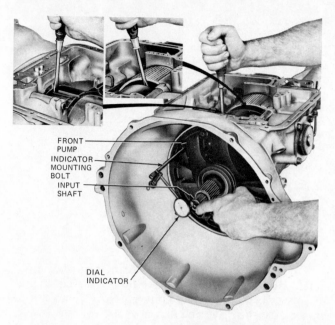

Fig. 13-25 Checking gear-train end play. (*Ford Motor Company*)

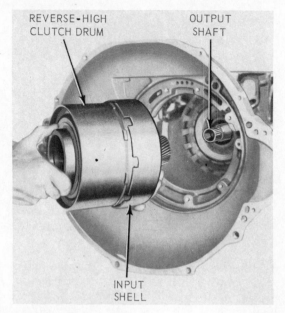

Fig. 13-27 Removing the forward part of the gear train. (*Ford Motor Company*)

planet carrier in the low-and-reverse clutch hub, as shown in Fig. 13-28. Lift the planet carrier from the drum.

10. Remove the snap ring that secures the reverse ring gear and hub to the output shaft. Slide the ring gear and hub off the shaft.
11. Rotate the low-and-reverse clutch hub in a clockwise direction and at the same time withdraw it from the case.
12. Remove the reverse-clutch snap ring from the case, and then remove the clutch disks, plates, and pressure plate from the case.
13. Remove the extension-housing attaching bolts and vent tube from the case. Remove the extension housing and gasket.
14. Slide the output-shaft assembly from the case.
15. Remove the distributor-sleeve attaching bolts and remove the sleeve, parking-pawl gear, and thrust washer. If the thrust washer is staked in place, use

a sharp chisel and cut off the metal from behind the washer. Clean up the rear of the case with air pressure and solvent to remove any metal particles.

16. Compress the reverse-clutch-piston spring with a compressor tool, as shown in Fig. 13-29. Remove the snap ring, tool, and spring retainer.
17. Remove the one-way-clutch inner-race attaching bolts and the inner race. Then remove the reverse-clutch piston from the case, using air pressure, as shown in Fig. 13-30.

✿ 13-22 Downshift and manual linkages The downshift and manual linkages are shown in the case in Fig. 13-31. Follow this illustration if the parts must be removed. If they are removed, remove the shaft seal from the case and install a new seal by using a special tool to drive it into place.

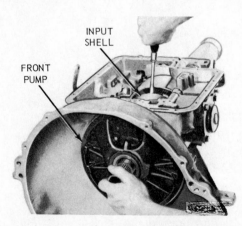

Fig. 13-26 Removing the pump. (*Ford Motor Company*)

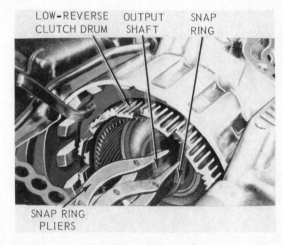

Fig. 13-28 Removing the reverse-ring-gear-hub retaining ring. (*Ford Motor Company*)

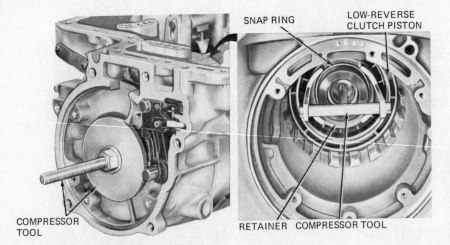

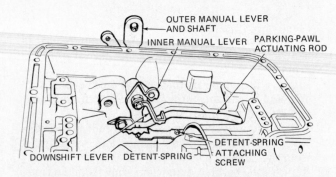

Fig. 13-29 Compressing the low-reverse-clutch springs. (*Ford Motor Company*)

✿ 13-23 Parking-pawl linkage The parking-pawl linkage is shown in Fig. 13-32. Refer to this illustration if the parts must be removed. Drill a ⅛-inch [3.175-mm] hole in the cupped plug and pull it out with a wire hook. Then, after the spring is lifted off the park-plate pin, thread a ¼-20 screw into the shaft and pull it out. You will need a new cupped plug to retain the shaft when you reassemble the parts.

✿ 13-24 Servo apply lever If the servo-apply lever requires replacement, drive the lever shaft from inside the case to drive out the cup plug. Then the shaft can be pulled out. Use a new plug on reinstallation, and coat it with a locking sealant, such as Loctite.

✿ 13-25 Valve body Figure 13-33 shows the upper and lower valve bodies separated. Figure 13-34 shows the upper valve body disassembled. Follow these illustrations to disassemble and reassemble the valve body. To remove the manual valve, you will need a special tool to depress the manual-valve detent spring, as shown in Fig. 13-35. Note the locations of the relief balls and springs, as shown in Fig. 13-36.

✿ 13-26 Pump Figure 13-37 shows the pump in disassembled view. To disassemble it, remove the selec-

tive thrust washer and the two seal rings, the large square-cut seal ring, and the five bolts that secure the pump support to the pump housing. Now, remove the support and the drive and driven gears from the housing.

Fig. 13-31 Downshift and manual linkage. (*Ford Motor Company*)

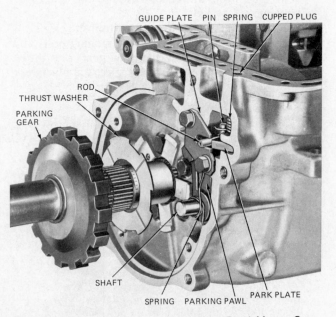

Fig. 13-30 Removing the low-reverse-clutch piston with compressed air. (*Ford Motor Company*)

Fig. 13-32 Parking-pawl mechanism. (*Ford Motor Company*)

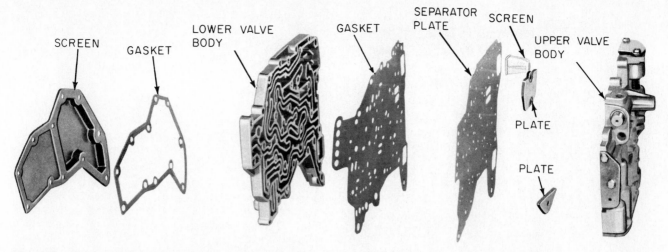

SCREEN   GASKET   LOWER VALVE BODY   GASKET   SEPARATOR PLATE   SCREEN   UPPER VALVE BODY   PLATE   PLATE

Fig. 13-33 Upper and lower valve bodies separated. (*Ford Motor Company*)

When reassembling the pump, make sure that the gears are installed with the identification marks facing the front of the pump housing. Also, make sure that the two locking seal rings on the support are locked. Use new seal rings on reinstallation.

Finally, make sure that the correct thickness of selective thrust washer is used to produce the correct gear-train end play. If the end play is not within the specifications given in the car shop manual, select a thrust washer of a different thickness to give the correct end play.

✿ 13-27 Reverse-and-high clutch Separate the gear train into its major component parts, as shown in Fig. 13-24. Then remove the pressure-plate snap ring, as shown in Fig. 13-38, so that the internal parts can be taken out of the clutch drum, as shown in Fig. 13-39. Next, use the special compressor tool, as shown in Fig. 13-40, to compress the springs so that the snap ring can be removed. Then the tool can be removed along with the spring retainer and springs. Compressed air can then be used to remove the piston, as shown in Fig. 13-41. The piston seals, shown in Fig. 13-39, should be removed from the piston and drum.

If the front bushing is worn, it can be removed by cutting along the bushing seam with a cape chisel. The rear bushing is removed with a removal tool. New bushings can then be pressed into place.

To reassemble the reverse-and-high clutch, dip new seals in transmission fluid and install one on the piston and one in the drum. Put the piston in the drum. Put the piston springs in the piston sockets and place the spring retainer on the springs. Compress the springs with the spring compressor, as shown in Fig. 13-40, and install the snap ring. Make sure the snap ring is inside the four snap-ring guides on the spring retainer.

Install the clutch plates, starting with a steel drive plate and alternating with the composition plates. Soak the composition plates in transmission fluid for 15 minutes before installing them. Then install the pressure plate and secure it with the pressure-plate snap ring.

Check the clearance between the pressure plate and snap ring with a thickness gauge, as shown in Fig. 13-42. If the clearance is not within specifications, replace the snap ring with another snap ring of the correct thickness to produce the proper clearance. Selective snap rings of three thicknesses are available.

✿ 13-28 Forward clutch The forward clutch is shown in disassembled view in Fig. 13-43. It is disassembled in the same way as the reverse-and-high clutch. The snap ring is removed so that the rear pressure plate, disks and forward pressure plate can be removed. After these parts are removed, the disk spring must be compressed by a compressor so that the snap ring and spring can be removed. Then the piston can be removed by air pressure applied to the passage in the cylinder.

On reassembly, clearance must be checked between the snap ring and the pressure plate, just as in the forward clutch. Also, there are snap rings of different thicknesses so that the correct one can be installed to provide the proper clearance.

✿ 13-29 Input shell and sun gear The sun gear is held in place in the input shell by two snap rings and a thrust washer, as shown in Fig. 13-44. Removal of the snap rings permits separation of the parts.

✿ 13-30 One-way clutch The one-way clutch is shown disassembled and separated from the low-and-reverse-clutch hub in Fig. 13-45. This clutch is disassembled by removing the snap rings. On reassembly, first install a snap ring in the forward-snap-ring groove of the clutch hub. Then put the hub forward end down, as shown in Fig. 13-46. Put the forward-clutch bushing into place against the snap ring with the flat side up. Install the spring retainer on top of the bushing. Be sure to install it in the hub so that the springs load the rollers in a counterclockwise direction when you are looking down at the unit (Fig. 13-46). Install a spring and roller into each of the spring-retainer compartments by slightly compressing each spring and positioning the roller between the spring and spring retainer. Install the rear bushing, flat side down, and install the snap ring to secure the assembly.

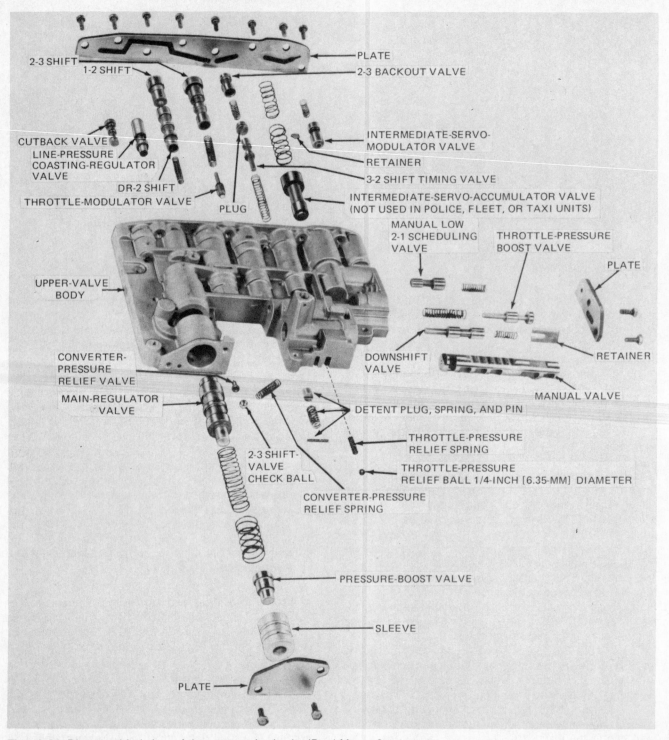

Fig. 13-34 Disassembled view of the upper valve body. (*Ford Motor Company*)

Within the figure, the following labels appear:

2-3 SHIFT

1-2 SHIFT

PLATE

2-3 BACKOUT VALVE

CUTBACK VALVE

LINE-PRESSURE COASTING-REGULATOR VALVE

DR-2 SHIFT

THROTTLE-MODULATOR VALVE

PLUG

INTERMEDIATE-SERVO-MODULATOR VALVE

RETAINER

3-2 SHIFT TIMING VALVE

INTERMEDIATE-SERVO-ACCUMULATOR VALVE (NOT USED IN POLICE, FLEET, OR TAXI UNITS)

UPPER-VALVE BODY

MANUAL LOW 2-1 SCHEDULING VALVE

THROTTLE-PRESSURE BOOST VALVE

PLATE

RETAINER

CONVERTER-PRESSURE RELIEF VALVE

MAIN-REGULATOR VALVE

DOWNSHIFT VALVE

MANUAL VALVE

2-3 SHIFT-VALVE CHECK BALL

DETENT PLUG, SPRING, AND PIN

THROTTLE-PRESSURE RELIEF SPRING

CONVERTER-PRESSURE RELIEF SPRING

THROTTLE-PRESSURE RELIEF BALL 1/4-INCH [6.35-MM] DIAMETER

PRESSURE-BOOST VALVE

SLEEVE

PLATE

● 13-31 Output shaft   The output shaft contains the governor and governor distributor. The governor is attached by bolts, and the distributor is held on the shaft by snap rings. Remove bolts and snap rings to remove these parts.

● 13-32 Converter checks   The converter is a welded assembly and cannot be disassembled for service. If it is defective, it must be replaced as a unit. It can be checked for leakage and for end play and stator-clutch action. To check for leakage, the tools shown in Fig. 13-47 are required. With these tools, the converter can be sealed. Then, 20 psi [138 kPa] of air can be applied to the converter through the special tire valve. Finally, the converter can be placed in a tank of water to check it for air leaks.

The end-play check requires a special tool that is inserted into the pump-drive hub and then lifted so that a dial indicator will measure the amount of end play. The stator-clutch check requires another special tool

214

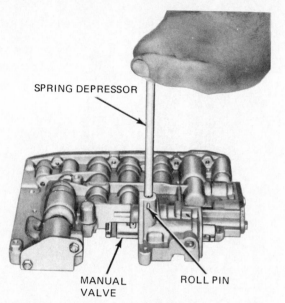

Fig. 13-35 Using a special spring depressor to remove the manual valve. (*Ford Motor Company*)

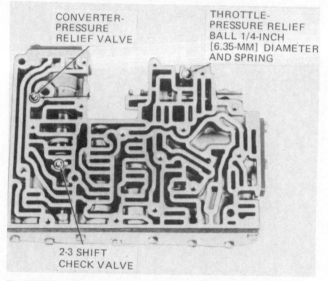

Fig. 13-36 Location of check balls in the valve body. (*Ford Motor Company*)

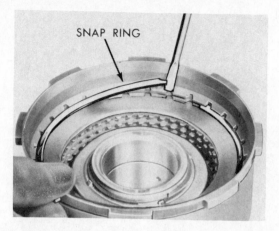

Fig. 13-38 Removing the reverse-high-clutch pressure-plate snap ring. (*Ford Motor Company*)

that permits application of torque to determine whether or not the clutch can hold against reverse rotation.

Any defect found in the converter requires converter replacement.

✿ 13-33 Transmission reassembly On reassembly, lubricate the parts with transmission fluid. Use petroleum jelly to hold washers in place during reassembly. Install the case in a repair fixture and proceed as follows:

1. Position the low-and-reverse-clutch piston so that the check ball is at 6 o'clock—at the bottom of the case—and tap it into place with a clean rubber hammer.
2. Hold the one-way-clutch inner race in position and install and torque the attaching bolts to specifications.
3. Install the low-and-reverse-clutch return springs into the pocket in the piston. Press the springs firmly into place so that they will not fall out.
4. Put the spring retainer over the springs and put the snap ring above it. Use a spring compressor to compress the springs and install the snap ring.
5. Put the case on the bench, front end down.

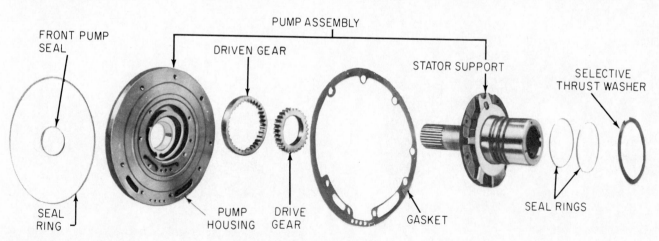

Fig. 13-37 Disassembled view of the pump. (*Ford Motor Company*)

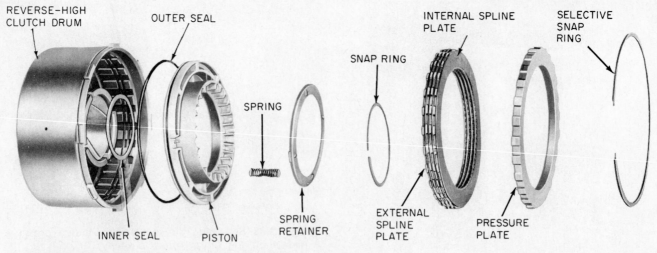

Fig. 13-39 Disassembled view of the reverse-high clutch. (*Ford Motor Company*)

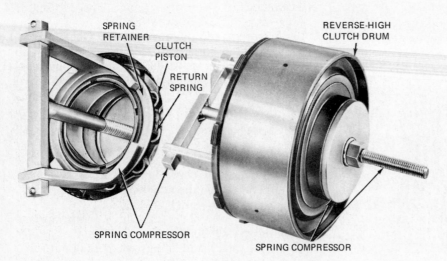

Fig. 13-40 Using a clutch-spring compressor to remove the clutch-piston snap ring. (*Ford Motor Company*)

Fig. 13-41 Removing the clutch piston with air pressure. (*Ford Motor Company*)

Fig. 13-42 Checking clearance between the snap ring and pressure plate with a thickness gauge. (*Ford Motor Company*)

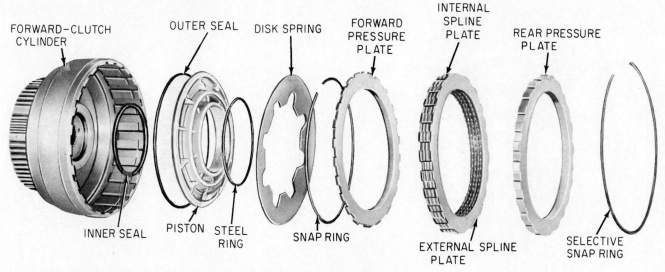

FORWARD-CLUTCH CYLINDER — OUTER SEAL — DISK SPRING — FORWARD PRESSURE PLATE — INTERNAL SPLINE PLATE — REAR PRESSURE PLATE

INNER SEAL — PISTON — STEEL RING — SNAP RING — EXTERNAL SPLINE PLATE — SELECTIVE SNAP RING

Fig. 13-43 Disassembled view of the forward clutch. (*Ford Motor Company*)

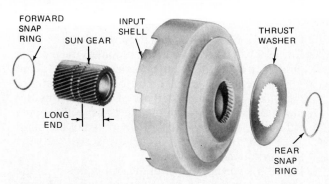

FORWARD SNAP RING — SUN GEAR — INPUT SHELL — THRUST WASHER — LONG END — REAR SNAP RING

Fig. 13-44 Sun gear separated from the input shell. (*Ford Motor Company*)

6. Put the parking-gear thrust washer and gear on the case. Do not stake the washer.
7. Position the oil distributor and tubes on the rear of the case. Then torque the attaching bolts to specifications.
8. Install the output shaft and governor.
9. Put a new gasket on the rear of the case and install the extension housing, torquing the bolts to specifications.
10. Put the case back in the repair fixture.
11. Align the low-and-reverse-clutch hub and one-way clutch with the inner race at the rear of the case. Rotate the low-and-reverse-clutch hub clockwise while applying pressure to seat it on the inner race.
12. Install the low-and-reverse-clutch plates, starting with a steel plate and following alternately with composition and steel plates. Retain them with petroleum jelly. New composition plates should be soaked in transmission fluid for 15 minutes before installation. Test the operation of the clutch with compressed air.

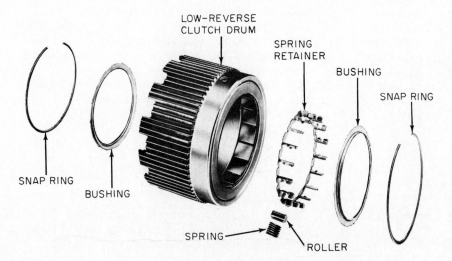

LOW-REVERSE CLUTCH DRUM — SPRING RETAINER — BUSHING — SNAP RING

SNAP RING — BUSHING — SPRING — ROLLER

Fig. 13-45 Disassembled view of the one-way clutch. (*Ford Motor Company*)

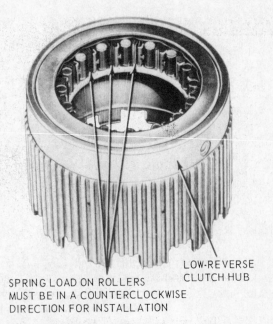

SPRING LOAD ON ROLLERS
MUST BE IN A COUNTERCLOCKWISE
DIRECTION FOR INSTALLATION

LOW-REVERSE
CLUTCH HUB

Fig. 13-46 Correct installed positions of the rollers and springs in the one-way clutch. (*Ford Motor Company*)

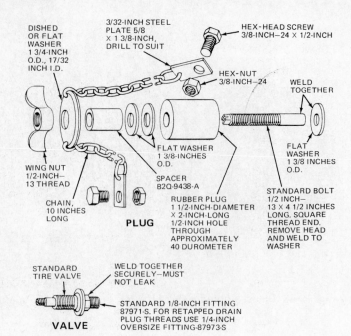

DISHED OR FLAT WASHER 1 3/4-INCH O.D., 17/32 INCH I.D.

3/32-INCH STEEL PLATE 5/8 X 1 3/8-INCH, DRILL TO SUIT

HEX-HEAD SCREW 3/8-INCH—24 X 1/2-INCH

HEX-NUT 3/8-INCH—24

WELD TOGETHER

WING NUT 1/2-INCH—13 THREAD

CHAIN, 10 INCHES LONG

FLAT WASHER 1 3/8-INCHES O.D.

SPACER B2Q-9438-A

RUBBER PLUG 1 1/2-INCH-DIAMETER X 2-INCH-LONG 1/2-INCH HOLE THROUGH APPROXIMATELY 40 DUROMETER

PLUG

FLAT WASHER 1 3/8 INCHES O.D.

STANDARD BOLT 1/2 INCH—13 X 4 1/2 INCHES LONG. SQUARE THREAD END. REMOVE HEAD AND WELD TO WASHER

STANDARD TIRE VALVE

WELD TOGETHER SECURELY—MUST NOT LEAK

STANDARD 1/8-INCH FITTING 87971-S. FOR RETAPPED DRAIN PLUG THREADS USE 1/4-INCH OVERSIZE FITTING-87973-S

VALVE

Fig. 13-47 Construction of a torque-converter leak detector. (*Ford Motor Company*)

13. Install the reverse-planet-ring-gear thrust washer and the ring-gear-and-hub assembly. Insert the snap ring in the groove in the output shaft.

14. Assemble the front and rear thrust washers onto the reverse-planet assembly. Retain them with petroleum jelly. Insert the assembly into the ring gear and secure with the snap ring.

15. Set the reverse-and-high clutch on the bench, front end facing down. Install the thrust washer on the rear end, retaining it with petroleum jelly. Insert the splined end of the forward clutch into the open end of the reverse-and-high clutch so that the splines engage the direct-clutch plates.

16. On the front end of the forward planet ring gear and hub, install the thrust washer and retain it with petroleum jelly. Insert the assembly into the ring gear. Install the input-shell-and-sun-gear assembly.

17. Install the reverse-and-high clutch, forward clutch, forward-planet assembly, and input shell and sun gear as an assembly into the transmission case.

18. Insert the intermediate band into the case around the direct-clutch drum. Install the struts and tighten the band-adjusting screw enough to hold the band.

19. Place a selective-thickness washer of the proper thickness as determined by the end-play check during disassembly (Fig. 13-25) on the shoulder of the stator support. This is the selective thrust washer

referred to in ✿ 13-26 and shown in Fig. 13-37. Retain the washer with petroleum jelly.

20. Lay a new gasket on the rear mounting face of the pump and bring the pump into place. Be careful not to damage the large seal on the outside diameter of the pump housing. Install six of the seven pump-mounting screws and torque them to specifications.

21. Adjust the intermediate band, and install the input shaft with the long splined end inserted into the forward-clutch assembly.

22. Install the special tool with the dial indicator in the seventh pump-bolt hole and recheck the end play. If the end play is correct, remove the tool and install the seventh bolt. If it is incorrect, remove the pump and install a selective thrust washer of the correct thickness.

23. Install the control-valve body in the case, making sure the levers engage the valves properly. Install the primary throttle valve, rod, and vacuum unit.

24. Install a new oil-pan gasket and pan.

25. Install the converter and install the transmission in the car. Attach and adjust linkages.

26. Add transmission fluid to bring the fluid up to the correct level.

27. Start the engine. Check the operation of the transmission in all selector-lever positions.

Chapter 13 review questions

Select the *one* correct, best, or most probable answer to each question. Then check your answers against the correct answers given at the end of the book.

1. The two linkage adjustments are to the:
   a. throttle linkage and intermediate band linkage
   b. throttle linkage and selector-lever linkage

c. selector-lever and band linkage

d. band linkage and clutch linkage

2. The C6 band can be adjusted:

a. only after removing the valve body

b. on the car

c. only after removing the transmission from the car

d. only after disassembling the transmission

3. The three diagnostic checks that can be made on the transmission include tests of the shift points, air checks of the clutches and servo, and tests of:

a. hydraulic control pressures

b. stall pressure

c. band pressure

d. output pressure

4. The Ford transmission tester contains three elements—a pressure gauge and a:

a. vacuum gauge and an ammeter

b. tachometer and rpm indicator

c. vacuum gauge and a voltmeter

d. tachometer and a vacuum gauge

5. If, on the stall test, the engine goes well above the specified rpm, then:

a. the band or clutches are slipping.

b. the engine needs a tune-up.

c. the converter one-way clutch is slipping.

d. none of the above.

6. If the engine does not reach the specified rpm on the stall test, then:

a. the clutches or band is slipping.

b. the engine needs a tune-up.

c. line pressure is too low.

d. line pressure is too high.

7. To make the air-pressure checks of the clutches and servo, you must:

a. remove the transmission from the car

b. disassemble the transmission to get to the clutches

c. remove the oil pan and valve body

d. operate the engine at about 1000 rpm

8. Mechanic A says that if the transmission fluid is contaminated, a new torque converter will be required. Mechanic B says the torque converter can be flushed out and reused. Who is right?

a. A only

b. B only

c. both A and B

d. neither A nor B

9. Mechanic A says that for normal service, the band or bands should be adjusted periodically—every 6 or 12 months. Mechanic B says that for normal service the bands do not need to be adjusted. Who is right?

a. A only

b. B only

c. both A and B

d. neither A nor B

10. Which of the following statements is correct?

a. the C6 has only one band.

b. the C4 has two bands.

c. the FMX has two bands.

d. all of the above.

11. For normal service, the recommendation on fluid changes is:

a. change every 12,000 miles [19,312 km].

b. change every 24,000 miles [38,624 km].

c. do not change fluid.

d. change every engine-oil change.

12. The band adjustment can be made:

a. on the car

b. only after the transmission is off the car

c. only after the valve body is removed

d. only when the transmission is disassembled

13. When using the Ford transmission tester to check shift points in the shop, you need:

a. a dynamometer

b. a source of vacuum

c. to disconnect the drive shaft

d. a source of pressure

14. When using the Ford transmission tester in the shop to check shift points, you simulate a high engine load or wide-open throttle by:

a. opening the carburetor throttle

b. reducing the vacuum to 0 to 2 inches [0 to 51 mm]

c. opening the kickdown valve

d. operating the engine at 2000 rpm

15. To remove a clutch piston, you must first compress the clutch springs so you can:

a. remove the clutch plates

b. loosen the screws

c. unfasten the clutch

d. remove the snap ring

16. Gear-train end play is adjusted by:

a. changing a snap ring

b. installing different pinion carriers

c. changing the selective thrust washer

d. installing different clutch plates

17. If the transmission slips on the stall-speed test, then:

a. release the throttle instantly.

b. open the throttle wide to see how high the speed will go.

c. shift to D and hold the throttle half open.

d. shift to N and race the engine to cool the transmission.

18. The control pressure checks are made with the:

a. engine running at about 2500 rpm

b. engine idling

c. engine not running

d. transmission removed from the car

19. The purpose of applying vacuum to the transmission is:

a. to simulate engine intake-manifold vacuum

b. to check shift points

c. to check car speeds at which shifts occur

d. all of the above

20. Air-pressure checks are made to check:

a. the operation of the hydraulic system

b. the condition of the clutches and servos

c. the shift points

d. none of the above

FORD AUTOMATIC OVERDRIVE
TRANSMISSION SERVICE

After studying this chapter, and with proper instruction and equipment, you should be able to:

1. Explain how to use the diagnostic charts to find the causes of troubles.
2. Describe the procedures for testing the transmission, including road and stall tests, and making line-pressure and air-pressure tests.
3. Discuss the procedures of making linkage adjustments.
4. List and describe the services that can be performed with the transmission in the car.
5. Explain how to remove and install the transmission.
6. Describe transmission disassembly and reassembly, including inspection and cleaning of parts.
7. Perform the above transmission services.

14-1 Introduction to Ford automatic overdrive transmission service The first part of this chapter describes the maintenance and trouble-diagnosis of the Ford automatic overdrive transmission. The latter part of the chapter describes the overhaul of this transmission. The previous chapter (✿ 13-18) covers the removal and installation of Ford automatic transmissions.

✿ 14-2 Diagnostic procedures Refer to ✿ 9-6 and 9-7 for the fundamentals of trouble diagnosis. The diagnosis sequence recommended by Ford on the automatic overdrive transmission is the same, in general, as that used on any automatic transmission:

1. Check fluid level and condition (✿ 13-3).
2. Check throttle-valve linkage for free movement and adjustment (✿ 14-7).
3. Check and adjust manual linkage (✿ 14-5).
4. Road- or dynamometer-test.
5. Make engine-idle throttle-valve adjustments.
6. Make stall test.
7. Make line-pressure test.
8. Make air-pressure tests.

✿ 14-3 Checking fluid level and condition The procedure for checking fluid level and condition has been described in ✿ 9-3 and 13-3. The dipstick for this transmission (Fig. 14-1) shows the full-level cold and

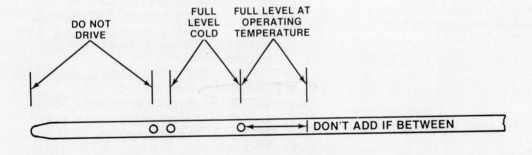

Fig. **14-1** Dipstick markings on the dipstick for the Ford automatic overdrive transmission. (*Ford Motor Company*)

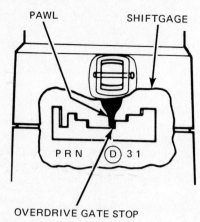

PAWL   SHIFTGAGE

OVERDRIVE GATE STOP

Fig. 14-2 Transmission selector-lever gate stops. (*Ford Motor Company*)

the full-level hot markings. Do not add fluid if the fluid level is within the arrow. Note that the type of fluid required is stamped on the dipstick.

NOTE: If the fluid level is low, check for leaks (✿ 13-3).

✿ **14-4 Throttle-valve-linkage checks** If the shifts are soft or mushy, suspect incorrect throttle-valve-linkage adjustment. Do not drive the car with a heavy throttle, because this could damage the transmission. Instead, check the fluid level (✿ 14-3), manual-linkage adjustment (✿ 14-5), throttle-valve-linkage adjustment (✿ 14-7), and TV pressure (✿ 14-8). After making necessary adjustments, road-test (✿ 14-6) and stall-test (✿ 14-9) the car.

Additional information on the condition of the transmission can be obtained from line-pressure and air-pressure tests (✿ 14-10 and 14-11).

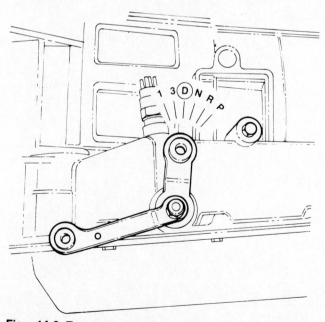

Fig. 14-3 Transmission-lever positions. (*Ford Motor Company*)

RANGE	CHECK FOR	CONDITION (OK OR NOT OK)
1	Engagement	
	No 1-2 upshift	
	Engine braking in 1st and 2d gear	
	Slipping	
3	Engagement and shift feel	
	1-2-3 automatic upshifts at specified speeds	
	Fourth gear lockout	
	Torque demand 3-2 only	
	WOT kickdown to second or first	
	Smooth coast-down	
	Slipping	
Ⓓ	Engagement and shift feel	
	1-2-3-4 automatic upshifts at specified speeds	
	Torque demand to third gear from overdrive	
	WOT kickdown to second or first	
	No fourth gear at wide open throttle	
	Smooth coast-down	
	Slipping	
R	Engagement	
	Backup without slip	
	Slipping	

Fig. 14-4 Road-test check chart. (*Ford Motor Company*)

✿ **14-5 Manual-linkage adjustment** If the manual valve is not centered in each position of the selector lever, proper pressure will not be delivered to the various operating members of the transmission. The result could be serious damage. Check the adjustment by lifting the selector lever toward you to ungate it (Fig. 14-2). Move the lever through the range from P to 1. Note whether or not the detents are centered with the

SYMPTOMS	CAUSE	CORRECTION
1. Early and/or soft upshift 2. Slip-bump feel on light throttle 3-4 or 4-3 3. No kickdown or wrong kickdown speeds	TV control linkage is set too short.	1. Adjust carburetor linkage. 2. If adjustment screw has insufficient travel, adjust transmission linkage.
1. Harsh idle engagement with engine hot (at curb idle) 2. Shift clunk in back-out shifts 3. Harsh 4-3 coasting shift	TV control linkage is set too long.	1. Adjust carburetor linkage. 2. If adjusting screw has insufficient travel, or condition is not corrected, adjust transmission linkage.
1. Very late and harsh upshifts, especially at moderate acceleration 2. Harsh idle engagement	Interference prevents TV rod return.	Correct interference.
	Binding grommets prevent TV return.	1. Check for bends or twists that cause misalignment. 2. Replace grommets if damaged. 3. Adjust TV linkage.
	TV rod disconnected (TV is at WOT).	1. Check for grommet damage. 2. Connect rod. 3. Adjust TV linkage.
	Loose clamping bolt on TV rod trunnion (TV is at WOT).	Adjust TV linkage.
	Linkage-lever return spring is broken or disconnected (TV is at WOT).	Repair or replace spring.

Fig. 14-5 Shift trouble-diagnosis chart. (*Ford Motor Company*)

markings on the selector. Shift to D and let the lever drop into position. Try to move the lever toward the 3 position without lifting (ungating) it. If there is free movement in the ⒹD position, or if the pawl is up on position 3 without ungating, adjustment is required.

Careful: Do not drive the car if the adjustment is not correct.

Adjustment is made as follows:

1. Loosen the linkage adjustment screw (see shop manual on the car you are servicing to find its location). Move the shift selector lever firmly against the ⒹD gate stop and hang a weight on the lever to hold it in place.
2. Locate the ⒹD detent in the transmission (Fig. 14-3). Set the transmission lever in this position. Tighten

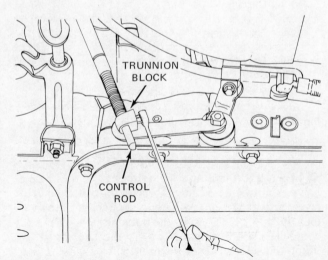

Fig. 14-6 Loosening the trunnion block on the TV control rod. (*Ford Motor Company*)

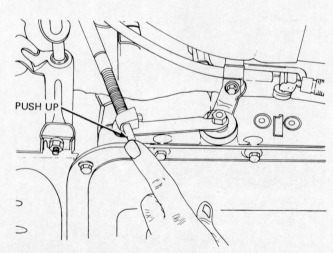

Fig. 14-7 Positioning the control rod. (*Ford Motor Company*)

the linkage-adjustment screw. The ⓓ detent and gate stop should now be synchronized.

3. Recheck the adjustment before driving the car.

✸ 14-6 Road test Road-testing the transmission is covered in ✸ 13-9. The purpose of the test is to check the shift points at various throttle openings and also check the quality of the shifts. The test can also be made in the shop with a vacuum pump (✸ 13-9). Figure 14-4 shows the conditions to check for. See also Fig. 7-22, which shows the holding and released members in each drive range. Figure 14-5 is a chart showing shift troubles and their possible causes and corrections.

✸ 14-7 Throttle-valve-linkage adjustment The engine hot-curb idle must be set correctly (see the decal in the engine compartment) before the TV linkage is adjusted. To make the TV linkage adjustment, change the length of the TV control rod at the transmission.

1. Set the screw at the TV linkage lever at its midpoint. Be sure the throttle is against the idle stop. Put the shift-selector lever in neutral and set the parking brake.
2. Loosen the bolt on the control-rod (Fig. 14-6) sliding trunnion block.
3. Make sure the trunnion slides freely on the rod. Push up on the end of the rod (Fig. 14-7) so the carburetor linkage lever is held firmly against the throttle lever.
4. Release the rod to make sure it stays in position. If it does not, the return spring at the carburetor is disconnected or damaged.
5. Push the transmission TV lever up against its internal stop and tighten the bolt on the trunnion block (Fig. 14-8). Verify that the throttle lever is at the idle stop.

✸ 14-8 Throttle-valve-pressure adjustment The throttle-valve linkage can also be set with a pressure gauge.

1. Connect a pressure gauge of the proper capacity to the TV limit pressure tap on the right side of the case (Fig. 14-9). The hose should be long enough that you can see the gauge from under the hood.
2. The throttle lever should be at the idle stop. Idle the engine in neutral and set the parking brake.
3. Put a $1/_{16}$-inch [1.6-mm] drill or gauge between the linkage-lever adjustment screw and the throttle lever (Fig. 14-10). The pressure should be 5 psi [35 kPa] or less. If it is high, the linkage is too long. Back out the screw as necessary to reduce the pressure.
4. Recheck the clearance with a $5/_{16}$-inch [8-mm] gauge. Throttle-valve pressure should now rise to at least 22 psi [152 kPa]. If it is too low, the control linkage is too short. Turn the screw in to bring the pressure up to specifications.

NOTE: If the pressure cannot be set at the linkage-lever-adjustment screw, adjust the length of the control rod (✸ 14-7).

✸ 14-9 Stall test Stall testing of the C6 and other Ford automatic transmissions is covered in ✸ 13-10. The test procedure for the overdrive automatic is sim-

ilar. However, there are some differences when testing the Ford automatic overdrive transmission. The stall test checks for clutch or band slippage, for engine performance, and for torque-converter operation. The test should be done only after the fluid level and linkages have been checked and corrected as necessary. Engine-coolant level should be correct and the engine and transmission fluid should be at normal operating temperature.

Each vehicle may have a different stall speed due to the engine and other equipment installed. Look up the stall speed for the vehicle you are checking in the Ford shop manual. Then mark this rpm on the dial of the tachometer with a grease pencil (Fig. 14-11).

1. Connect the tachometer to the engine and position it so you can read it from the driver's seat (Fig. 14-11).
2. In each driving range, push the accelerator to the floor and hold it just long enough to let the engine reach maximum rpm.

Careful: If the needle goes past the grease-pencil mark, the transmission is slipping. Release the throttle at once! See the other "Careful" and "Caution" notes in ✸ 13-10.

3. Record the engine rpm in each range.
4. Run the engine at fast idle in neutral between each pair of tests to cool the fluid.
5. Figure 14-12 shows what the results mean and what should be done if the stall speed is high or low.

✸ 14-10 Control-pressure test To make the control-pressure test, connect a transmission pressure gauge to the line-pressure port (Fig. 14-13). Check the specifications for the vehicle you are servicing and write them in a chart such as the one shown in Fig. 14-14. The pressures are checked at idle (zero throttle-valve pressure) and at wide-open throttle (full throttle-valve pressure). The reverse pressure should be higher than any other pressure.

1. The pressure gauge is connected to the line-pressure port on the left side of the case, just above the control levers (Fig. 14-13).
2. Run the engine to bring it up to normal operating temperature. Idle pressure must be read with the throttle off the fast-idle cam.
3. Throttle-valve pressure must be properly adjusted (✸ 14-8).
4. Apply service and parking brakes firmly and shift through all ranges. Record the line pressures in Fig. 14-14 and compare them with the specifications.

Careful: Wide-open-throttle (WOT) readings are made at full stall. Run the engine at fast idle in neutral between tests to cool the fluid. See also the "Careful" and "Caution" notes in ✸ 13-10.

5. Check the chart in Fig. 14-15 to determine possible causes if the pressure readings are not within the manufacturer's specifications.
6. Additional information on possible causes of the trouble can be learned by the air-pressure tests (✸ 14-11).

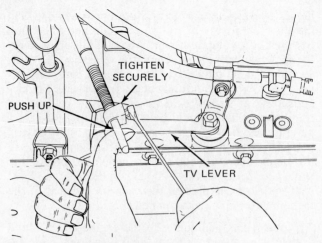

Fig. 14-8 Holding the lever and tightening the bolt on the trunnion block. (*Ford Motor Company*)

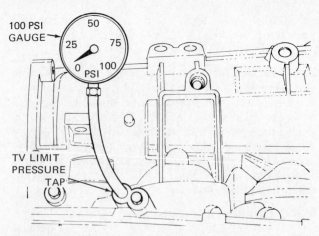

Fig. 14-9 Throttle-valve limit pressure gauge check. (*Ford Motor Company*)

✿ **14-11 Air-pressure tests** Air-pressure tests check the operation of the clutches and band servos. Figure 14-16 shows the transmission case with the oil pan and valve body removed. A nozzle such as the one shown to the upper right should be used. A rubber tip on the end of the nozzle will help make a tight fit between the nozzle and the port being checked. Or use clean, lint-free cloth to form a tight seal between the nozzle and port. The air pressure should be regulated to about 25 psi [172 kPa]. The air should be filtered and dry.

1. Clutches should engage with a dull thud when air pressure is applied to their pressure ports. A hissing sound and failure to engage indicates a leak.

2. Servos   The low-reverse servo application should be heard as the piston bottoms when air pressure is applied. When the air pressure is removed, you should feel the piston move back against the cover.

For the overdrive-servo test, first pressurize the apply side. Hold the pressure and pressurize the release side. You should hear the piston stroke to apply the band and also hear it bottom in the cover when release pressure is applied.

A hissing sound indicates a leak.

3. Air blown into the line-to-governor port should cause a whistle. If there is no noise, the governor may be stuck.

INSERT DRILL BIT HERE TO CHECK TV PRESSURE

Fig. 14-10 Positioning the linkage lever for a 5 psi (35 kPa) check. (*Ford Motor Company*)

GREASE PENCIL MARK

TACHOMETER

Fig. 14-11 Checking the tachometer while performing a stall test. (*Ford Motor Company*)

✻ 14-12 Band adjustments Periodic band adjustments are not required for normal driving. However, for severe service, such as in taxicabs, police cars, trailer hauling, and door-to-door delivery operation, the bands will require periodic checking and adjusting. The manufacturer's service manual contains the time or distance specifications for the vehicle.

✻ 14-13 Trouble diagnosis The trouble-diagnosis chart in ✻ 13-14 applies, in general, to the automatic overdrive transmission. Figure 13-20, which diagrams a checking procedure, may also be helpful in locating trouble causes.

✻ 14-14 On-the-car-repairs The transmission valve body, governor and extension-housing assembly, and servos can be removed and installed with the transmission in the vehicle. The procedures for these services are covered in ✻ 13-15 to 13-17.

✻ 14-15 Removing and installing the transmission The procedure for removing and installing the transmission is the same as for other Ford automatic transmissions (✻ 13-18). After the transmission is removed, any remaining oil should be drained from it. The outside should be cleaned and then the transmission installed in a holding fixture. Remove the pan to prevent any excess debris from falling into the transmission. Then turn the transmission upside down (Fig. 14-17). Now transmission disassembly can begin.

✻ 14-16 Transmission-service precautions Cleanliness is very important in transmission service work. Transmission-service precautions are discussed in detail in ✻ 9-10 and 13-19.

✻ 14-17 Cleaning and inspecting transmission parts During and after disassembly, clean and inspect all parts for wear or other damage. Refer to ✻ 9-11 for details. If the transmission shows signs of internal damage and there is evidence that foreign material has been circulating with the transmission fluid, then all parts, including the oil cooler and lines, must be flushed out. A new filter will be required. Also, a new torque converter must be installed. A contaminated converter cannot be flushed out satisfactorily in the field.

✻ 14-18 Transmission disassembly Refer to Figs. 7-1 to 7-20, which show the locations of all the operating components in the transmission. These illustrations will be your guides as you service the transmission. With the transmission on a holding fixture (Fig. 14-17), remove the following:

1. Torque converter.
2. Pan and pan gasket (14 bolts).
3. Transmission filter (3 bolts).
4. Transmission-filter grommet and gasket. Discard gasket.
5. Detent-spring bolt and spring.
6. Twenty-four valve-body bolts. Note differences in length. Four front, one center, and three rear bolts are shorter.
7. Valve body.

SELECTOR POSITIONS	STALL SPEED HIGH	STALL SPEED LOW
Ⓓ and 3	Slipping planetary one-way clutch	
Ⓓ, 3, and 1	Slipping forward clutch	(1) Check engine for tune-up. If OK, (2) remove torque converter and bench-test for a slipping stator clutch.
All driving ranges	1. Check TV adjustment. 2. Perform control-pressure test.	
R only	Reverse clutch or low-reverse band or slipping	

Fig. 14-12 Stall-test results chart. (*Ford Motor Company*)

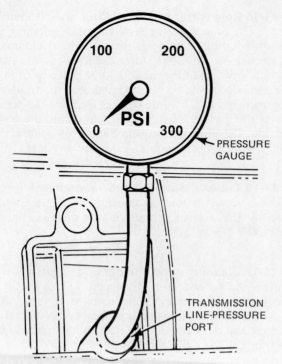

Fig. 14-13 Pressure gauge connected to make the control-pressure test. (*Ford Motor Company*)

Continue disassembly as follows:

1. Use a screwdriver to remove the snap ring from the overdrive servo cover (Fig. 14-18).
2. Position the servo-piston remover over the overdrive-servo-release passage. Figure 14-16 shows

LINE PRESSURE (psi)

Idle (closed throttle)

POSITION	SPECIFIED	ACTUAL
P		
R		
N		
D		
3		
1		

Wide-open Throttle (full TV)

POSITION	SPECIFIED	ACTUAL
D		
3		
1		
R		

Fig. 14-14 Chart to record specifications and results of control- (line-) pressure test. (*Ford Motor Company*)

CONTROL-PRESSURE CONDITION	POSSIBLE CAUSE(S)
Low in P	Valve body, low-reverse servo
Low in R	Reverse clutch, low-reverse servo
Low in N	Valve body
Low in ⒟	Forward clutch, overdrive servo, valve body
Low in 3	Forward clutch
Low in 1	Forward clutch, low-reverse servo
Low at idle in all ranges	Low fluid level, restricted inlet screen, loose valve-body bolts, pump leakage, case leakage, valve body, excessively low engine idle, fluid too hot.
High at idle in all ranges	TV linkage, valve body
Okay at idle but low at WOT	Internal leakage, pump leakage, restricted inlet screen, TV linkage, valve body (TV or TV limit valve sticking)

Fig. 14-15 Chart showing possible causes of abnormal conditions indicated by the control-pressure test results. (*Ford Motor Company*)

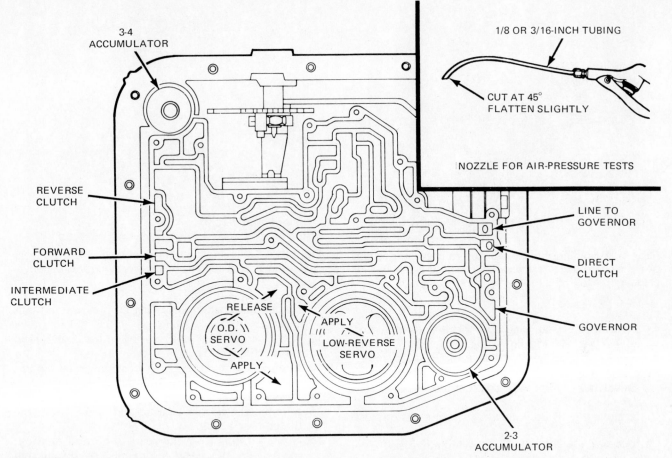

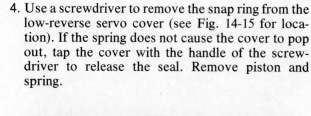

Fig. 14-16 Points at which to apply air pressure to check the clutch and servo actions. (*Ford Motor Company*)

the location of this passage. Use air pressure to remove the servo cover and piston. Figure 14-19 shows the servo-piston remover in place with air pressure being applied.

3. Remove the overdrive-servo return spring after the cover and piston are out.

4. Use a screwdriver to remove the snap ring from the low-reverse servo cover (see Fig. 14-15 for location). If the spring does not cause the cover to pop out, tap the cover with the handle of the screwdriver to release the seal. Remove piston and spring.

Fig. 14-17 Ford automatic overdrive transmission mounted in holding fixture in readiness for service. (*Ford Motor Company*)

Fig. 14-18 Using a screwdriver to remove the overdrive-servo-cover snap ring. (*Ford Motor Company*)

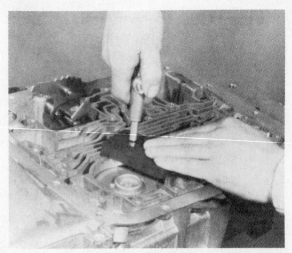

Fig. 14-19 Holding the servo-piston remover in position over the overdrive-servo-release passage and applying air pressure through the tool. (*Ford Motor Company*)

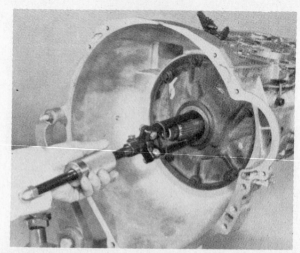

Fig. 14-21 Removing the front-pump seal with the special seal remover and slide hammer. (*Ford Motor Company*)

5. Remove the snap ring and the 3-4 accumulator cover (Fig. 14-15 shows the location). Use the servo-piston remover and air pressure to force the piston from the transmission case. Figure 14-20 shows the piston out. Some transmissions may have a spring in the bore.
6. Remove the 2-3 accumulator snap ring, piston, and spring.
7. Now, working from the front end, remove the front-pump seal (Fig. 14-21) and discard it.
8. Rotate the transmission in the holding fixture so that the front end is up. Then remove the bolts from the front pump. Now use slide hammers to remove the pump.
9. Remove the pump gasket and the direct-drive shaft.
10. Turn the transmission to horizontal. Grasp the turbine shaft and remove as an assembly the intermediate clutch, overdrive band, forward-and-reverse-clutch assembly, and forward-and-reverse sun-gear shell (Fig. 14-22).

11. Remove the center snap ring by pressing the tabs together.
12. Remove the "anticlunk" spring with a small screwdriver (Fig. 14-23).
13. Remove the planetary gearset with center support and take off the direct-clutch hub (Fig. 14-24).
14. Remove the low-reverse band.
15. Remove the needle bearing and spacer from the direct-clutch drum.
16. Now, working from the rear of the transmission, remove the bolts and the extension housing and gasket. Remove the extension-housing seal.
17. Remove the governor snap ring, and the governor. Remove the drive ball from the shaft.
18. Now, from the front, remove the output shaft, ring gear, and direct-clutch assembly (Fig. 14-25). Remove the output-shaft needle bearing.

Fig. 14-22 Pulling out the assembly by grasping the turbine shaft. The parts that come out include the intermediate-clutch pack, overdrive band, forward-and-reverse-clutch assembly, and forward-and-reverse sun-gear-shell assembly. (*Ford Motor Company*)

Fig. 14-20 Removing the 3-4 accumulator piston from the case. (*Ford Motor Company*)

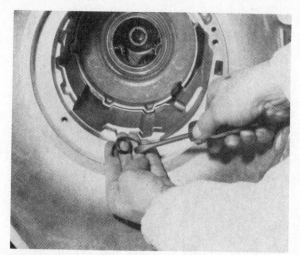

Fig. 14-23 Removing the center-support anticlunk spring with a small screwdriver. (*Ford Motor Company*)

Fig. 14-24 Top, removing the center-support low-reverse planetary gearset. Bottom, taking the direct-clutch hub from the assembly. (*Ford Motor Company*)

Careful: Do not damage the seal-ring grooves and mating surface in the case.

19. If the bushing in the rear of the case needs replacement, remove the old seal with a slide hammer. Install a new bushing with the replacer tool and slide hammer.
20. To service the internal and external linkage, first study the locations of all parts so you can restore them to their correct positions. About the only time you need to take these parts off would be if something were broken or if you had to replace the throttle-lever seal.
21. To remove the outer throttle lever, hold it steady and remove the retaining nut. Then remove the seal from the counterbore in the outer manual lever.
22. Remove the manual-lever locating pin (Fig. 14-26). Now hold the outer manual lever and loosen and remove the attaching nut from the outer-manual-lever shaft. Remove the outer manual lever.
23. Remove the inner throttle lever and spring (Fig. 14-27).
24. Remove the inner detent manual lever and park-pawl actuating rod. Disconnect the actuating rod from the inner detent manual lever.
25. Remove the park-pawl shaft and remove pawl and spring.
26. To replace the manual-lever seal, remove it with a slide hammer. Install a new seal with the proper seal replacer.

✿ 14-19 Internal and external linkages To replace the linkages, connect the park-pawl actuating rod to the inner detent manual lever and install the assembly in the park-rod-guide cup. Install park pawl, spring, and shaft. The location of the park pawl must be in the same position as before removal.

Install the inner throttle lever, nut, and spring (Fig. 14-27). Install the outer manual lever on the inner throttle-lever shaft, making sure it is in the same position as on removal. Hold the outer manual lever and tighten the nut on the manual-lever shaft. Install the manual-lever locating pin. Figure 14-26 shows its removal. Use a soft hammer to tap it in if necessary.

Install a new throttle-lever seal in the outer manual-lever counterbore. Identification numbers on the seal must face out. Then install the retaining nut, lock washer, and outer throttle lever on the inner-lever shaft.

✿ 14-20 Front-pump and intermediate-clutch-piston service Figure 14-28 is a disassembled view of the pump. To aid in servicing this assembly, cut a 2.5-inch [63.5-mm] hole in the work bench, 6 inches [152 mm] from the edge. Proceed as follows.

1. The intermediate-clutch piston is secured to the inside of the pump with a retainer (Fig. 14-29). Note, in Fig. 14-29, that the lower end of the stator support goes through the hole in the workbench. Open the tabs with a screwdriver to release the retainer and return spring. Note the position of the piston bleed hole at the top of the case.

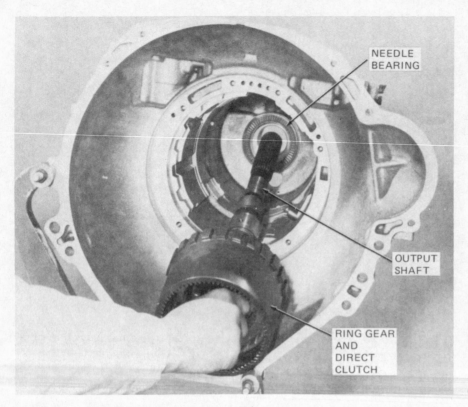

Fig. 14-25 Removing the output shaft with ring gear and direct-clutch assembly. Do not damage the seal-ring grooves and mating surfaces in the case. (*Ford Motor Company*)

2. Remove the clutch piston. Discard the inner and outer piston-lip seals.
3. Remove the stator support by removing the attaching bolts. Remove the selective thrust washer and the four sealing rings on the stator support (Fig. 14-30).
4. Lift off the stator support. Then remove the drive and driven gears from the pump body (Fig. 14-31). If the oil-pump bushing is worn, replace it. The oil groove in the bushing faces the stator support.
5. Inspect all parts and replace those that are worn or damaged.

6. To reassemble, lubricate the pump-gear pocket with petroleum jelly. Put the gears in place. The chamfer on the drive gear faces down. The triangle on the driven gear also faces down.
7. Install the stator support (Fig. 14-31). Secure it with five bolts.
8. Install the selective thrust washer. This washer controls the transmission end play. There are five thicknesses available.
9. Remove the old O ring from and install a new one on the pump body (Fig. 14-30).
10. Install new inner and outer clutch-piston-lip seals.

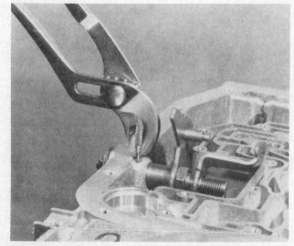

Fig. 14-26 Using pliers to remove the manual-lever locating pin. (*Ford Motor Company*)

Fig. 14-27 Removing the inner throttle lever and spring. (*Ford Motor Company*)

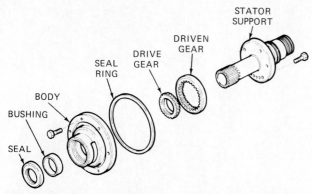

Fig. 14-28 Disassembled pump. (*Ford Motor Company*)

Fig. 14-29 Using a screwdriver to open the tabs holding the intermediate-clutch-piston return-spring retainer. (*Ford Motor Company*)

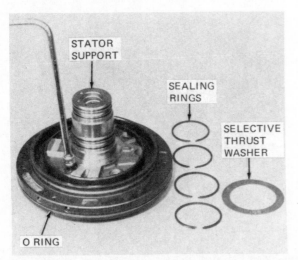

Fig. 14-30 Take out the bolts attaching the stator support. Remove the selective thrust washer and the four sealing rings. New sealing rings should be used on reassembly. (*Ford Motor Company*)

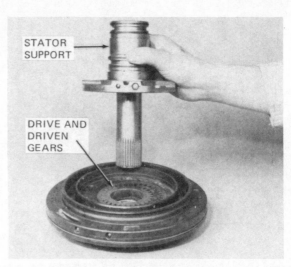

Fig. 14-31 Drive and driven gears in the pump. (*Ford Motor Company*)

11. Install the piston on the pump, using a lip-seal protector. Be sure the piston bleed hole is at 12 o'clock with reference to the top of the case.
12. Press the clutch-piston-spring retainer down on the piston so it snaps into place on the pump housing.

✿ 14-21 Reverse and intermediate one-way clutch service Figure 7-4 shows the location of the reverse and intermediate one-way clutches. Figure 14-32 shows the subassembly on the bench. The technician is removing the reverse-clutch thrust washer and clutch-pack-retaining snap ring. With these out, you can remove the three steel and four friction plates, and the top and bottom pressure plates.

Then remove the waved snap ring that retains the Belleville spring. Remove the Belleville spring and thrust ring. Then turn the assembly over and remove the clutch piston with air pressure (Fig. 14-33). Block the opposing air hole with a finger.

Remove and discard the inner and outer piston seals.

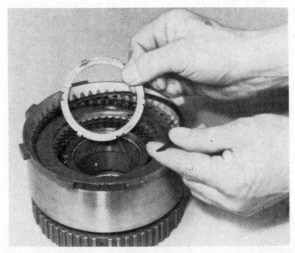

Fig. 14-32 Removing reverse-clutch thrust washer and clutch-pack-retaining snap ring. (*Ford Motor Company*)

Fig. 14-33 Using air pressure at the clutch-drum apply passage to remove the clutch piston. (*Ford Motor Company*)

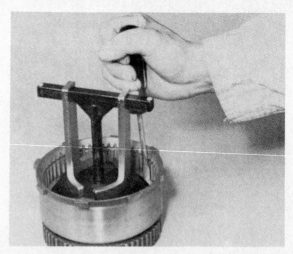

Fig. 14-35 To remove the reverse-clutch plates, use the clutch-spring compressor and pry out the waved snap ring. (*Ford Motor Company*)

Install new seals after lubricating them with transmission fluid.

Remove the snap ring, retaining plate, and outer race of the one-way clutch (Fig. 14-34). Turn the outer race counterclockwise to remove it. Then remove the roller-clutch assembly. If a roller is damaged or lost, the entire assembly must be replaced.

To reassemble, install the one-way roller clutch on the clutch drum, followed by the outer race, chamfer side facing the clutch drum. Then install the retaining plate and snap ring.

Turn the assembly over and install the clutch piston in the drum. Put the thrust ring and the Belleville spring over the clutch piston. Position the waved snap ring, with points down, over the spring. Compress with a special compressor (Fig. 14-35). Install the waved snap ring in the drum groove.

Install the clutch pack and pressure plates. Then install the selective snap ring, positioning the three tabs on the reverse-clutch washer so they seat in mating

cutouts in the clutch drum. Insert a thickness gauge between the selective snap ring and the top pressure plate. If the thickness is not correct, use a different snap ring. Apply air pressure (Fig. 14-33) to check if the clutch is free to move as it should.

✿ 14-22 Forward-clutch and turbine-shaft service The forward-clutch-and-turbine-shaft assembly is shown in Fig. 7-5. It is disassembled by removing the clutch plates and piston, as follows. Remove the forward-clutch hub and needle bearing. Then remove the snap ring and the clutch pack, which consists of a wave spring, five friction plates, five steel plates, and the top pressure plate.

Compress the piston-return spring with the clutch-spring compressor (Fig. 14-36), and remove the snap ring, retainer, clutch-piston return spring, and piston. Remove and discard the inner and outer seals on the piston. Install new seals.

Fig. 14-34 To remove the outer race from the clutch assembly, rotate it counterclockwise and lift off. (*Ford Motor Company*)

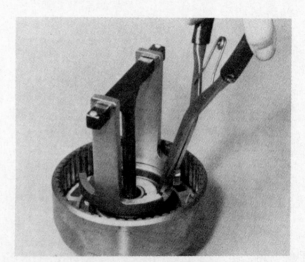

Fig. 14-36 To remove the forward-clutch plates, use the clutch-spring compressor and take out the snap ring. (*Ford Motor Company*)

Fig. 14-37 Installing the forward clutch in the reverse-clutch drum. (*Ford Motor Company*)

Fig. 14-38 Lubricate the outer race of the needle bearing with petroleum jelly and install it in the drive shell. Tabs on the bearing should face toward the drive shell. (*Ford Motor Company*)

Make sure the two ball checks in the clutch drum are free.

To reassemble the clutch, lubricate the new piston seals with transmission fluid and install the piston, using seal protectors to protect both the inner and outer seals. Then install the clutch-piston return spring and retainer. Compress the spring and install the snap ring (Fig. 14-36).

Install the clutch pack and secure with the selective snap ring. Use a thickness gauge to measure the clearance between the snap ring and the top pressure plate. If it is not correct, install another snap ring of a different thickness.

Apply air pressure to make sure the piston is working properly. Cover the outer race of the needle bearing with petroleum jelly and install it in the clutch hub. Then install the forward clutch in the reverse-clutch drum (Fig. 14-37).

Fig. 14-39 Lubricate the inner race of the needle bearing with petroleum jelly and stick it on the forward sun-gear shaft. Bearing tabs should face away from the gear. (*Ford Motor Company*)

Lubricate the outer race of the needle bearing with petroleum jelly. Install it in the drive shell (Fig. 14-38). Tabs on the bearing face inward toward the drive shell.

Lubricate the inner race of the needle bearing with petroleum jelly and install it on the forward sun-gear shaft (Fig. 14-39). Bearing tabs should face away from the gear. Insert the forward sun-gear shaft in the drive shell.

✿ 14-23 Center-support and low-reverse planetary-gearset service The center support and low-reverse planetary gearset are shown to the lower right in Fig. 7-6. There are only two parts that can be separated from the planetary gearset—the center support and the roller-clutch assembly (Fig. 14-40). The center support must be turned counterclockwise to remove it.

Fig. 14-40 To remove the center support from the low one-way roller clutch, turn it counterclockwise and lift it off. (*Ford Motor Company*)

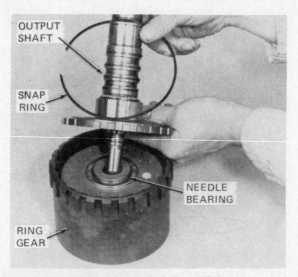

Fig. 14-41 To remove the direct-clutch output shaft from the ring gear, remove the waved snap ring. (*Ford Motor Company*)

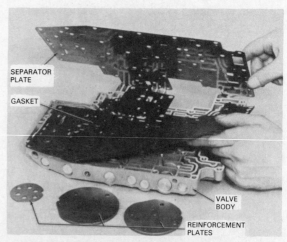

Fig. 14-42 Removing the reinforcement plates, gasket, and separator plate from the valve body. (*Ford Motor Company*)

✿ 14-24 Output-shaft and direct-clutch service
The output shaft and direct clutch are shown in Fig. 7-5. To disassemble, first remove the waved snap ring and direct-clutch output shaft from the ring gear (Fig. 14-41). Remove and discard the six output-shaft steel sealing rings. Install new sealing rings.

Remove the needle bearing and lift the direct-clutch assembly from the ring gear. Remove the selective snap ring and the clutch pack (five steel and five friction plates with top pressure plate). Use a clutch-spring compressor (Fig. 14-36) to compress the spring retainer. Remove the snap ring and clutch-piston-return-spring-and-retainer assembly. Remove the piston. Discard the piston seals and install new ones. Make sure the ball check in the clutch piston is free.

To reassemble the direct clutch, install the inner piston seal in the direct-clutch drum. The seal lip faces into the drum. Lubricate the outer piston seal with transmission fluid and install it on the piston with lip-seal protector. Then install the direct-clutch return spring and retainer. Use a spring compressor to install the snap ring (Fig. 14-36).

Install the clutch pack and selective snap ring. Check clearance between the snap ring and the top pressure plate. If it is not correct, use a snap ring of a different thickness. Then install the clutch assembly in the ring gear and put the needle bearing on the clutch.

Block one apply hole and apply air pressure to the other to make sure the clutch works. Then install the output shaft in the ring gear (Fig. 14-41). Turn the clutch drum over and install the needle bearing with the tabs facing away from the clutch drum. Install the direct-clutch hub.

✿ 14-25 Valve-body service Figure 7-24 shows the valve body completely disassembled. To disassemble the valve body, remove the 11 bolts holding the three reinforcement plates. Then separate the plates, separator plate, and gasket from the valve body (Fig. 14-

42). Note the locations of the eight shuttle balls in the valve body (Fig. 14-43). Ball 1 is orange, while the other seven balls are black. The orange ball is not interchangeable with the black balls.

Follow Fig. 7-24 if valves or springs need replacement. When installing the separator and reinforcement plates, use a new gasket.

✿ 14-26 Transmission reassembly Lubricate the race side of the needle bearing that goes in the rear end of the transmission case (Fig. 14-25). Then install the output shaft and ring-gear-and-direct clutch assembly. (Do not damage the seal-ring grooves and case mating surfaces.) Then proceed as follows.

1. Install the drive ball and the governor assembly on the output shaft. Retain the governor by installing the snap ring.
2. Use a new extension-housing seal and install the extension-housing gasket and housing. Secure with six housing bolts, tightened to specifications.
3. Install the low-reverse band in correct relationship to the anchor pins.
4. Install the center-support low-reverse planetary gearset (Fig. 14-24). Apply forward pressure to make sure the assembly meshes and fully seats with the ring gear and the direct-clutch hub.
5. Install the anticlunk spring (Fig. 14-23). To make sure that the spring is properly installed, apply thumb pressure to the center support in a clockwise direction to verify spring tension.
6. Install the large center-support snap ring in the case groove next to the low-reverse planetary gearset.
7. Install the direct drive shaft. Then install, as a unit, the forward-and-reverse sun-gear shell assembly and the low-and-reverse clutch pack (Fig. 14-22).
8. Rotate the transmission to a vertical position. Install the overdrive band on the anchor pin.
9. Install the intermediate-clutch pack. Install the pressure plate first, followed with three friction plates and three steel plates alternately installed. The outer steel plate is selective.

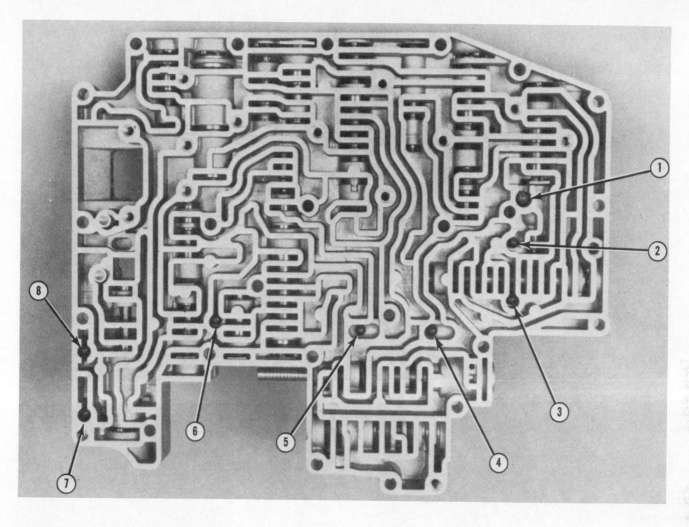

Fig. 14-43 Locations of the shuttle balls in the valve body. All are black except number 1, which is orange. It is a different size from the other balls. (*Ford Motor Company*)

Fig. 14-44 Using a depth micrometer to measure the intermediate-clutch clearance. Measurement is taken through a hole in a steel bar. It should be taken at opposing points and the results averaged. (*Ford Motor Company*)

Fig. 14-45 Using a depth gauge to check the end play. Check at opposing points and average the results. (*Ford Motor Company*)

Fig. 14-46 Installing the selective thrust washer. Lubricate the tab side with petroleum jelly so it will stick in place during reassembly. (*Ford Motor Company*)

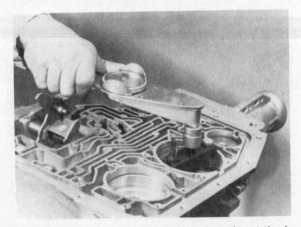

Fig. 14-47 Checking the reverse-band stroke at the low-reverse piston. Tighten to the specified torque. (*Ford Motor Company*)

Fig. 14-48 Dial indicator setup to measure piston-rod-assembly movement. (*Ford Motor Company*)

Fig. 14-49 Use a screwdriver to center the overdrive band under the overdrive-servo-piston hole before installing the piston. Tap piston cover firmly in place with a hammer. (*Ford Motor Company*)

10. Measure the intermediate-clutch clearance (Fig. 14-44) with a depth micrometer and an end-play checking tool. Check the clearance at two points 180° apart. Push down on the clutch pack while making the measurement. If it is not correct, install another selective steel plate of a different thickness.

11. Check the transmission end play at the reverse-clutch-drum thrust face (Fig. 14-45). The same tools are used for this measurement as for the clutch-clearance measurement. Check the measurement at two places 180° apart. If it is not correct, select a thrust washer of the correct thickness.

12. Install the selective thrust washer (Fig. 14-46). Lubricate the tab side with petroleum jelly and install the washer, tab down, on the pump support.

13. Install the front-pump gasket and the pump with seven bolts, tightened to specifications.

14. Install a new seal in the front pump with a special driver tool and hammer.

15. Rotate the transmission to a horizontal position. Use new seals and install the accumulator pistons. Secure them with the covers and snap rings.

16. Before installing the low-reverse piston and spring, make sure the band is aligned. Then install the piston and spring. Check the piston stroke with the servo-piston selection tool (Fig. 14-47). Tighten the tool center bolt to specifications. Then set up a dial indicator as shown in Fig. 14-48. Loosen the tool center bolt until all torque is relieved. If the dial indicator does not read within the specifications, install another piston-rod assembly (three lengths are available). Secure with cover and snap ring.

17. Install new seals on the overdrive piston-and-piston-cover assembly. Use a screwdriver as shown in Fig. 14-49 to center the overdrive band under the overdrive-servo-piston hole. Then install the piston spring and piston-cover assembly. Hold the assembly in place. Remove the screwdriver. Tap the piston cover firmly with a soft hammer and install the snap ring.

18. Install two valve-body guide pins (Fig. 14-50) on

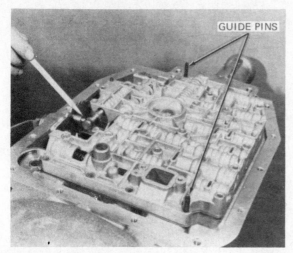

Fig. 14-50 How the manual valve must be mated with the inner manual shift-lever linkage. (*Ford Motor Company*)

the transmission case. Position the gasket and valve body over the pins. The manual valve must be mated with the inner manual-lever shift linkage,

and the TV plunger must be mated with the inner TV lever.

19. Secure the valve body with 22 bolts tightened to specifications. The four front, one center, and three rear bolts are shorter than the other bolts. Remove the guide pins and install the last two bolts.

20. Reconnect the inner linkage spring to the separator-plate notch.

21. Install the detent spring and bolt (Fig. 12-18), tightened to specifications.

22. Coat a new filter seal and gasket with petroleum jelly and position on a new transmission filter. Install the transmission filter and tighten the three filter bolts to specifications.

23. Use a new pan gasket and install the pan with 14 bolts, tightened to specifications.

24. Install the torque converter. Make sure it aligns on the drive shaft, turbine shaft, and pump support and is fully seated in the pump drive gear.

25. After installing the transmission on the car, add fluid until the dipstick shows the proper level (✿ 14-3).

26. Adjust the linkages and check the TV control pressure (✿ 14-4 to 14-10).

Chapter 14 review questions

Select the *one* correct, best, or most probable answer to each question. Then check your answers against the correct answers given at the end of the book.

1. Mechanic A says the two linkage adjustments for the Ford overdrive automatic transmission are to the throttle-valve linkage and manual linkage. Mechanic B says the two linkage adjustments are to the lockup clutch and the direct band. Who is right?
   a. A only
   b. B only
   c. both A and B
   d. neither A nor B

2. If the shifts are soft and mushy, the most likely cause is:
   a. sticking manual valve
   b. sticking clutch
   c. incorrect throttle-valve adjustment
   d. all of the above

3. Mechanic A says if the shifts are soft and mushy, it is better to drive with a heavy throttle to improve the shifts. Mechanic B says you should avoid a heavy throttle. Who is right?
   a. A only
   b. B only
   c. both A and B
   d. neither A nor B

4. To adjust the TV linkage, you:
   a. change the length of the control cable at the carburetor
   b. change the length of the control rod at the transmission

   c. change the engine idle speed
   d. change the length of the manual-valve linkage

5. Mechanic A says that if the engine overspeeds on the stall test, the transmission is slipping. Mechanic B says that if this happens, release the throttle at once. Who is right?
   a. A only
   b. B only
   c. both A and B
   d. neither A nor B

6. The purpose of the air-pressure test is to check the operation of:
   a. the clutches
   b. the servos
   c. the bands
   d. all of the above

7. If the engine does not reach specified rpm on the stall test, the trouble probably is that the:
   a. bands are slipping
   b. clutches are slipping
   c. stator clutch is locked up
   d. engine needs a tune-up

8. Mechanic A says that you can tell when the clutches engage during the air-pressure test because they hiss when engaged. Mechanic B says cluches engage with a dull thud when air pressure is applied. Who is right?
   a. A only
   b. B only
   c. both A and B
   d. neither A nor B

9. Mechanic A says band adjustment is not required for normal driving. Mechanic B says band adjustment is required if the car is in severe service. Who is right?
   a. A only
   b. B only
   c. both A and B
   d. neither A nor B
10. To make the air-pressure tests of the clutches and servos, you must:
    a. operate the engine at fast idle
    b. remove the transmission from the vehicle
    c. disassemble the transmission to get to the clutches and servos
    d. remove the oil pan and valve body
11. The purpose of having a 2.5-inch [63.5-mm] hole in the workbench to service the front-pump and intermediate-clutch piston assembly is to:
    a. accommodate the stator support
    b. allow the fluid to drain
    c. permit the clutch piston to move down out of the assembly
    d. allow you to measure transmission end play
12. To remove the piston from the reverse and intermediate one-way clutch, use:
    a. a piston puller
    b. a screwdriver
    c. snap-ring pliers
    d. air pressure
13. When disassembling the forward clutch and turbine shaft assembly, you find that the clutch has:
    a. no return spring
    b. a series of small coil springs
    c. one heavy coil spring
    d. a Belleville spring
14. During reassembly, the forward clutch is installed in the:
    a. forward-clutch drum
    b. reverse planetary gearset
    c. reverse-clutch drum
    d. front-band drum
15. Mechanic A says the valve body has eight shuttle balls. Mechanic B says one of the balls is orange and the other seven are black. Who is right?
    a. A only
    b. B only
    c. both A and B
    d. neither A nor B

CHRYSLER TORQUEFLITE AUTOMATIC TRANSAXLE

After studying this chapter, you should be able to:

1. Point out the similarities between the compound planetary gearset used in the TorqueFlite automatic transaxle and the one used in the standard TorqueFlite.

2. Identify and name the transaxle parts when they are laid out on the bench.

3. Explain why the transaxle requires a differential and how the differential works.

4. Describe the six shift-lever positions and explain the power flow through the transmission with each.

15-1 Introduction to the TorqueFlite transaxle This chapter describes the construction and operation of the Chrysler model A-404 TorqueFlite automatic transaxle. Transaxles are mounted on the engine (Fig. 15-1) and the engine is mounted sideways, or transversely, in the vehicle (Fig. 15-2). Figure 15-3 is a sectional view of the Chrysler TorqueFlite transaxle. This transaxle includes an automatic transmission based on the TorqueFlites discussed in Chap. 12. It also includes a differential and two couplings for the drive shafts to the two front wheels. Figures 15-4 to 15-6 show right, front, and rear views of the transaxle detached from the engine.

15-2 TorqueFlite transaxle construction The transaxle includes a torque converter, a three-speed automatic transmission, final-drive gearing, and a differential.

NOTE: This transaxle is basically a metric design. Parts and specifications are sized to the metric measuring system.

Compare Fig. 15-3 with Figs. 5-1 to 5-3. All TorqueFlite automatic transmissions have similar components. These include two multiple-disk clutches, two bands with servos, an overrunning clutch, and a compound planetary gearset.

The transaxle automatic transmission (Fig. 15-3) is similar to the other TorqueFlites except that it has been cut off in back of the planetary gearset and a transfer shaft has been added. A gear on the end of the output shaft is meshed with a gear on the end of the transfer shaft. The transfer shaft carries the governor. The other end of the transfer shaft, next to the torque converter, has a pinion gear that is meshed with the ring gear on the differential.

The ring gear drives the differential case and the differential works the same as other differentials described in ✿ 1-14.

✿ 15-3 Hydraulic system The hydraulic system for the transaxle TorqueFlite is similar to those for other TorqueFlites discussed in Chap. 5. Compare Plates 9 to 13 for the standard TorqueFlite with Plates 24 to 27

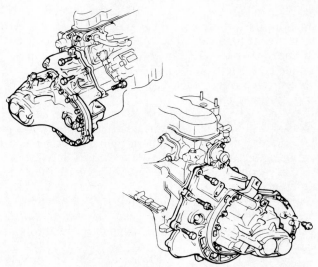

Fig. 15-1 Attachment of the transaxle to the engine. (*Chrysler Corporation*)

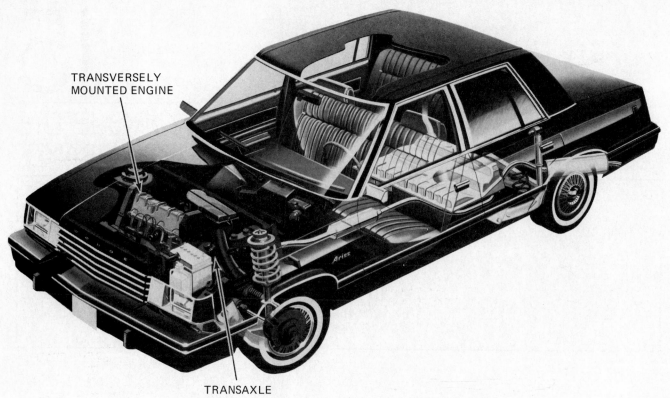

TRANSVERSELY MOUNTED ENGINE

TRANSAXLE

Fig. 15-2 Phantom view of a car with a transversely mounted engine. (*Chrysler Corporation*)

for the transaxle TorqueFlite in the *Color Plates of Automatic-Transmission Hydraulic Circuits,* sixth edition. The systems look different at first, but both have many of the same basic parts and valves. For example, the low-and-reverse servo (LOW & REV SERVO) is the same as the rear servo. The KD SERVO (kickdown servo) is the same as the front servo. The accumulator, 2-3 shift valve, shuttle valve, 1-2 shift valve, manual valve, and throttle and kickdown valves are all alike.

The governor valve is shown differently in the two sets of plates because it has a different construction in the transaxle. Missing from the transaxle hydraulic circuits are the limit valve, the 1-2 shift control valve, and the lockup valves. The transaxle does not have a lockup torque converter. The limit valve and 1-2 shift control valve are not required, because the transaxle is used with smaller engines.

✿15-4 Transaxle operation   The engine torque enters the transaxle through the torque converter. The various gear positions and power flows are the same as in the standard TorqueFlite (Figs. 5-4 to 5-10). The difference is that in the transaxle there is an output-shaft gear on the end of the output shaft. This gear is meshed with the transfer-shaft gear. The transfer-shaft pinion gear drives the differential pinion carrier through the ring gear.

To review the operating sequence briefly, the driver moves the manual valve to the desired operating position. Then, car speed and throttle position take over to produce whatever shift pattern has been selected. ✿ 5-5 to 5-12 cover the complete range of operating conditions for the standard TorqueFlite. The operating conditions for the transaxle are similar, but the transaxle hydraulic system (✿ 15-3) is simpler than in the standard TorqueFlite (✿ 5-4). A study of Plates 24 to 27 while you review Figs. 5-4 to 5-10 will show that the operating sequences are essentially the same.

✿15-5 Differential action   Engine power enters the differential through the transfer-shaft pinion gear and the differential ring gear. Figure 15-7 shows in simplified form the construction of the differential. Figure 15-8 shows the actual differential removed from the transaxle. Figure 15-9 shows the differential side gears, pinion gears, and related parts removed from the differential.

The drive shafts which connect the differential to the two front wheels are splined into the two side gears. Figure 15-10 is an underside view from the front of the car, showing the locations of the transaxle and the two front-wheel drive shafts.

The drive shafts have universal joints and slip joints (✿ 1-11) that permit them to carry rotary motion through angles. They also allow the effective length of the drive shafts to change. These joint actions allow the front wheels to move up and down as they meet irregularities in the road. Also, the joint actions permit the front wheels to be swung from side to side for steering.

The differential allows the two front wheels to rotate

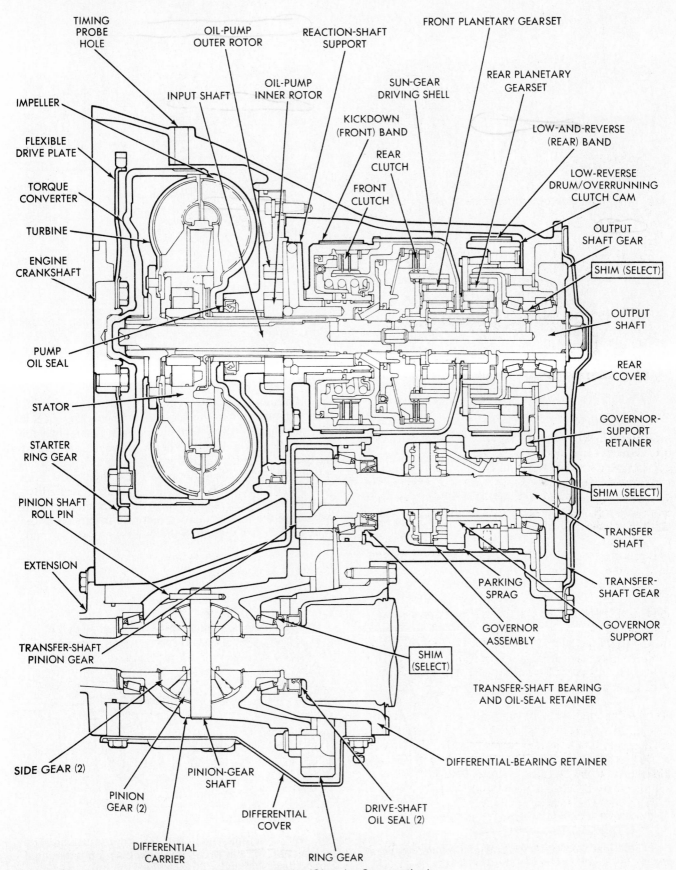

TIMING
PROBE
HOLE

OIL-PUMP
OUTER ROTOR

REACTION-SHAFT
SUPPORT

FRONT PLANETARY GEARSET

INPUT SHAFT

OIL-PUMP
INNER ROTOR

SUN-GEAR
DRIVING SHELL

REAR PLANETARY
GEARSET

IMPELLER

KICKDOWN
(FRONT) BAND

LOW-AND-REVERSE
(REAR) BAND

FLEXIBLE
DRIVE PLATE

REAR
CLUTCH

LOW-REVERSE
DRUM/OVERRUNNING
CLUTCH CAM

TORQUE
CONVERTER

FRONT
CLUTCH

OUTPUT
SHAFT GEAR

TURBINE

SHIM (SELECT)

ENGINE
CRANKSHAFT

OUTPUT
SHAFT

PUMP
OIL SEAL

REAR
COVER

STATOR

GOVERNOR-
SUPPORT
RETAINER

STARTER
RING GEAR

SHIM (SELECT)

PINION SHAFT
ROLL PIN

TRANSFER
SHAFT

EXTENSION

TRANSFER-
SHAFT GEAR

PARKING
SPRAG

GOVERNOR
SUPPORT

TRANSFER-SHAFT
PINION GEAR

GOVERNOR
ASSEMBLY

SHIM
(SELECT)

TRANSFER-SHAFT BEARING
AND OIL-SEAL RETAINER

DIFFERENTIAL-BEARING RETAINER

SIDE GEAR (2)

PINION-GEAR
SHAFT

PINION
GEAR (2)

DIFFERENTIAL
COVER

DRIVE-SHAFT
OIL SEAL (2)

DIFFERENTIAL
CARRIER

RING GEAR

Fig. 15-3 Sectional view of the TorqueFlite transaxle. (*Chrysler Corporation*)

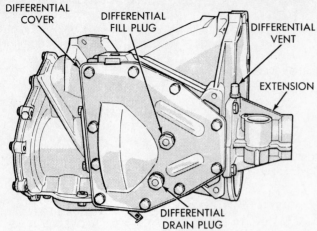

Fig. 15-4 Right side of the transaxle. (*Chrysler Corporation*)

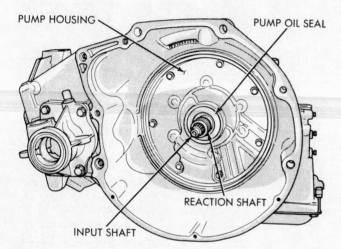

Fig. 15-5 Front end of the transaxle. (*Chrysler Corporation*)

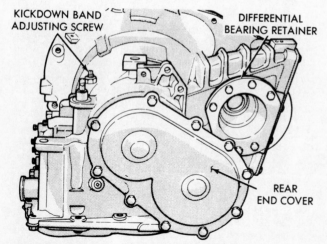

Fig. 15-6 Rear view of the transaxle. (*Chrysler Corporation*)

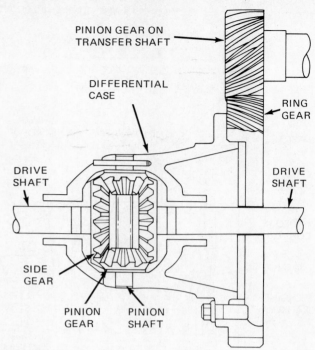

Fig. 15-7 Basic construction of the differential. (*Chrysler Corporation*)

at different speeds as the car moves around a curve or makes a turn. When this happens, the outside wheel in the turn has to travel farther and turn faster than the inside wheel (Fig. 15-11). When both wheels are turning at the same speed, the differential acts as a solid coupling (✿ 1-14). The differential case then turns both drive shafts and both wheels at the same speed. However, when the car makes a turn, the differential pinions rotate on the pinion shaft to increase the speed of the side gear that is splined to the outer-wheel drive shaft.

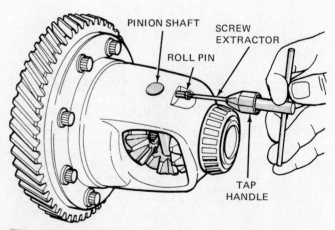

Fig. 15-8 Differential removed from the transaxle. The pinion-shaft roll pin is being removed so the differential can be disassembled. (*Chrysler Corporation*)

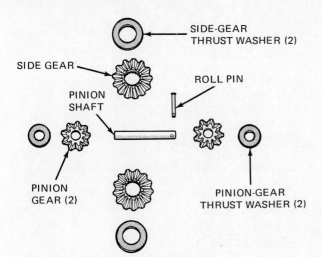

SIDE-GEAR
THRUST WASHER (2)

SIDE GEAR

PINION
SHAFT

ROLL PIN

PINION
GEAR (2)

PINION-GEAR
THRUST WASHER (2)

Fig. 15-9 Gears and related parts used in the differential. (*Chrysler Corporation*)

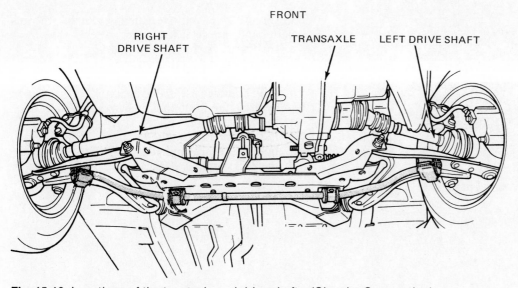

FRONT

RIGHT
DRIVE SHAFT

TRANSAXLE   LEFT DRIVE SHAFT

Fig. 15-10 Locations of the transaxle and drive shafts. (*Chrysler Corporation*)

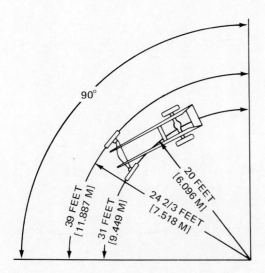

90°

39 FEET
[11.887 M]

31 FEET
[9.449 M]

24 2/3 FEET
[7.518 M]

20 FEET
[6.096 M]

Fig. 15-11 When the car goes around a curve, the outside wheel travels farther than the inside wheel.

Chapter 15 review questions

Select the *one* correct, best, or most probable answer to each question. Then check your answers against the correct answers given at the end of the book.

1. The transfer shaft is geared to the:
   - *a.* torque-converter shaft and ring gear
   - *b.* differential ring gear and differential side gears
   - *c.* output shaft and ring gear
   - *d.* sun gear and ring gear
2. Counting all the active control members in the transaxle TorqueFlite transmission, including the clutches and bands, there are:
   - *a.* three
   - *b.* four
   - *c.* five
   - *d.* six
3. Counting the transfer-shaft pinion gear, the differential has:
   - *a.* three gears
   - *b.* four gears
   - *c.* five gears
   - *d.* six gears
4. The compound planetary gearset has two sets of planet pinions, two ring gears, and:
   - *a.* one sun gear
   - *b.* two sun gears
   - *c.* three sun gears
   - *d.* no sun gears
5. The drive shafts are:
   - *a.* meshed with the ring gear
   - *b.* splined to the differential pinions
   - *c.* splined to the differential side gears
   - *d.* splined to the transfer-shaft pinion
6. The two types of joints in the drive shafts are:
   - *a.* universal and drive
   - *b.* elbow and slip
   - *c.* universal and slip
   - *d.* slip and drive
7. The Chrysler model A-404 transaxle manual valve has:
   - *a.* four positions
   - *b.* two positions
   - *c.* five positions
   - *d.* six positions
8. When the lift-foot shift is made, the shift is from:
   - *a.* first to second
   - *b.* first to third
   - *c.* second to third
   - *d.* third to fourth
9. In direct drive, the overrunning clutch overruns, both bands are released, and:
   - *a.* the front clutch is disengaged.
   - *b.* the rear clutch is disengaged.
   - *c.* both clutches are engaged.
   - *d.* both clutches are disengaged.
10. In reverse, the front clutch is engaged, and:
    - *a.* the front band is applied.
    - *b.* the rear band is applied.
    - *c.* both bands are applied.
    - *d.* the rear clutch is engaged.

CHRYSLER TORQUEFLITE AUTOMATIC TRANSAXLE SERVICE

After studying this chapter, and with proper instruction and equipment, you should be able to:

1. Explain how to use the diagnostic charts to find causes of trouble.
2. Discuss the manual-linkage and throttle-linkage adjustments.
3. Describe the procedures for making hydraulic- and air-pressure checks.
4. Describe the procedure of removing and reinstalling the transaxle.
5. Discuss the disassembling and reassembling of the transmission and the servicing procedures on the subassemblies.
6. Perform the services listed above.

16-1 Introduction to TorqueFlite transaxle service This chapter describes the trouble diagnosis, checks, adjustments, in-car service, removal, overhaul, and installation of the TorqueFlite automatic transaxle. The transaxle is essentially the same as the TorqueFlite transmissions described in Chaps. 5 and 12. However, it includes a differential, and the components are arranged so the transaxle will mount on a transversely positioned engine (Figs. 15-1 and 15-2). The transaxle has a torque converter, clutches, bands, planetary gearsets, and a hydraulic system of the same basic design as the other TorqueFlites. Chapter 15 describes and illustrates the construction and operation of the TorqueFlite transaxle.

In this chapter the service procedures that apply especially to the transaxle are covered. Many of the service procedures for the transaxle are the same as for the TorqueFlites covered in Chap. 12. Refer to the appropriate sections in Chap. 12 for details of procedures that are similar on both the standard and the transaxle TorqueFlites.

16-2 Operating cautions The operating cautions are the same as for the standard TorqueFlite and are covered in ✿ 12-2.

16-3 Transaxle trouble diagnosis The basic theory of trouble diagnosis is covered in ✿ 9-6. Figures 16-1 to 16-4 are trouble-diagnosis charts for the TorqueFlite transaxle.

✿16-4 TorqueFlite transaxle diagnostic tests Diagnosis of TorqueFlite transaxle trouble should start with checks of the fluid level and condition and linkage adjustments. Then the car should be road-tested to determine if the trouble has been eliminated. If not, pressure and stall tests should be made to pinpoint the cause of trouble.

✿16-5 Fluid level and condition Fluid-checking procedures are covered in ✿ 9-3 and 12-5. Figure 16-5 shows the dipstick for the transaxle. This dipstick has a hollow tube which serves as a vent. Also, the dipstick has only ADD and FULL marks. A properly filled transaxle will read near the ADD mark when cold and near the FULL mark when hot.

NOTE: The differential has a separate lubricant sump. It is entirely independent of the transmission-section lubricating and hydraulic system.

✿16-6 Gearshift-linkage adjustment Figure 16-6 shows the linkage. To adjust, make sure the adjustable swivel block is free to slide on the shift cable. If you must disconnect linkages that have plastic grommets, the grommets must be pried and cut out and new grommets must be snapped in with pliers.

1. Put the gearshift lever in P.
2. With all linkages assembled and the adjustable swivel-lock bolt loose, move the shift lever on the transmission all the way to the rear detent position (PARK). Tighten the bolt.

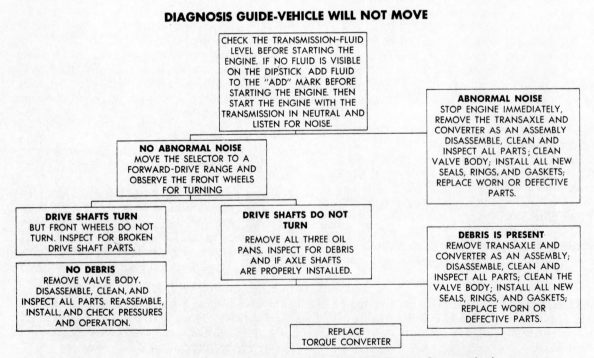

DIAGNOSIS GUIDE-ABNORMAL NOISE

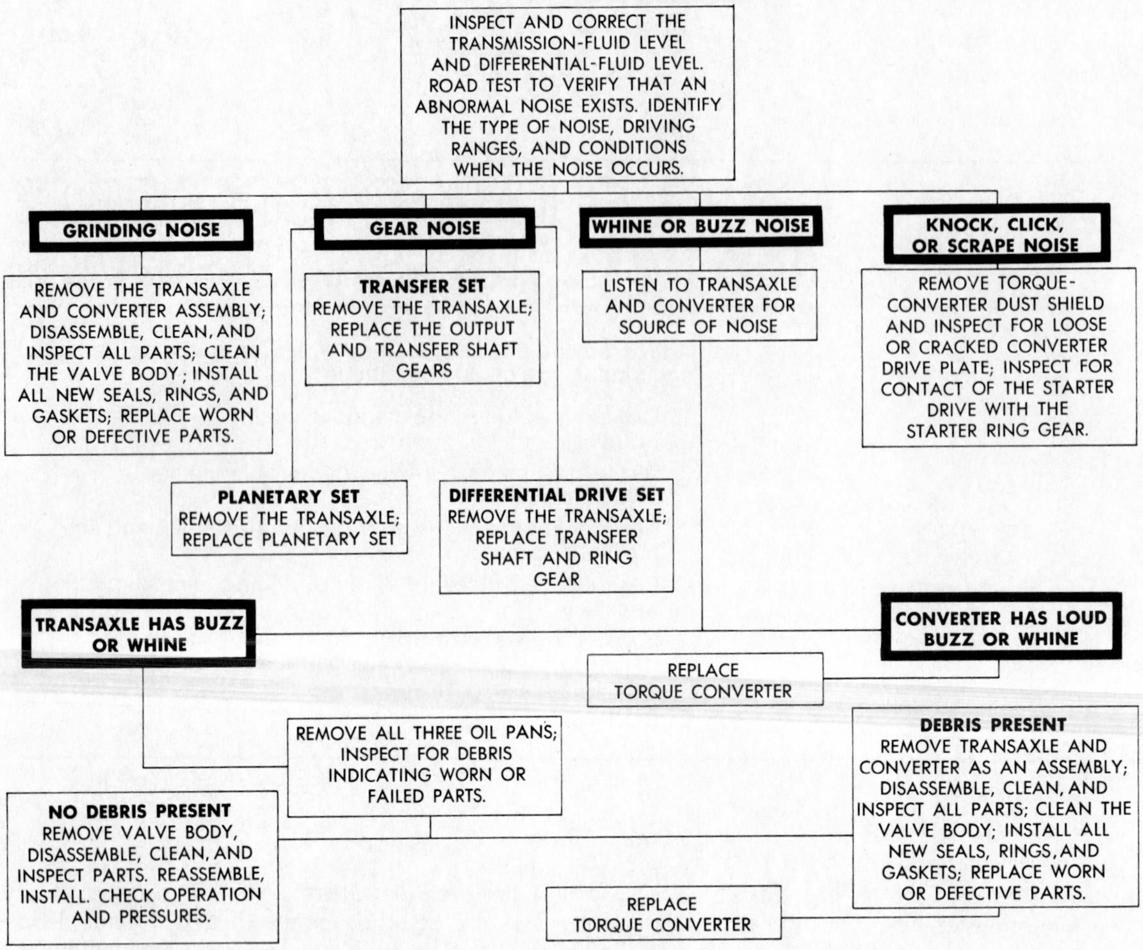

INSPECT AND CORRECT THE TRANSMISSION-FLUID LEVEL AND DIFFERENTIAL-FLUID LEVEL. ROAD TEST TO VERIFY THAT AN ABNORMAL NOISE EXISTS. IDENTIFY THE TYPE OF NOISE, DRIVING RANGES, AND CONDITIONS WHEN THE NOISE OCCURS.

GRINDING NOISE

REMOVE THE TRANSAXLE AND CONVERTER ASSEMBLY; DISASSEMBLE, CLEAN, AND INSPECT ALL PARTS; CLEAN THE VALVE BODY; INSTALL ALL NEW SEALS, RINGS, AND GASKETS; REPLACE WORN OR DEFECTIVE PARTS.

GEAR NOISE

TRANSFER SET
REMOVE THE TRANSAXLE; REPLACE THE OUTPUT AND TRANSFER SHAFT GEARS

WHINE OR BUZZ NOISE

LISTEN TO TRANSAXLE AND CONVERTER FOR SOURCE OF NOISE

KNOCK, CLICK, OR SCRAPE NOISE

REMOVE TORQUE-CONVERTER DUST SHIELD AND INSPECT FOR LOOSE OR CRACKED CONVERTER DRIVE PLATE; INSPECT FOR CONTACT OF THE STARTER DRIVE WITH THE STARTER RING GEAR.

PLANETARY SET
REMOVE THE TRANSAXLE; REPLACE PLANETARY SET

DIFFERENTIAL DRIVE SET
REMOVE THE TRANSAXLE; REPLACE TRANSFER SHAFT AND RING GEAR

TRANSAXLE HAS BUZZ OR WHINE

CONVERTER HAS LOUD BUZZ OR WHINE

REPLACE TORQUE CONVERTER

REMOVE ALL THREE OIL PANS; INSPECT FOR DEBRIS INDICATING WORN OR FAILED PARTS.

DEBRIS PRESENT
REMOVE TRANSAXLE AND CONVERTER AS AN ASSEMBLY; DISASSEMBLE, CLEAN, AND INSPECT ALL PARTS; CLEAN THE VALVE BODY; INSTALL ALL NEW SEALS, RINGS, AND GASKETS; REPLACE WORN OR DEFECTIVE PARTS.

NO DEBRIS PRESENT
REMOVE VALVE BODY, DISASSEMBLE, CLEAN, AND INSPECT PARTS. REASSEMBLE, INSTALL. CHECK OPERATION AND PRESSURES.

REPLACE TORQUE CONVERTER

Fig. 16-1 Transaxle noise trouble-diagnosis chart. (*Chrysler Corporation*)

DIAGNOSIS GUIDE-VEHICLE WILL NOT MOVE

CHECK THE TRANSMISSION-FLUID LEVEL BEFORE STARTING THE ENGINE. IF NO FLUID IS VISIBLE ON THE DIPSTICK ADD FLUID TO THE "ADD" MARK BEFORE STARTING THE ENGINE. THEN START THE ENGINE WITH THE TRANSMISSION IN NEUTRAL AND LISTEN FOR NOISE.

ABNORMAL NOISE
STOP ENGINE IMMEDIATELY, REMOVE THE TRANSAXLE AND CONVERTER AS AN ASSEMBLY DISASSEMBLE, CLEAN AND INSPECT ALL PARTS; CLEAN VALVE BODY; INSTALL ALL NEW SEALS, RINGS, AND GASKETS; REPLACE WORN OR DEFECTIVE PARTS.

NO ABNORMAL NOISE
MOVE THE SELECTOR TO A FORWARD-DRIVE RANGE AND OBSERVE THE FRONT WHEELS FOR TURNING

DRIVE SHAFTS TURN
BUT FRONT WHEELS DO NOT TURN. INSPECT FOR BROKEN DRIVE SHAFT PARTS.

DRIVE SHAFTS DO NOT TURN
REMOVE ALL THREE OIL PANS. INSPECT FOR DEBRIS AND IF AXLE SHAFTS ARE PROPERLY INSTALLED.

DEBRIS IS PRESENT
REMOVE TRANSAXLE AND CONVERTER AS AN ASSEMBLY; DISASSEMBLE, CLEAN AND INSPECT ALL PARTS; CLEAN THE VALVE BODY; INSTALL ALL NEW SEALS, RINGS, AND GASKETS; REPLACE WORN OR DEFECTIVE PARTS.

NO DEBRIS
REMOVE VALVE BODY. DISASSEMBLE, CLEAN, AND INSPECT ALL PARTS. REASSEMBLE, INSTALL, AND CHECK PRESSURES AND OPERATION.

REPLACE TORQUE CONVERTER

Fig. 16-2 Transaxle diagnosis guide for a vehicle that will not move. (*Chrysler Corporation*)

DIAGNOSIS GUIDE-FLUID LEAKS

VISUALLY INSPECT FOR SOURCE OF LEAK. IF THE SOURCE OF LEAK CANNOT BE READILY DETERMINED, CLEAN THE EXTERIOR OF THE TRANSMISSION. CHECK TRANSMISSION-FLUID LEVEL. CORRECT IF NECESSARY.

THE FOLLOWING LEAKS MAY BE CORRECTED WITHOUT REMOVING THE TRANSAXLE:

MANUAL LEVER SHAFT OIL SEAL
PRESSURE-GAUGE PLUGS
NEUTRAL START SWITCH
OIL PAN RTV
OIL-COOLER FITTINGS
EXTENSION-HOUSING-TO-CASE BOLTS
SPEEDOMETER ADAPTER O RING
FRONT-BAND ADJUSTING SCREW
EXTENSION-HOUSING AXLE SEAL
DIFFERENTIAL-BEARING-RETAINER AXLE SEAL
REAR-END COVER RTV
DIFFERENTIAL COVER RTV
EXTENSION HOUSING O RING
DIFFERENTIAL-BEARING-RETAINER RTV

THE FOLLOWING LEAKS REQUIRE REMOVAL OF THE TRANSAXLE AND TORQUE CONVERTER FOR CORRECTION:

TRANSMISSION FLUID LEAKING FROM THE LOWER EDGE OF THE CONVERTER HOUSING; CAUSED BY FRONT PUMP SEAL, PUMP TO CASE SEAL, OR TORQUE CONVERTER WELD.

CRACKED OR POROUS TRANSMISSION CASE.

Fig. 16-3 Transaxle leakage trouble-diagnosis chart. (*Chrysler Corporation*)

Harsh engagement from neutral to D or R	Delayed engagement from neutral to D or R	Runaway upshift	No upshift	3-2 kickdown runaway	No kickdown or normal downshift	Shifts erratic	Slips in forward-drive positions	Slips in reverse only	Slips in all positions	No drive in any position	No drive in reverse	No drive in forward-drive positions	Drives in neutral	Drags or locks	Grating, scraping, growling noise	Buzzing noise	Hard to fill, oil blows out filler hole	Transmission overheats	Harsh upshift	Delayed upshift	POSSIBLE CAUSE
															X						Overrunning-clutch inner race damaged
							X				X			X	X						Overrunning clutch worn, broken, or seized
										X	X	X		X	X						Planetary gearsets broken or seized
													X								Rear clutch dragging
X	X						X				X	X	X								Worn or faulty rear clutch
													X					X			Insufficient clutch-plate clearance
																		X			Faulty cooling system
														X				X			Kickdown-band adjustment too tight
X																			X		Hydraulic pressure too high
																	X				High fluid level
	X	X	X	X		X		X				X								X	Worn or faulty front clutch
		X	X	X	X	X														X	Kickdown-servo band or linkage malfunction
			X		X	X														X	Governor malfunction
	X	X	X			X		X				X								X	Worn or broken reaction-shaft-support seal rings
			X	X		X														X	Governor-support seal rings broken or worn
															X						Drive-shaft bushing(s) damaged

Fig. 16-4 Chart relating transaxle troubles with possible causes. (*Chrysler Corporation*) (Figure 16-4 continued on page 248)

Harsh engagement from neutral to D or R	Delayed engagement from neutral to D or R	Runaway upshift	No upshift	3-2 kickdown runaway	No kickdown or normal downshift	Shifts erratic	Slips in forward-drive positions	Slips in reverse only	Slips in all positions	No drive in any position	No drive in forward-drive positions	No drive in reverse	Drives in neutral	Drags or locks	Grating, scraping, growling noise	Buzzing noise	Hard to fill, oil blows out filler hole	Transmission overheats	Harsh upshift	Delayed upshift	POSSIBLE CAUSE
				X				X													Overrunning clutch not holding
		X										X						X	X		Kickdown band out of adjustment
	X	X	X	X	X	X												X	X		Incorrect throttle-linkage adjustment
X																					Engine idle speed too low
X	X		X		X	X	X	X							X	X					Aerated fluid
X					X		X		X												Worn or broken input-shaft seal rings
X				X	X	X	X	X								X					Faulty oil pump
X	X			X	X		X	X							X						Oil filter clogged
X		X		X	X	X					X	X				X					Incorrect gearshift-control linkage adjustment
X	X	X	X		X	X	X	X	X	X	X				X		X				Low fluid level
X					X							X									Low-reverse servo, band, or linkage malfunction
X	X	X	X	X	X	X	X	X	X	X	X	X	X	X		X					Valve-body malfunction or leakage
							X					X		X	X						Low-reverse band worn out
X	X	X	X		X	X	X	X	X	X	X				X	X					Hydraulic pressures too low
X																X					Engine idle speed too high
																X					Stuck switch valve

Fig. 16-4 (*Continued*)

3. Check by making sure the detent positions for neutral and drive are within the limits of the hand-lever gate stops. Also, key starts must occur only when the shift lever is in P or N.

✿ 16-7 Throttle-cable adjustment The linkage is shown in Fig. 16-7. With the engine at operating temperature and carburetor off fast idle, adjust the idle speed of the engine using a tachometer. Set the engine speed to the specified rpm. Loosen the adjustment-bracket lock screw. The bracket must be free to slide on its slot. Hold the transmission lever firmly to the rear against its internal stop and tighten the lock screw to specifications.

✿ 16-8 Band adjustments The front, or kickdown, band is adjusted in the same way as the front band for the standard TorqueFlite (✿ 12-8). The low-reverse band is not adjustable in the transaxle. If the band is worn, it should be replaced.

✿ 16-9 Hydraulic-pressure tests The pressure tests with the selector lever in 1, 2, D, and R, and for governor and throttle pressures, are the same as for the standard TorqueFlite (✿ 12-10 to 12-17). Figures 16-8 and 16-9 show the ports where connections are made to make the pressure checks.

✿ 16-10 Line-pressure and throttle-pressure adjustments Refer to ✿ 12-20 for details of the line-pressure and throttle-pressure adjustments.

✿ 16-11 Stall test The stall test tests the torque-converter stator clutch and the holding ability of the clutches. The procedure is described in ✿ 12-18.

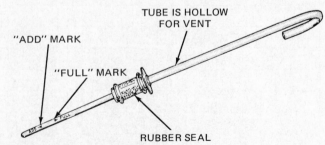

Fig. 16-5 Transaxle dipstick and vent. (*Chrysler Corporation*)

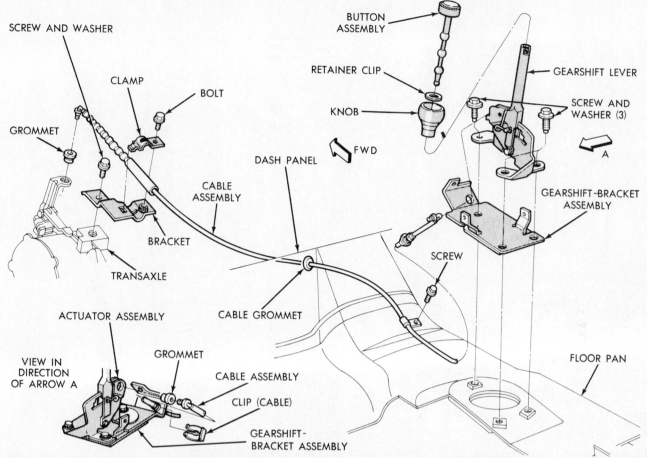

Fig. 16-6 Cable type of gearshift (manual-valve) linkage for a console mounted shift. (*Chrysler Corporation*)

✿ 16-12 Air-pressure tests Applying air pressure to the various circuits in the transmission causes the clutches and band servos to operate, providing they are in good condition. The procedure is outlined in ✿ 12-19. Figure 16-9 shows the ports to which air pressure is applied.

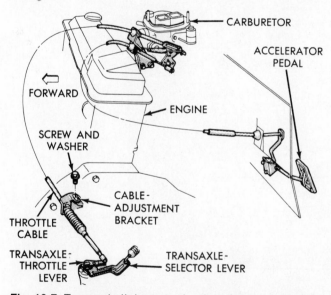

Fig. 16-7 Transaxle linkage to the throttle or accelerator pedal. (*Chrysler Corporation*)

✿ 16-13 On-the-car service In addition to adjusting the linkages, and checking and adjusting fluid level, some other services can be performed on the transaxle in the vehicle. These include removing and installing the following: valve body, transfer oil seal, parking sprag, speedometer pinion gear, and the neutral-starting and backup-light switch. Procedures for servicing these components are similar to those covered previously for the TorqueFlite (Chap. 12).

The procedures of removing, overhauling, and installing the transaxle are covered in the remainder of the chapter. Refer to ✿ 9-10 for the precautions to take when servicing the transaxle. Refer to ✿ 9-11 for cleaning and inspecting procedures during disassembly.

It is not necessary to remove the engine when removing the transaxle. Part of the general procedure outlined in ✿ 12-32 for removing the standard TorqueFlite also applies to the transaxle.

Transaxle removal Disconnect the grounded battery cable. Disconnect the throttle and shift linkage from the transaxle. Install the engine support fixture (Fig. 16-10). Remove the three upper-housing bolts (Fig. 16-11). Remove the hub locknuts and cotter pins at the front wheels. Raise the vehicle on a lift. Drain the transaxle. Then:

1. Remove the left splash shield.

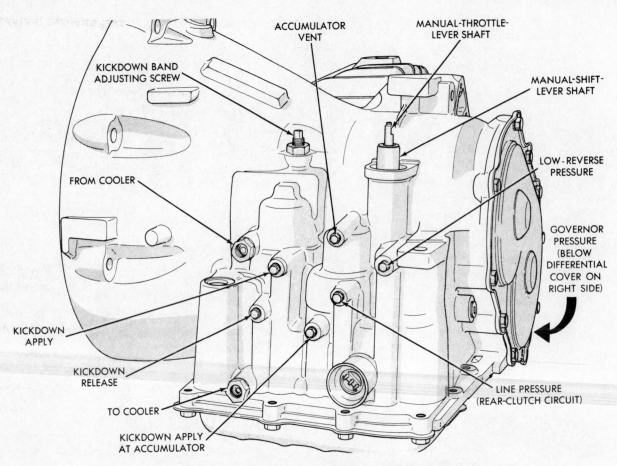

Fig. 16-8 Left side of transaxle, showing various pressure taps. (*Chrysler Corporation*)

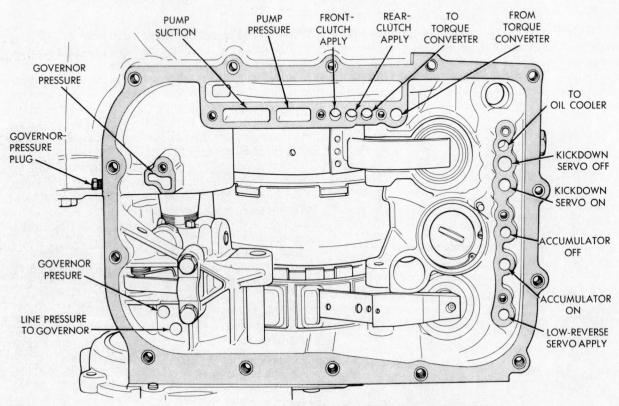

Fig. 16-9 Points at which to apply air pressure to check clutch and band action. (*Chrysler Corporation*)

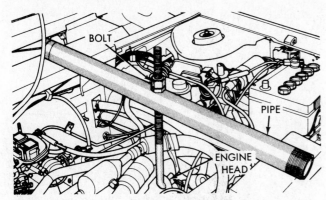

Fig. 16-10 Engine support installed. (*Chrysler Corporation*)

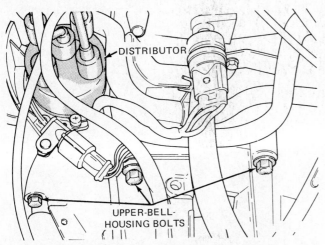

Fig. 16-11 Upper bell-housing bolts. (*Chrysler Corporation*)

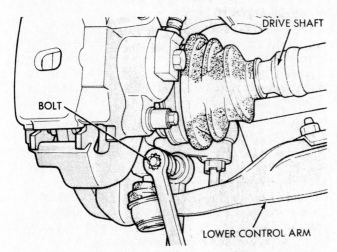

Fig. 16-12 Removing lower ball-joint-to-steering-knuckle bolt. (*Chrysler Corporation*)

2. Drain the differential and remove the differential cover.
3. Remove the speedometer adapter, cable, and pinion as an assembly.
4. Remove the sway bar.

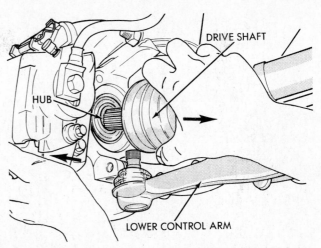

Fig. 16-13 Removing drive shaft from wheel hub. (*Chrysler Corporation*)

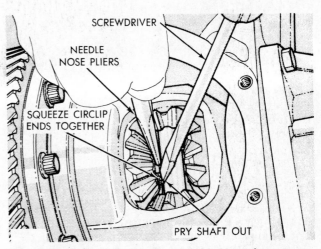

Fig. 16-14 Prying drive shaft out of side gear after squeezing the Circlip ends together. (*Chrysler Corporation*)

5. Remove the lower ball-joint-to-steering-knuckle bolts (Fig. 16-12) and pry the lower ball joints from the steering knuckles.
6. Remove the drive shafts from the hubs (Fig. 16-13).
7. Use needle-nose pliers to squeeze the Circlip ends and pry the drive shafts out of the side gears (Fig. 16-14).
8. Remove both drive shafts (Fig. 16-15).
9. Mark the torque converter and drive plate so they can be reconnected in the same relative positions. Then remove the converter-mounting bolts. Rotate the engine to get to the bolts. Use a wrench through the access plug in the right splash shield to turn the engine (Fig. 16-16).
10. Remove the lower cooler tube and disconnect the wire to the neutral-park safety switch.
11. Remove the engine-mount bracket from the front crossmember.
12. Remove the front mount insulator through-bolt and bell-housing bolts.
13. Position the transmission jack under the transaxle (Fig. 16-17).

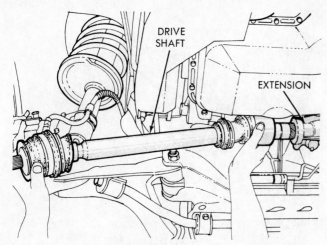

Fig. 16-15 Removing drive shaft. (*Chrysler Corporation*)

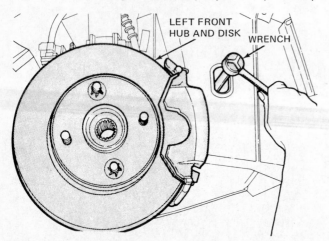

Fig. 16-16 Using wrench through the right splash shield to turn the engine crankshaft. (*Chrysler Corporation*)

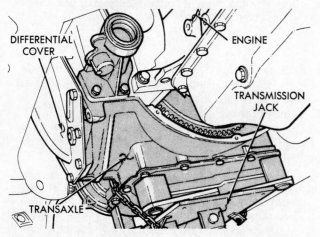

Fig. 16-17 Positioning the transmission jack under the transaxle. (*Chrysler Corporation*)

14. Remove the left engine mount and the long bolt through the mount.
15. Lower the transaxle (Fig. 16-18). Pry against the engine to get enough clearance. Do not let the torque converter slide off during transaxle removal.
16. Install the transaxle in a holding fixture.

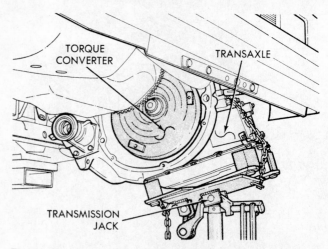

Fig. 16-18 Lowering the transaxle. (*Chrysler Corporation*)

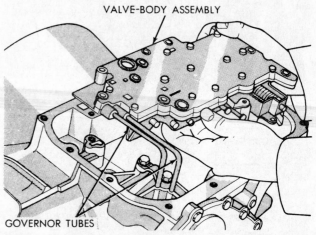

Fig. 16-19 Removing valve body with governor tubes. (*Chrysler Corporation*)

❇ 16-14 Disassembly Before disassembling the transaxle, plug all openings and clean the exterior. Note the service precautions about cleanliness and care in handling parts in ❇ 9-10. Also, reread the instructions on cleaning and inspecting parts (❇ 9-11).

1. Remove the torque converter and the transmission oil pan, filter, and parking rod.
2. Remove the valve body and governor tubes (Fig. 16-19). Lay the valve body aside on a clean surface for later inspection and overhaul.
3. Measure the input-shaft end play (Fig. 16-20). If it is not correct, a new thrust washer of a different thickness will be required. This washer is located between the input and output shafts.
4. Remove the oil pump by taking out the seven attaching bolts and using two slide hammers (Fig. 16-21).
5. Loosen the kickdown-band adjusting screw (top center in Fig. 16-8), and remove the kickdown band and strut.
6. Remove the front-clutch assembly (Fig. 16-22), large thrust washer, and rear-clutch assembly. Then remove the small thrust washer in back of the rear-clutch assembly.

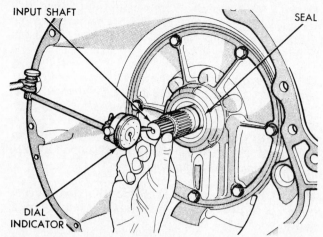

Fig. 16-20 Checking input-shaft end play. (*Chrysler Corporation*)

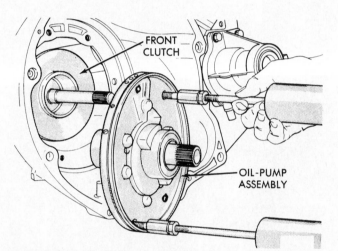

Fig. 16-21 Removing oil pump with slide hammers. (*Chrysler Corporation*)

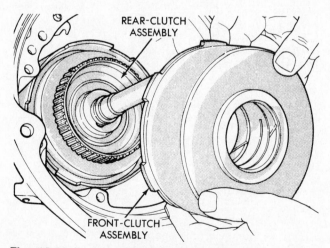

Fig. 16-22 Removing front-clutch assembly. (*Chrysler Corporation*)

7. Remove the snap ring and the front planetary-gear assembly and the large thrust washer (with four tangs) under it.
8. Remove the sun-gear driving shell and the large thrust washer under it.
9. Remove the rear planetary-gear assembly and the large tanged thrust washer under it.
10. Remove the overrunning-clutch cam assembly.
11. Remove the low-reverse band and strut and the thrust washer under it (Fig. 16-23).

❀ **16-15 Servicing the pump** Disassemble the pump by taking out the bolts and the two gears (Fig. 16-24). Check the clearance of the outer gear with the housing and the side clearance of the gears. To check the side clearance, lay a straightedge on the housing and measure the clearance between it and the gears with a thickness gauge.

❀ **16-16 Front clutch** Remove the large waved snap ring and take out the thick steel plate, the three clutch plates, and the two driving disks (Fig. 16-25). Then use the compressor tool and snap-ring pliers (Fig. 16-26) to compress the return spring, and remove the snap ring, retainer, spring, and piston.

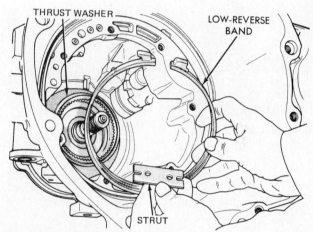

Fig. 16-23 Removing the low-reverse band and strut. (*Chrysler Corporation*)

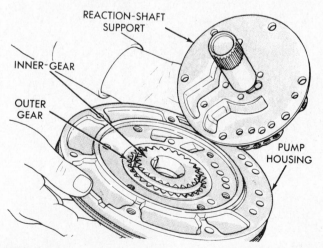

Fig. 16-24 Removing the reaction-shaft support from the pump body. (*Chrysler Corporation*)

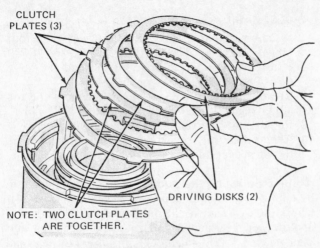

Fig. 16-25 Removing the clutch plates from the clutch housing. (*Chrysler Corporation*)

❋ 16-17 Rear clutch Remove the outer snap ring and the clutch plates (Fig. 16-27). Then remove the clutch-piston-spring snap ring and take out the piston spring and piston.

❋ 16-18 Front planetary gearset The front planetary gearset can be separated into its component parts by removing the snap ring and small thrust washer from the front so the gear assembly can be separated from the annulus (ring) gear (Fig. 16-28). The annulus gear has a gear support at the back which can be removed by removing a large snap ring.

❋ 16-19 Low-reverse (rear) servo Remove the snap ring (Fig. 16-29) and take out the retainer, return spring, and servo.

❋ 16-20 Accumulator Remove the snap ring and take out the accumulator plate, spring, and piston. The accumulator plate is shown to the left in Fig. 16-29.

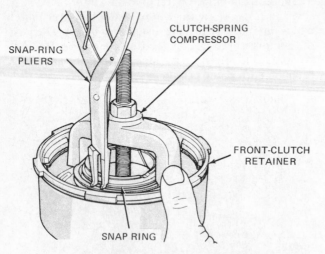

Fig. 16-26 Using the clutch-spring compressor to compress the return spring so the snap ring can be removed. This allows removal of the clutch piston. (*Chrysler Corporation*)

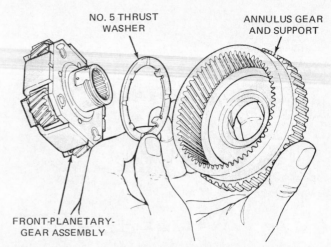

Fig. 16-28 Removing the front planetary gearset and thrust washer. (*Chrysler Corporation*)

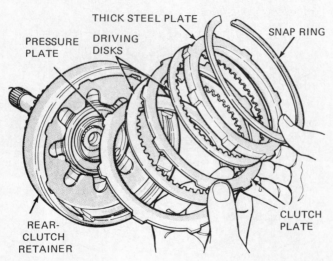

Fig. 16-27 Removing the rear-clutch clutch plates. (*Chrysler Corporation*)

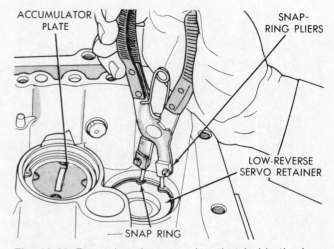

Fig. 16-29 Removing the snap ring that holds the low-reverse servo assembly in the case. (*Chrysler Corporation*)

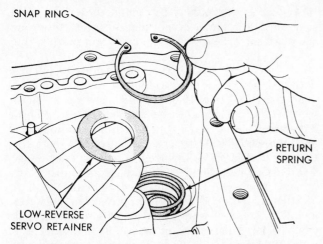

Fig. 16-30 Removing the kickdown piston. (*Chrysler Corporation*)

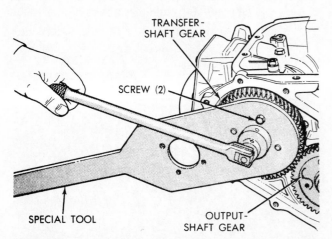

Fig. 16-31 Using a special tool, bolted to the transfer-shaft gear, to hold it while the gear-attaching nut is loosened. (*Chrysler Corporation*)

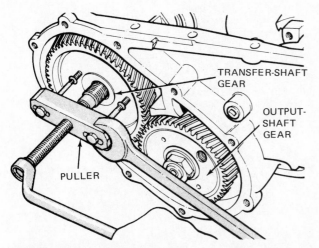

Fig. 16-32 Using a gear puller to pull the transfer-shaft gear. The wrench holds the puller steady when the screw is turned. (*Chrysler Corporation*)

✿**16-21 Kickdown servo** Use two screwdrivers to remove the kickdown-servo snap ring. Then remove the piston-rod guide, return spring, and piston rod. Use snap-ring pliers to lift the kickdown piston out (Fig. 16-30).

✿**16-22 Transfer shaft** Note the position of the transfer shaft in the transaxle (Fig. 15-3). Its purpose is to pick up the power output from the transmission (at right in Fig. 15-3) and transfer it to the differential (lower left). Servicing the transfer shaft follows.

1. Remove the transfer shaft by first removing the rear cover. Then use the tools as shown in Fig. 16-31 to remove the gear nut. Use the puller as shown in Fig. 16-32 to pull the gear. There is a select shim next to the bearing cone, which will come off with the gear. If the bearing cone needs replacement, use the special tools to remove the old and install the new cone.
2. Remove the governor-support retainer (Fig. 16-33). Replace the retainer-bearing cup with special tools if required.
3. Remove the low-reverse-band anchor pin.
4. Remove the governor assembly.
5. Remove the snap ring (Fig. 16-34) and use the slide hammer to remove the transfer shaft (Fig. 16-35).

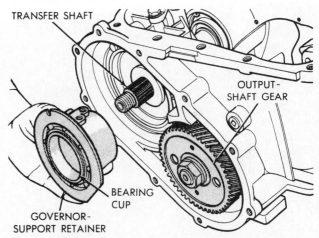

Fig. 16-33 Removing the governor-support retainer. (*Chrysler Corporation*)

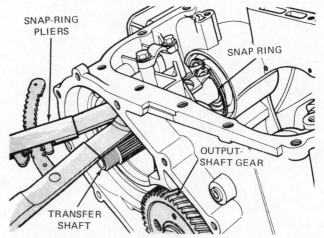

Fig. 16-34 Removing the transfer-shaft-retainer snap ring. (*Chrysler Corporation*)

6. The retainer with the outer bearing race can be slipped off (Fig. 16-36). If the shaft-bearing cone requires replacement, use the special tools to remove the old and install the new cone.

7. After reassembly, check the transfer-shaft end play (Fig. 16-37). If it is not correct, use a shim of a different thickness back of the transfer-shaft-gear bearing cone.

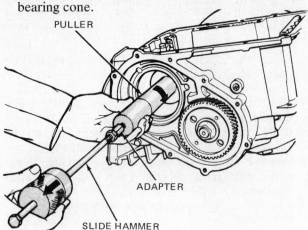

Fig. 16-35 Using a special puller with a slide hammer to remove the transfer shaft and retainer assembly. (*Chrysler Corporation*)

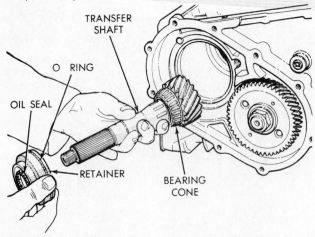

Fig. 16-36 Transfer shaft and retainer removed from case. (*Chrysler Corporation*)

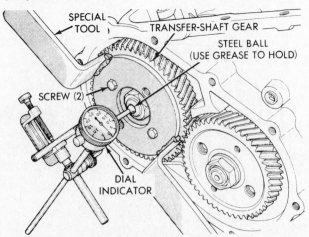

Fig. 16-37 Dial indicator installed to measure transfer-shaft end play. (*Chrysler Corporation*)

✿ 16-23 Parking pawl To remove the parking pawl, take out two retainer bolts, take off the retainer, pull the pivot shaft, and lift out the pawl and return spring.

✿ 16-24 Output shaft Remove the output-shaft gear with the same tools used to remove the transfer-shaft gear (Figs. 16-31 and 16-32). If the bearing cone on the gear needs replacement, use special tools to pull the old and install the new cone. Take out the output shaft and planetary annulus gear. Press the output shaft from the gear if necessary. Replace the bearing if necessary. The bearing cup remains in the case and must be removed, if necessary, by using a special puller tool.

After the output shaft has been pressed into the gear, install the shim, holding it in place with petroleum jelly (Fig. 16-38). Then install the assembly and tighten the gear-attaching nut to specifications. Check end play (Fig. 16-39). If it is not correct, use a different thickness of shim.

✿ 16-25 Differential repair The oil seal in the end of the extension housing can be removed and installed

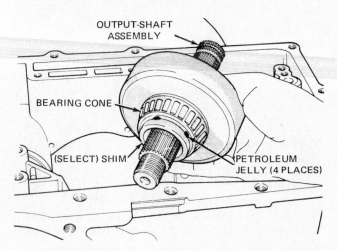

Fig. 16-38 Installing the select shim in place on the rear planetary ring gear, using petroleum jelly to hold the shim in place. (*Chrysler Corporation*)

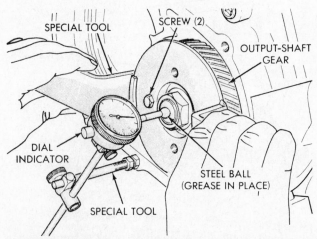

Fig. 16-39 Using dial indicator to check the output-shaft end play. (*Chrysler Corporation*)

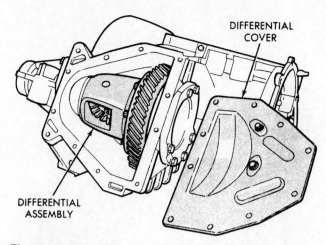

Fig. 16-40 Differential cover removed from the transaxle. (*Chrysler Corporation*)

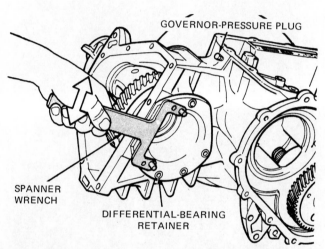

Fig. 16-41 Using a spanner wrench to gently rotate the differential-gearing retainer and remove it. (*Chrysler Corporation*)

without removing the housing. The housing does come off, however, when the differential is removed.

1. Remove the 10 differential-cover bolts and the cover (Fig. 16-40).
2. Remove the bearing-retainer bolts. Use the spanner wrench tool shown in Fig. 16-41 to gently rotate the bearing retainer and remove it.
3. Remove four extension-housing bolts and use the special tool to gently rotate the housing and remove it. Now, the differential assembly can be removed (Fig. 16-42).

Careful: Hold onto the differential while removing the extension housing to keep it from rolling out of the case.

4. If the two bearing cones that support the differential need replacement, use the special tools to remove the old cones and install the new cones.
5. To disassemble the differential, use a screw extractor to pull the pinion-shaft roll pin (Fig. 15-8). Then tap the pinion shaft loose with a brass drift punch

(Fig. 16-43). Now, the two side gears, two pinion gears, and thrust washers can be removed (Fig. 15-9).

6. If the ring gear needs replacement, remove the eight attaching bolts.
7. Replace the differential-bearing oil seal.
8. Remove the bearing cup from the bearing retainer (Fig. 16-44). There is a shim under this cup which determines the differential end play. Figure 16-45 shows the check of the end play. Turning torque of the differential is also checked to determine bearing preload (Fig. 16-46).

✿ **16-26 Valve-body service** Figure 16-19 shows the valve body and governor tubes being lifted from the transaxle. Figures 16-47 to 16-54 show the locations of the steel balls and valves. Follow these illustrations to service the valve body. Avoid mixing parts. Clean the parts and lay them out on the bench in the same relationship as they occupy in the valve body.

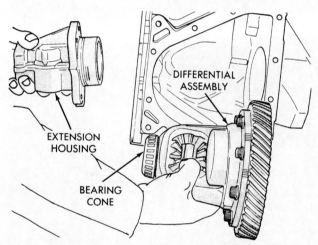

Fig. 16-42 Removing the extension housing and the differential assembly. (*Chrysler Corporation*)

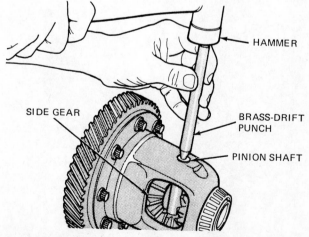

Fig. 16-43 Using a brass drift punch to drive the pinion shaft out after the roll pin has been removed. (*Chrysler Corporation*)

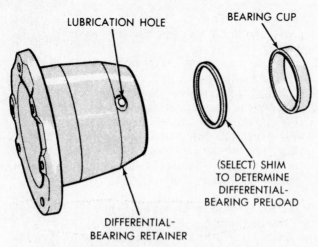

Fig. 16-44 Location of the select shim. (*Chrysler Corporation*)

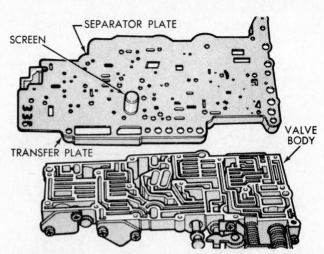

Fig. 16-47 Removing the transfer plate from the valve body. (*Chrysler Corporation*)

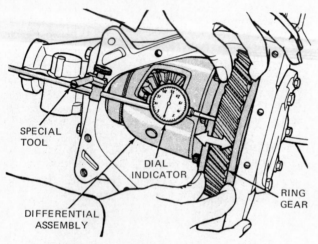

Fig. 16-45 Checking differential end play. (*Chrysler Corporation*)

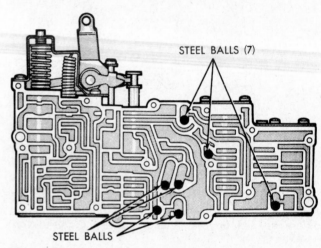

Fig. 16-48 Steel-ball locations. (*Chrysler Corporation*)

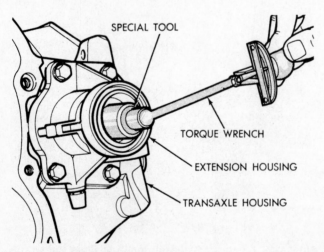

Fig. 16-46 Checking differential-bearing turning torque. (*Chrysler Corporation*)

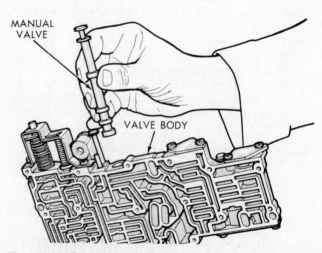

Fig. 16-49 Removing the manual valve from the valve body. (*Chrysler Corporation*)

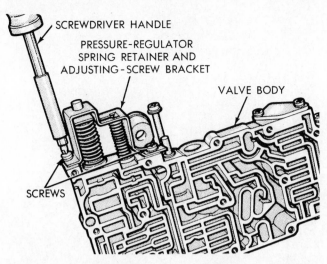

Fig. 16-50 Removing the pressure-regulator-and-adjusting-screw bracket. (*Chrysler Corporation*)

✿ 16-27 Reassembling the transaxle Generally, the reassembly procedure is the reverse of disassembly. Install the differential, extension housing, and cover. Then:

1. Install the output shaft as already explained (✿ 16-24).
2. Install the parking pawl.
3. Install the transfer shaft as already explained (✿ 16-22).
4. Install other parts that were removed—kickdown and low servos, accumulator, low-and-reverse band and strut, overrunning-clutch cam assembly, rear planetary-gear assembly, sun-gear driving shell,

front planetary-gear assembly, rear clutch, front clutch, kickdown band and strut, and oil pump. Reinstall the valve body, filter, oil pan, and parking rod.

Be sure to install the thrust washers and snap washers in their correct positions as you found them on disassembly.

✿ 16-28 Reinstalling the transaxle To reinstall the transaxle, put it on a transmission jack (Fig. 16-17). Raise it up into position. Reattach the insulators, mounting bracket, cooler tube, and switch wire.

1. Note the marks on the torque converter and drive plate and position the converter so they line up. Attach the torque converter with mounting bolts. Rotate the engine as necessary (Fig. 16-16) to get at all the bolts.
2. Install the drive shafts (Fig. 16-15), pushing them into the differential and securing the ends with Circlips (Fig. 16-14).
3. Insert the outer ends of the drive shafts into the wheel hubs (Fig. 16-13).
4. Attach the ball-joint-to-steering-knuckle bolts (Fig. 16-12) and attach the lower ball joints to the steering knuckles.
5. Install the sway bar.
6. Install the speedometer adapter, cable, and pinion.
7. Fill the differential with the specified type and amount of lubricant.
8. Add the correct amount and type of transmission fluid to the transaxle.
9. Adjust the linkages and the kickdown band.
10. Road-test the vehicle to make sure the transaxle is working properly before returning the car to the customer.

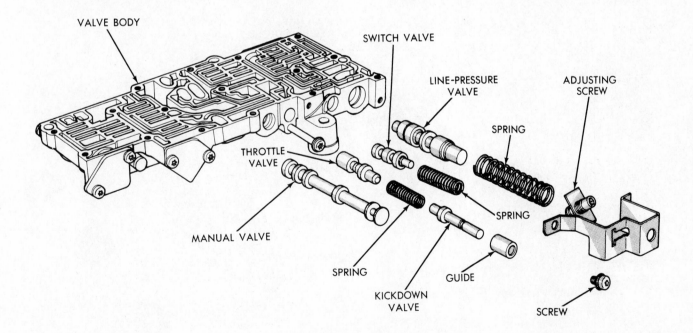

Fig. 16-51 Pressure regulators and manual controls. (*Chrysler Corporation*)

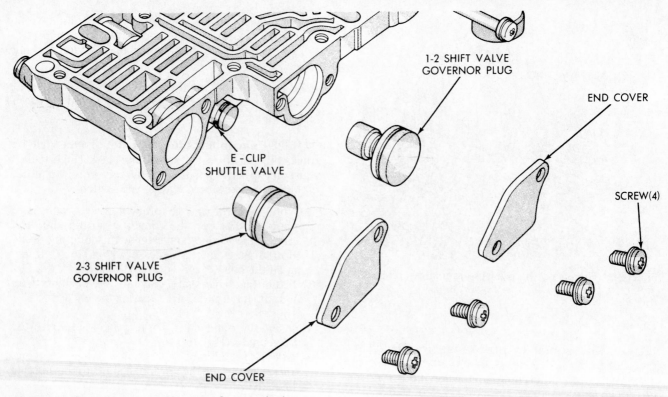

Fig. 16-52 Governor plugs. (*Chrysler Corporation*)

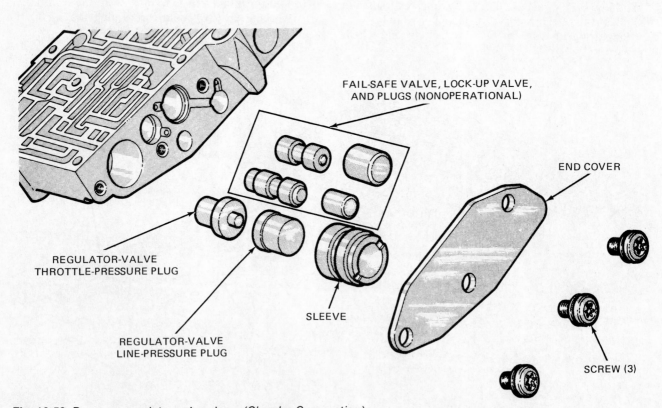

Fig. 16-53 Pressure-regulator-valve plugs. (*Chrysler Corporation*)

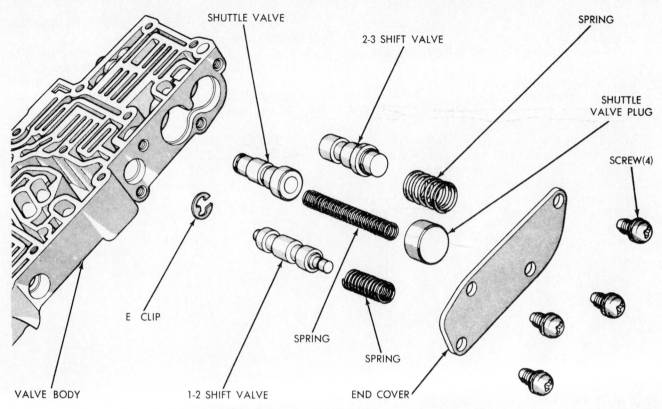

Fig. 16-54 Shift valves and shuttle valve. (*Chrysler Corporation*)

Labels in figure:
SHUTTLE VALVE
2-3 SHIFT VALVE
SPRING
SHUTTLE VALVE PLUG
SCREW(4)
E CLIP
VALVE BODY
1-2 SHIFT VALVE
SPRING
SPRING
END COVER

Chapter 16 review questions

Select the *one* correct, best, or most probable answer to each question. Then check your answers against the correct answers given at the end of the book.

1. Diagnosis of transaxle trouble should start with checks of the:
   a. clutches and bands
   b. stalling ability
   c. fluid and linkages
   d. torque converter and clutches
2. Mechanic A says that in the Chrysler transaxle the differential uses the same lubricant as the transaxle. Mechanic B says the differential is sealed off from the transaxle and uses a different lubricant. Who is right?
   a. A only
   b. B only
   c. both A and B
   d. neither A nor B
3. Mechanic A says the two linkage adjustments to be made are to the gearshift and throttle linkages. Mechanic B says the adjustments are to the linkages to the manual valve and the kickdown valve. Who is right?
   a. A only
   b. B only

c. both A and B
d. neither A nor B
4. The throttle-cable adjustment is made with the:
   a. engine hot and not running
   b. engine cold and not running
   c. engine hot and operating at medium speed
   d. engine hot and idling
5. Mechanic A says both bands in the transaxle can be adjusted. Mechanic B says only the front, or kickdown, band can be adjusted. Who is right?
   a. A only
   b. B only
   c. both A and B
   d. neither A nor B
6. The purpose of making air-pressure tests is to check the:
   a. operation of the band servos and clutches
   b. pressures in the hydraulic system in N
   c. hydraulic system for leaks
   d. torque converter for freeness
7. Mechanic A says you do not have to remove the engine from the vehicle when removing the transaxle. Mechanic B says you must remove the engine and transaxle together. Who is right?
   a. A only
   b. B only

c. both A and B

d. neither A nor B

8. To remove the transaxle, the drive shafts must be disconnected:

   a. from the transaxle only

   b. from the wheel hubs only

   c. from both the transaxle and the wheel hubs

   d. from the universal joints only

9. Which of the four statements that follow is correct?

   a. A clutch-spring compressor is required to disassemble both the front and rear clutches.

   b. Neither clutch requires a spring compressor to disassemble it.

   c. Only the front clutch requires a spring compressor to disassemble it.

   d. Only the rear clutch requires a spring compressor to disassemble it.

10. Mechanic A says the transfer shaft is between the transmission output shaft and the differential. Mechanic B says the transfer shaft is geared to the output-shaft gear and to the differential ring gear. Who is right?

   a. A only

   b. B only

   c. both A and B

   d. neither A nor B

CHAPTER 17

GENERAL MOTORS 125 AUTOMATIC TRANSAXLE

After studying this chapter, you should be able to:

1. Describe the similarities between the General Motors 125 transaxle and the General Motors 200C automatic transmission.
2. Identify and name the parts of a disassembled 125 transaxle.
3. Explain why the transaxle requires a differential and how the differential works.
4. Identify which transaxle members are holding in the various operating ranges.

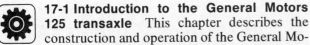

17-1 Introduction to the General Motors 125 transaxle This chapter describes the construction and operation of the General Motors 125 automatic transaxle. Transaxles are mounted on the engine (Fig. 15-1). The engine is mounted sideways, or transversely, in the vehicle (Fig. 15-2). Figure 17-1 is a partial cutaway view of the 125 automatic transaxle. This transaxle includes an automatic transmission based on the General Motors type 200 automatic transmission, described in Chap. 8. The transaxle includes a differential. Two drive shafts connect the differential to the front-wheel hubs (Fig. 15-10). Figure 17-2 shows the transaxle with the two drive shafts disconnected from the transaxle.

17-2 General Motors 125 transaxle construction The transaxle includes a torque converter, a three-speed automatic transmission, and a differential. Compare Fig. 17-1 with Fig. 8-1, which is a cutaway view of the 200 transmission. Note that the transaxle and the 200C transmission are very similar. Both have three multiple-disk clutches, a roller clutch, and a band to provide the three forward speeds and reverse.

The transaxle torque converter has, in effect, been cut off the end of the transmission and mounted up above it. The transmission turbine shaft has a drive sprocket on it. This drive sprocket is connected by a drive chain to a driven sprocket on the transmission output shaft (Fig. 17-3). The power flows from the turbine shaft to the output shaft, through the transmission and into the differential. The differential sends the power to the front wheels through the two drive shafts.

17-3 Hydraulic system The hydraulic system (Fig. 17-4) has the usual valves, servo, accumulator, and governor. This hydraulic system is very similar to that for the 200 hydraulic system (Figs. 8-2 and 8-3). The two hydraulic systems look different, but both have accumulator pistons, pressure-regulator valves, intermediate servo, 1-2 and 2-3 shift valves, manual valve, clutches, and band.

The pump is different for the two hydraulic systems. The 200C uses a gear-type pump. The 125 transaxle uses a vane-type pump. The 125 transaxle hydraulic circuit does not have an actuator valve, solenoid, or governor pressure switch because it does not have the torque-converter clutch, which the 200C does have. In the *Color Plates of Automatic-Transmission Hydraulic Circuits*, sixth edition, compare Plates 21 to 23, which show the 200C hydraulic circuit, with Plates 28 to 30, which show the 125 hydraulic circuits.

17-4 General Motors 125 transaxle operation Engine torque enters the torque converter and drives the turbine shaft. The turbine shaft, connected by sprockets and chain to the output shaft of the transmission, drives the transmission. The transmission, responding to the positions of the selector lever and manual valve, and to engine and car speed, shifts into the required gear. Figure 17-5 shows in chart form which members of the transmission are holding (applied or engaged) and which are released or disengaged.

For a detailed discussion of how the valves and other components work in the various selector-lever positions and operating modes, refer to Chap. 8. Note that

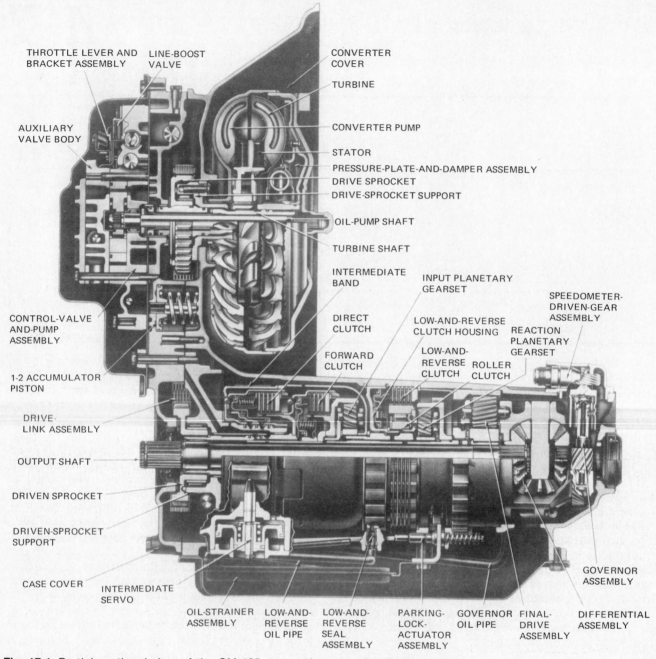

THROTTLE LEVER AND BRACKET ASSEMBLY
LINE-BOOST VALVE
CONVERTER COVER
TURBINE
AUXILIARY VALVE BODY
CONVERTER PUMP
STATOR
PRESSURE-PLATE-AND-DAMPER ASSEMBLY
DRIVE SPROCKET
DRIVE-SPROCKET SUPPORT
OIL-PUMP SHAFT
TURBINE SHAFT
CONTROL-VALVE AND-PUMP ASSEMBLY
INTERMEDIATE BAND
INPUT PLANETARY GEARSET
SPEEDOMETER-DRIVEN-GEAR ASSEMBLY
DIRECT CLUTCH
LOW-AND-REVERSE CLUTCH HOUSING
REACTION PLANETARY GEARSET
FORWARD CLUTCH
LOW-AND-REVERSE CLUTCH
ROLLER CLUTCH
1-2 ACCUMULATOR PISTON
DRIVE-LINK ASSEMBLY
OUTPUT SHAFT
DRIVEN SPROCKET
DRIVEN-SPROCKET SUPPORT
GOVERNOR ASSEMBLY
CASE COVER
INTERMEDIATE SERVO
OIL-STRAINER ASSEMBLY
LOW-AND-REVERSE OIL PIPE
LOW-AND-REVERSE SEAL ASSEMBLY
PARKING-LOCK-ACTUATOR ASSEMBLY
GOVERNOR OIL PIPE
FINAL-DRIVE ASSEMBLY
DIFFERENTIAL ASSEMBLY

Fig. 17-1 Partial sectional view of the GM 125 automatic transaxle. (*Pontiac Motor Division of General Motors Corporation*)

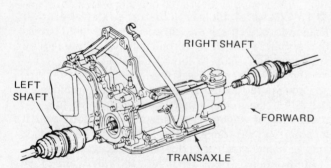

RIGHT SHAFT
LEFT SHAFT
FORWARD
TRANSAXLE

Fig. 17-2 Transaxle with both drive shafts disconnected. (*Pontiac Motor Division of General Motors Corporation*)

DRIVE SPROCKET
DRIVEN SPROCKET
LINK ASSEMBLY

Fig. 17-3 Drive and driven sprockets and drive chain being lifted from the case. (*Pontiac Motor Division of General Motors Corporation*)

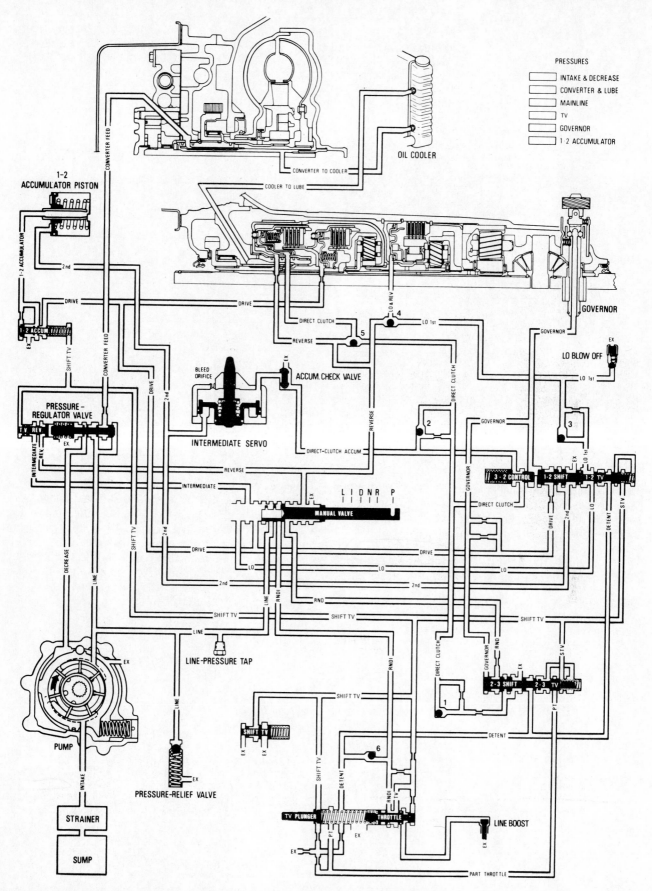

Fig. 17-4 Hydraulic system of the GM 125 transaxle, shown in Park. (*Pontiac Motor Division of General Motors Corporation*)

	INTERMEDIATE BAND	DIRECT CLUTCH	FORWARD CLUTCH	ROLLER CLUTCH	LOW-AND-REVERSE CLUTCH
Park (engine on)	R	D	D		D
N	R	D	D		D
D1			E	L	
D2	A		E		
D3		E	E		
Part-throttle 3-2 downshift	A		E		
Detent downshift	A		E		
Intermediate range			E		
Low			E		E
R		E			E

R—released
A—applied
E—engaged
D—disengaged
L—locked

Fig. 17-5 Clutch-engagement and band-application chart for the GM 125 transaxle. (*Pontiac Motor Division of General Motors Corporation*)

the 200C automatic transmission covered in Chap. 8 has a torque-converter clutch while the 125 transaxle does not. Otherwise, the operating elements of the 200C and the 125 work in the same way.

⚙ 17-5 Differential action The differential and final drive are shown disassembled in Fig. 17-6. After passing through the transmission, the engine torque is delivered to the differential. The differential works as explained in Chap. 1 (⚙ 1-14) and illustrated in Figs. 1-36 to 1-38. The differential allows the two driving wheels to turn at different speeds when the car goes around a turn. When this happens, the outside wheel has to travel farther and turn faster than the inside

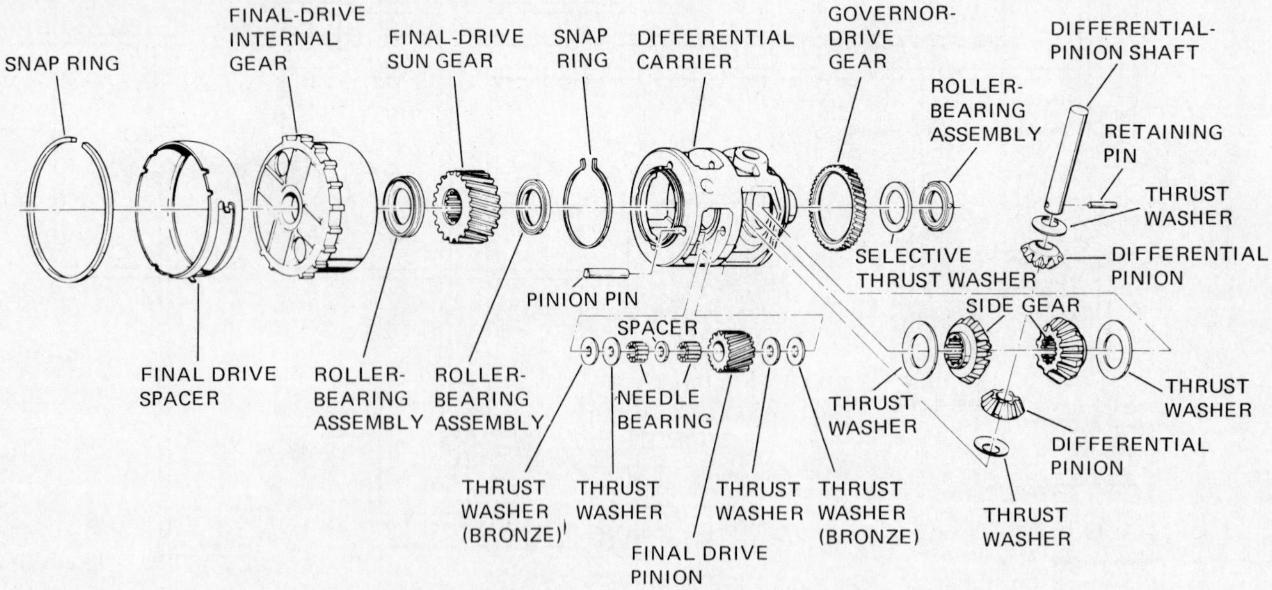

Fig. 17-6 Disassembled differential and final drive of the GM 125 transaxle. (*Chevrolet Motor Division of General Motors Corporation*)

wheel (Fig. 15-11). The differential permits this. When the car is moving straight ahead, the differential turns both drive shafts and both front wheels at the same speed. But when the car makes a turn, the differential pinions rotate on the pinion shaft to increase the speed of the side gear that is splined to the outer-wheel drive shaft.

Chapter 17 review questions

Select the *one* correct, best, or most probable answer to each question. Then check your answers against the correct answers given at the end of the book.

1. The General Motors 125 automatic transaxle includes a compound planetary gearset, a roller clutch, and:
   a. two multiple-disk clutches and two bands
   b. three multiple-disk clutches and two bands
   c. three multiple-disk clutches and one band
   d. four multiple-disk clutches and no bands
2. Mechanic A says the major difference between the 125 automatic transaxle and the 200 automatic transmission is that the torque converter has been repositioned. Mechanic B says the turbine shaft on the 125 transaxle drives the output shaft through a drive chain. Who is right?
   a. A only
   b. B only
   c. both A and B
   d. neither A nor B
3. When the selector lever is moved to R, the following hold:
   a. direct and intermediate clutches
   b. forward and direct clutches
   c. low-and-reverse clutch and intermediate band
   d. low-and-reverse and direct clutches
4. During a detent downshift to second, the following hold:
   a. direct and forward clutches
   b. forward clutch and intermediate band
   c. forward and roller clutches
   d. forward and reverse clutches
5. In drive range, third gear, the following are holding:
   a. forward and direct clutches

   b. direct clutch and intermediate band
   c. direct and reverse clutches
   d. direct and roller clutches
6. In drive range, second gear, the following are holding:
   a. forward clutch and reverse clutch
   b. forward clutch and intermediate band
   c. forward and intermediate clutches
   d. forward and roller clutches
7. In drive range, first gear, the following are holding:
   a. forward and reverse clutches
   b. roller clutch and band
   c. forward and intermediate clutches
   d. forward clutch and roller clutch
8. Mechanic A says the 1-2 accumulator piston smooths the 1-2 upshift. Mechanic B says the servo piston acts as a 2-3 accumulator to smooth the 2-3 upshift. Who is right?
   a. A only
   b. B only
   c. both A and B
   d. neither A nor B
9. Opposing the governor pressure working on the 2-3 shift valve is the:
   a. 2-3 throttle valve
   b. manual valve
   c. regulator valve
   d. clutch valve
10. After a part-throttle 3-2 downshift, the following are holding:
    a. direct and forward clutches
    b. intermediate band and forward clutch
    c. forward clutch and low-and-reverse clutch
    d. roller and direct clutch

GENERAL MOTORS 125
AUTOMATIC TRANSAXLE SERVICE

After studying this chapter, and with proper instruction and equipment, you should be able to:

1. Explain how to use the trouble-diagnosis chart to find causes of trouble.
2. Discuss manual- and throttle-linkage adjustments.
3. Describe the procedure for checking the oil pressure.
4. Explain how to remove and reinstall the transaxle.
5. Describe the procedure of disassembling and reassembling the transaxle and of servicing the subassemblies.
6. Perform the services listed above.

 18-1 Introduction to General Motors 125 transaxle service This chapter describes the trouble diagnosis, checks, adjustments, removal, overhaul, and installation of the General Motors 125 automatic transaxle. Chapter 17 describes the construction and operation of this transaxle. The transaxle has a differential and torque converter, clutches, band, planetary gearset, and a hydraulic system that controls the actions of the transaxle shifts. Figure 17-1 is a partial cutaway of the transaxle.

In this chapter, the service procedures that apply to the 125 transaxle are covered. Many of these services are the same as or similar to those for other transmissions discussed earlier. Reference is made to earlier chapters for procedures that apply to both the transaxle and other transmissions.

✿ **18-2 Transaxle trouble diagnosis** The chart in ✿ 18-9 lists various transaxle problems, possible causes, and checks or corrections. The sequence for trouble diagnosis recommended by General Motors follows.

1. Check and correct the fluid level.
2. Check and adjust the throttle-valve (TV) cable if necessary.
3. Check and correct the manual linkage if necessary.
4. Check engine condition. It should be in good operating condition. If it lacks power, the transaxle will not function properly.
5. Install oil-pressure gauge and tachometer to check fluid pressures.

6. Road-test the car, noting oil pressures in shifts in all ranges. Note the quality of the shifts. The fluid pressure can also be checked in the shop (✿ 18-6).

✿ **18-3 Fluid level and condition** Refer to ✿ 9-3 for a discussion of how to check fluid level and condition. The dipstick for the 125 transaxle is different from others. The COLD reading is *above* the FULL mark (Fig. 18-1). The shape of the filler tube may make the fluid-level reading misleading. Look carefully for the fluid ring on both sides of the dipstick. Here is the procedure:

1. Shift to P (park) and leave the selector lever in that position. Do not move the lever through the ranges.
2. The car must be on a level surface and the brakes should be applied.
3. Start and idle the engine.
4. Check fluid level on the dipstick. If it is low, check for leaks (✿ 18-5).
5. Check also the condition of the fluid. Checking fluid level and condition is covered in ✿ 9-3. If the fluid

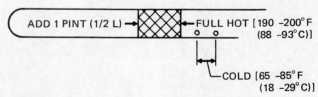

Fig. 18-1 Dipstick for the 125 transaxle. Note that the COLD reading is *above* the FULL mark. (*Pontiac Motor Division of General Motors Corporation*)

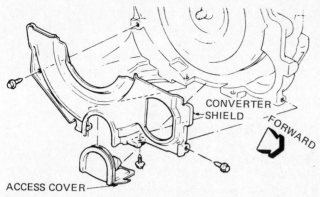

Fig. 18-2 Converter shield for the V-6 engine. (*Pontiac Motor Division of General Motors Corporation*)

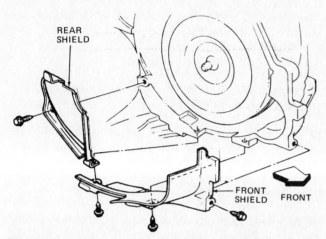

Fig. 18-3 Converter shield for the four-cylinder engine. (*Pontiac Motor Division of General Motors Corporation*)

is contaminated, the oil cooler and lines must be flushed out, as well as the transmission itself. Also, a new torque converter will be required because a contaminated torque converter cannot be flushed out satisfactorily in the field.

6. At room temperature of 70°F [21°C] the fluid should be about ¼ inch [7 mm] above the FULL mark or between the two dimples on the dipstick. At operating temperature of 200°F [90°C] the fluid level should be between the ADD and the FULL marks.

7. If fluid is needed, do not overfill. It takes only 1 pint [0.47 L] to raise the level from ADD to FULL with a hot transaxle. Overfilling can cause foaming, which will cause fluid loss through the vent, and poor clutch and band action.

☼ 18-4 Changing fluid   In normal operation, change the fluid and replace the strainer every 100,000 miles [161,000 km]. In severe service, such as city traffic (stop-and-go), pulling a trailer, taxicabs, police cars, or door-to-door delivery service, change the fluid and install a new strainer every 15,000 miles [24,140 km].

Draining the fluid and adding new fluid is the same as for other transmissions previously discussed. Loosen the rear oil pan and pry it down so the fluid drains into a container. Then remove the oil pan and gasket, discarding the gasket. Clean the pan. Remove and discard the strainer and O-ring seal. Install a new strainer and O-ring seal. Use a new gasket and install the oil pan, tightening the bolts to specifications.

Add the specified amount and type of fluid to the transaxle. The correct fluid for this transaxle is Dexron II automatic-transmission fluid.

Recheck the fluid level (☼ 18-3).

TV CONTROL CABLE ADJUSTMENT

AFTER INSTALLATION INTO TRANSMISSION, INSTALL CABLE FITTING INTO ENGINE BRACKET. CAREFUL: CABLE MUST NOT BE ADJUSTED BEFORE OR DURING ASSEMBLY INTO BRACKET.

1. INSTALL CABLE TERMINAL TO THROTTLE-IDLER LEVER.
2. ROTATE THROTTLE-IDLER LEVER TO "FULL TRAVEL STOP" POSITION TO SET AUTOMATIC CABLE ADJUSTER ON CABLE TO CORRECT SETTING.
3. RELEASE THROTTLE-IDLER LEVER.

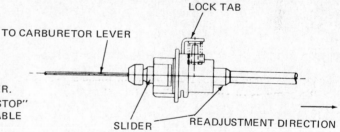

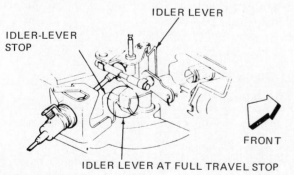

TV CONTROL CABLE READJUSTMENT

IN CASE READJUSTMENT IS NECESSARY BECAUSE OF INADVERTANT ADJUSTMENT BEFORE OR DURING ASSEMBLY, OR FOR DOWN LINE REPAIR.

PERFORM FOLLOWING:

1. DEPRESS AND HOLD METAL LOCK TAB.
2. MOVE SLIDER BACK THROUGH FITTING IN DIRECTION AWAY FROM THROTTLE-IDLER LEVER UNTIL SLIDER STOPS AGAINST FITTING.
3. RELEASE METAL LOCK TAB.
4. REPEAT STEPS 2 AND 3 OF ADJUSTMENT.

Fig. 18-4 Throttle-valve (TV) cable adjustment. (*Pontiac Motor Division of General Motors Corporation*)

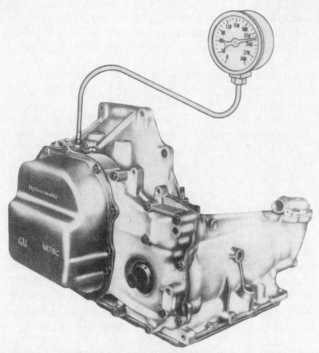

Fig. 18-5 Oil-pressure gauge attached, ready for pressure checks. (*Pontiac Motor Division of General Motors Corporation*)

✸ 18-5 Checking for fluid leaks If fluid has been lost, check as follows to locate the source. The fluid is red and this will help pinpoint where the leak occurs.

1. Clean the suspected leak area with solvent.
2. Remove the converter shield (Figs. 18-2 and 18-3).
3. After cleaning the suspected area, spray it with a pressure can of white foot powder. This will show up the red fluid at the leak point.
4. Start and run the engine at high idle speed.
5. Repair the leak by replacing gaskets or seals, or tightening the attaching bolts.

✸ 18-6 Adjusting the throttle-valve (TV) control cable Check for freeness by pulling out on the upper end of the cable. It should return to the zero position when released.

Make the adjustment as shown in Fig. 18-4. This illustration shows how to adjust the cable. The adjustment is checked with an oil-pressure gauge (Fig. 18-5) and a tachometer. Take the pressure readings at the engine speed shown in Fig. 18-6, as follows:

1. With the brakes applied, take the line pressure readings in the ranges and at the engine speed indicated.
2. To get maximum TV pressure readings without overspeeding the engine, tie or hold the TV cable at full travel. With the brakes applied and the engine speed as shown in Fig. 18-6, take the pressure readings. Note that items starred (*) should be checked in less than 2 minutes. Longer will overheat and possibly damage the transaxle.

✸ 18-7 Shifter-cable adjustment Figures 18-7 and 18-8 show the shifter-cable adjustment. Remove the lock pin from the pivot pin. Remove the lock clip that attaches the cable to the mounting bracket. Then proceed as follows:

1. Shift to N. Put the transaxle lever in N by moving the lever to L and then move it back (counterclockwise) through three detent positions (L, S, and D) to neutral (N).
2. Attach the shift cable to the pin.
3. Adjust the shift selector lever in N. Guide the clip on the edge of the shift bowl to the central position (N). Push clip into bowl.

✸ 18-8 Road test Road-test the car to check the upshifts and downshifts. Note their quality and the speeds and throttle openings at which they occur. Figure 18-9 shows the clutch engagement and band application in the different ranges.

✸ 18-9 Transaxle trouble-diagnosis chart Pressure checks and a road test provide information as to the source of trouble in the transaxle. Following is a trouble-diagnosis chart that lists various complaints that the driver might report to you, or problems that you may have found during testing. To the right of the troubles are listed possible causes and the checks or corrections to be made.

When analyzing a trouble, refer to Fig. 18-9, which shows the clutches and bands that should engage or apply in each range. For example, if there is slipping in drive, third gear, the direct clutch or the forward clutch, both of which should engage in drive, third gear, may be slipping.

NOTE: If the transaxle has been overhauled, trouble could be due to missing, misplaced, or incorrectly installed parts such as check balls, clutch plates, seals, or snap rings.

GENERAL MOTORS TYPE 125 AUTOMATIC TRANSAXLE TROUBLE-DIAGNOSIS CHART

PROBLEM	POSSIBLE CAUSE	CHECK OR CORRECTION
1. No drive in D (drive range)	a. Low fluid level	Leaks—porous casting, damaged seals, loose bolts.
	b. Manual linkage misadjusted	Readjust.
	c. Low fluid pressure	Strainer plugged or restricted, strainer O ring damaged, pressure-regulator valve stuck, pump damaged, gasket damaged or out of position.

PROBLEM	POSSIBLE CAUSE	CHECK OR CORRECTION
	d. Forward clutch	Does not engage, because of internal damage, leak in feed circuit, stuck ball check, plugged feed orifice in input shaft.
	e. One-way clutch	Not holding, because of internal damage—rollers galled or missing, springs broken or missing.
2. High or low fluid pressure (Fig. 18-6)	*a.* Trouble in TV linkage	Cable damaged or loose, throttle lever damaged or loose.
	b. Trouble in valve body	Valves sticking—check TV valve or plunger, shift TV valve, line-boost valve, TV boost valve, reverse-boost valve, pressure-regulator valve, etc.
	c. Manual valve unhooked	Reattach; adjust linkage.
	d. Pump damaged	Remove for repair.
3. 1-2 shift full throttle only	*a.* TV linkage trouble	Cable or throttle lever damaged or loose, TV valve or plunger binding.
4. First speed only, no 1-2 upshift	*a.* Governor and feed passages	Remove governor for check. Check seals, fluid passages.
	b. 1-2 shift train	1-2 shift valve or 1-2 throttle valve stuck in downshift.
	c. Case and case cover	Defect in case—porosity or undrilled 2d speed holes. Leakage at intermediate-band apply pin. Band defective.
	d. Intermediate servo	Oil-seal ring missing or damaged; wrong apply pin.
5. First and second speeds only; no 2-3 shift	*a.* 2-3 shift train	2-3 shift valve or 2-3 throttle valve stuck in downshift.
	b. Direct clutch	Feed line blocked, oil seals damaged, clutch plates damaged, snap ring out of groove.
	c. Intermediate servo	Oil seal damaged.
	d. Internal parts misplaced or missing	This would occur only after an overhaul. Check for placement of balls in valve assembly, missing plates, misplaced snap ring, missing or damaged seals, etc.
6. Third speed only	*a.* 2-3 shift valve	Stuck in upshift.
	b. Governor	Oil passages or pipe plugged.
7. Drive in neutral	*a.* Manual linkage	Misadjusted or disconnected.
	b. Forward clutch	Stuck closed because check ball is missing or plates are burned together.
	c. Case and cover	Cross-leakage to forward-clutch passages.
8. No drive in reverse or slips in reverse	*a.* Throttle valve or manual valve linkage	Readjust, check operation.
	b. Throttle valve, shift TV or reverse-boost valve stuck in low	Disassemble control-valve body for service.
	c. Low-and-reverse or direct clutch not engaging, or slipping	Plates, piston, seals damaged or missing. Feed lines plugged or leaking.
9. Slips on 1-2 shift	*a.* Low fluid level	Add fluid, check for leaks.
	b. 1-2 accumulator piston	Sticking, weak, or broken spring.
	c. 1-2 accumulator valve	Sticking, weak spring.
	d. Intermediate band	Worn or burned.
	e. Intermediate servo	Seal missing or damaged.
	f. Throttle-valve linkage	Readjust.
	g. Throttle or shift TV valve	Sticking.

PROBLEM	POSSIBLE CAUSE	CHECK OR CORRECTION
10. Rough 1-2 shift	*a.* Throttle-valve linkage *b.* Throttle-valve and plunger *c.* Shift TV or 1-2 accumulator valve *d.* 1-2 accumulator *e.* Intermediate servo	Misadjusted or binding. Binding. Binding. Stuck piston, damaged seal, broken spring. Piston seal damaged.
11. Slips in 2-3 shift	*a.* Low fluid level *b.* Throttle-valve linkage *c.* Throttle valve *d.* Intermediate servo *e.* Direct clutch	Add fluid, check for leaks. Misadjusted. Sticking. Fluid seal damaged. Seal rings damaged, plates burned, feed passages blocked.
12. Rough 2-3 shift	*a.* Throttle-valve linkage *b.* Throttle valve and plunger *c.* Shift throttle valve	Misadjusted or binding. Binding. Binding.
13. No engine braking—intermediate range, second gear	*a.* Intermediate servo *b.* Intermediate band	Seal ring missing or damaged. Mispositioned or burned.
14. No engine braking—low, first gear	*a.* Low-reverse clutch	Plates burned, seal leaking, piston damaged.
15. No part-throttle downshifts	*a.* Throttle plunger *b.* 2-3 throttle valve *c.* Throttle-valve linkage	Passage not open—check bushing. Passage not open—check bushing. Misadjusted.
16. High or low shift points (Fig. 18-6)	*a.* Throttle-valve linkage *b.* Throttle valve, shift TV valve, line-boost valve, throttle-valve plunger, 1-2 or 2-3 throttle valve *c.* Governor	Misadjusted. Binding of any of these valves can affect the shift point. Seal ring defective; governor malfunctioning.
17. Won't hold in park	*a.* Manual linkage *b.* Internal linkage	Misadjusted. Binding, damaged, disconnected.
18. Noisy	*a.* Pump noise *b.* Gear or bearing noise	Fluid low: Add fluid, check for leaks. Plugged line to pump. Water (coolant) in fluid. Transaxle mount worn or loose, allowing transaxle to touch frame or body part. Worn or defective gears or bearings. If noise occurs on turns only, differential is probably at fault.

❋**18-10 Transmission-service precautions** Cleanliness is very important in automatic-transmission service work. Service precautions are discussed in ❋ 9-10.

❋**18-11 Parts cleaning and inspection** During and after disassembly, clean and inspect all parts for wear and other damage. Refer to ❋ 9-11 for details. If the transaxle shows signs of internal damage and there is evidence that foreign material has been circulating with the transmission fluid, then all parts, including the oil cooler and lines, must be flushed out. A new filter will be required. Also, a new or exchange torque converter may be required because there is no satisfactory way to flush out the converter in the field.

Do not use solvents on neoprene seals, composition-faced clutch plates, or thrust washers. This could damage the parts. Discard old O rings, gaskets, and other seals on disassembly.

❋**18-12 On-car service** Several services can be performed with the transaxle in the car, in addition to checking fluid level and condition, and adjusting linkages. These include removing and replacing the governor, speedometer driven gear, oil pan and gasket, and valve-body assembly. The removal and replace-

MODEL	RANGE	NORMAL OIL PRESSURE AT MINIMUM TV		NORMAL OIL PRESSURE AT FULL TV	
		kPa	psi	kPa	psi
PZ,CV	Park at 1000 rpm	450–515	65–75	No TV pressure in park. Line pressure is the same as park at minimum TV.	
PZ CV	*Reverse at 1000 rpm	825–895 825–895	120–130 120–130	1275–1450 1275–1825	185–210 185–265
PZ CV	Neutral at 1000 rpm	450–515 450–515	65–75 65–75	825–895 965–1035	120–130 140–150
PZ CV	*Drive at 1000 rpm	450–515 450–515	65–75 65–75	825–895 965–1035	120–130 140–150
PZ, CV	*Intermediate at 1000 rpm	790–895	115–130	790–895	115–130
PZ, CV	*Low at 1000 rpm	790–895	115–130	No TV pressure in low range. Line pressure is the same as intermediate at minimum TV.	

Fig. 18-6 Chart of correct oil pressures for various driving ranges. Total running time for tests shown with an asterisk (*) must not exceed 2 minutes. (*Pontiac Motor Division of General Motors Corporation*)

ment of these parts is covered in the following sections, which describe disassembly and reassembly of the transaxle.

✿ 18-13 Transaxle removal Basically, the removal procedure requires detachment of all related parts and then lowering the transaxle from the car.

1. Disconnect the battery ground cable and tie it up out of the way on the upper radiator hose.
2. Remove the detent cable at the carburetor and take out the detent-cable attaching screw at the transaxle (Figs. 18-10 and 18-11). Then pull up on the detent-cable cover to expose the cable. Disconnect the cable from the rod.

3. If the vehicle has transaxle strut brackets (Fig. 18-12), remove the bolts attaching each bracket to the transaxle.
4. Remove all engine-to-transaxle bolts except the one by the starting motor (Fig. 18-13). Break this bolt loose but do not remove it yet. The bolt nearest the cowl is installed from the engine side with a short-handled box or ratchet wrench.
5. Disconnect the speedometer cable at the upper and lower cable couplings (Fig. 18-14).
6. On cars with cruise control, remove the speedometer cable at the cruise transducer (Fig. 18-15).
7. Remove the retainer clip and washer from the transaxle linkage and the two shift-linkage-bracket bolts (Figs. 18-16 and 18-17).

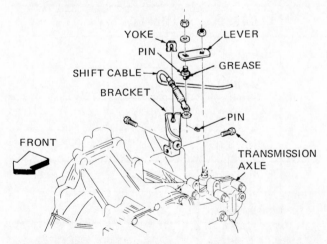

Fig. 18-7 Transaxle shift-cable mounting. (*Chevrolet Motor Division of General Motors Corporation*)

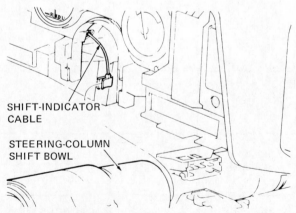

Fig. 18-8 Shift-indicator cable. (*Chevrolet Motor Division of General Motors Corporation*)

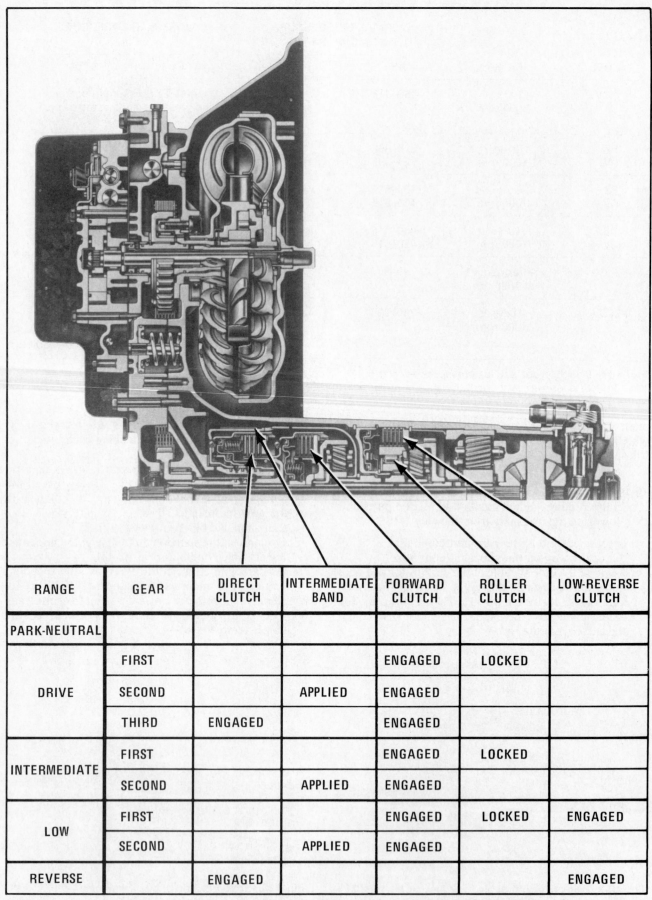

RANGE	GEAR	DIRECT CLUTCH	INTERMEDIATE BAND	FORWARD CLUTCH	ROLLER CLUTCH	LOW-REVERSE CLUTCH
PARK-NEUTRAL						
DRIVE	FIRST			ENGAGED	LOCKED	
	SECOND		APPLIED	ENGAGED		
	THIRD	ENGAGED		ENGAGED		
INTERMEDIATE	FIRST			ENGAGED	LOCKED	
	SECOND		APPLIED	ENGAGED		
LOW	FIRST			ENGAGED	LOCKED	ENGAGED
	SECOND		APPLIED	ENGAGED		
REVERSE		ENGAGED				ENGAGED

Fig. 18-9 Clutch-engagement and band-application chart. (*Pontiac Motor Division of General Motors Corporation*)

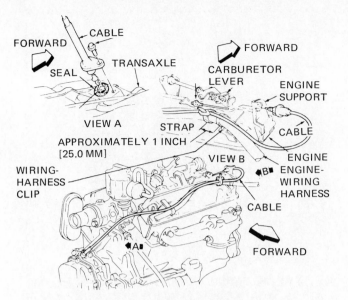

Fig. 18-10 Throttle-valve (TV) control cable on V-6 engine. (*Pontiac Motor Division of General Motors Corporation*)

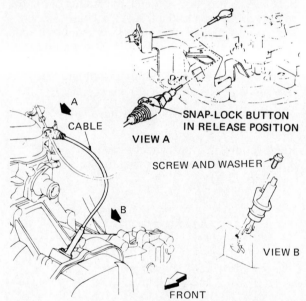

Fig. 18-11 Throttle-valve control cable on four-cylinder engine. (*Pontiac Motor Division of General Motors Corporation*)

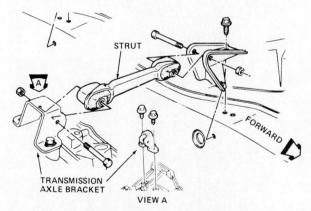

Fig. 18-12 Transaxle strut. (*Pontiac Motor Division of General Motors Corporation*)

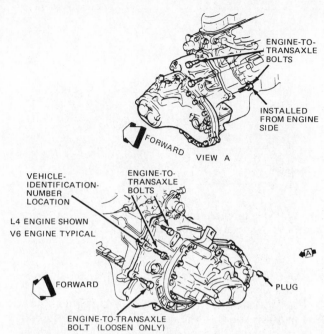

Fig. 18-13 Transaxle-to-engine mounting. (*Pontiac Motor Division of General Motors Corporation*)

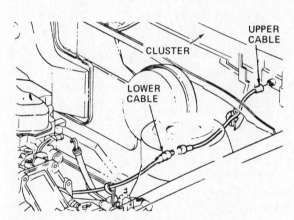

Fig. 18-14 Upper and lower speedometer cable. (*Pontiac Motor Division of General Motors Corporation*)

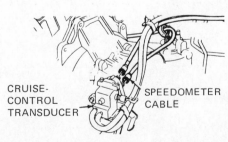

Fig. 18-15 Speedometer-cable connection at cruise-control transducer. (*Pontiac Motor Division of General Motors Corporation*)

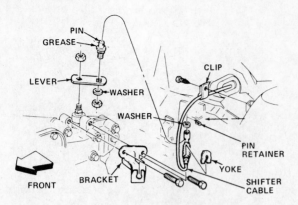

Fig. 18-16 Column-shift cable mounting. (*Pontiac Motor Division of General Motors Corporation*)

8. Disconnect the transaxle cooler lines at the transaxle. Use a ½-inch wrench with an $^{11}/_{16}$-inch backup wrench so you do not twist the cooler lines.
9. Install an engine-holding fixture and attach the fixture hook to the engine-lift ring (Fig. 16-10). Raise the engine just enough to take the weight off the engine mounts.
10. Unlock the steering column.
11. Raise the car on a lift.
12. Remove the two nuts attaching the stabilizer bar to the driver-side lower control arm (Fig. 18-18).
13. Remove the bolts attaching the retainer plate holding the driver-side stabilizer bar to the cradle (Fig. 18-19).
14. Loosen the four bolts holding the stabilizer bracket on the passenger side of the cradle.
15. Pull the stabilizer bar down on the driver's side.

NOTE: As an aid in the installation of the stabilizer bar, a pry hole has been put in the cradle.

16. Disconnect the front and rear transaxle mounts at the cradle (Fig. 18-20).
17. Remove the two rear-center cross-member bolts (Fig. 18-21).
18. Remove the passenger-side front-cradle attaching

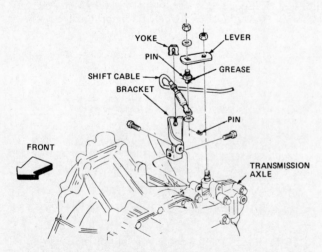

Fig. 18-17 Console-shift cable mounting. (*Pontiac Motor Division of General Motors Corporation*)

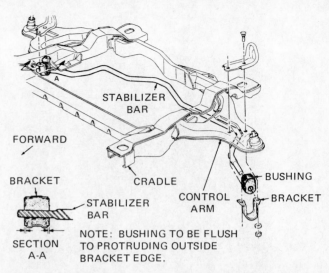

Fig. 18-18 Stabilizer-to-control-arm mounting. (*Pontiac Motor Division of General Motors Corporation*)

bolts. The nuts are reached by pulling back the splash shield next to the frame rail (Fig. 18-21).
19. Remove the top bolt from the lower front transaxle damper (shock absorber) on V-6 engines (if so equipped) (Fig. 18-22).
20. From the driver's side, remove front and rear cradle-to-body bolts (Fig. 18-21).

NOTE: The cradle-to-body bolts are installed before any of the stabilizer bolts.

21. Remove the left-front wheel assembly.
22. Attach the axle-shaft-removing tool to a slide hammer (Fig. 18-23).

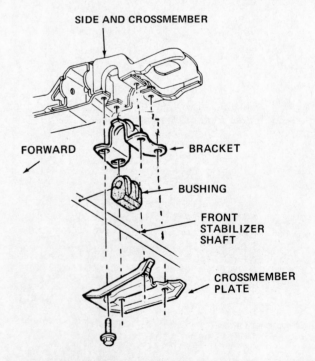

Fig. 18-19 Stabilizer-to-cradle mounting. (*Pontiac Motor Division of General Motors Corporation*)

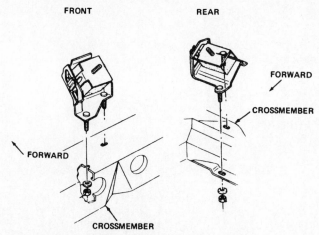

Fig. 18-20 Transaxle mounts. (*Pontiac Motor Division of General Motors Corporation*)

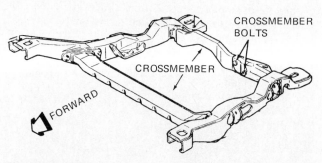

Fig. 18-21 Split cradle. (*Pontiac Motor Division of General Motors Corporation*)

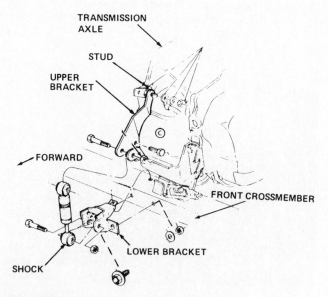

Fig. 18-22 Engine-damper mounting. (*Pontiac Motor Division of General Motors Corporation*)

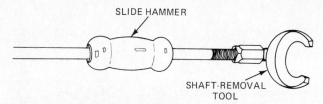

Fig. 18-23 Axle- (drive-) shaft-removing tool. (*Chevrolet Motor Division of General Motors Corporation*)

23. Place the removing tool behind the axle-shaft cones and pull the cones out of the way from the transaxle (Fig. 18-24). Set the shaft out of the way and plug the bore to minimize fluid leakage.
24. Swing the partial cradle (Fig. 18-21) to the driver's side and wire it securely outboard of the fender well.
25. Remove the four converter-and-starter shield bolts (Figs. 18-2 and 18-3).
26. Remove the transaxle extension bolts from the engine-to-transaxle bracket.
27. Position the transmission jack under the transaxle. Raise it to support the transaxle and attach the transaxle securely to the jack (Fig. 16-17).
28. Remove the three converter-to-flywheel bolts.
29. Remove the remaining transaxle-to-engine bolt located near the starting motor (Fig. 18-13). Attach a C clamp or converter-retaining strap to keep the converter from sliding off.
30. Remove the transaxle by sliding it to the driver's side, away from the engine.

❀ **18-14 Installing the transaxle** Basically, the transaxle is installed by reversing the removal procedure. The following points should be observed.

1. As the transaxle is being installed, slide the right axle shaft into the transaxle case.
2. Install the cradle-to-body bolts before attaching the stabilizer bar.
3. There is a pry hole in the cradle to assist in installing the stabilizer bar.

❀ **18-15 Transaxle disassembly** Remove the retaining strap or C clamp and slide the converter off. Install the transaxle in a holding fixture. Drain the fluid by turning the right-hand axle end down.

NOTE: If oil seals are damaged, replace them. The full-circle type must be cut off. Make sure the ring grooves are free of burrs or other damage. When installing angle-cut seal rings, make sure the tapered edges face each other (Fig. 18-25). Be sure the rings are seated in their grooves. Retain them with petroleum jelly. New seal rings may appear to be distorted after they are installed. However, when they are exposed to normal transaxle temperatures, the new seals will return to their proper shapes.

❀ **18-16 Removing external parts** Remove external parts as follows.

1. Remove the speedometer driven gear and sleeve assembly. Its location is shown in Fig. 17-3.

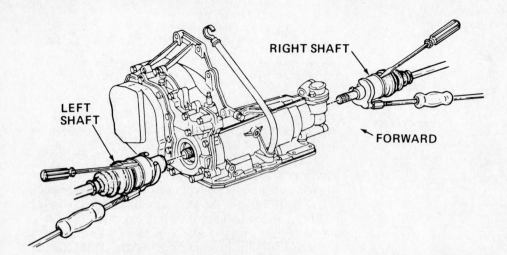

RIGHT SHAFT

LEFT SHAFT

FORWARD

Fig. 18-24 Using tools to remove drive shafts from transaxle (*Chevrolet Motor Division of General Motors Corporation*)

2. Remove governor cover.
3. Remove the speedometer drive gear and governor.
4. Rotate the transaxle so the oil-pan side is up. Remove the oil pan and discard the gasket.
5. Remove the oil strainer. Discard the O ring.
6. Remove the reverse oil-pipe brackets, intermediate-servo cover, and gasket (Fig. 18-26). Now lift out the intermediate-servo assembly.
7. If it is necessary to check the intermediate-band apply pin for length, you need a special gauge, which is installed with two intermediate-servo-cover screws. Then put a gauge collar on the pin and install it in the gauge. Next, apply the specified torque with a torque wrench. If you can see the white line in the gauge window, the pin is the correct length. If not, select a pin having a different length.
8. Remove the third accumulator check valve (Fig. 18-27). Early-production transaxles had a cup plug to hold the check valve in place. If you find a cup plug, cut ¼ inch [6 mm] off a No. 6 Ezy-Out and use it to back out the cup plug.
9. Remove the reverse oil pipe, backup ring, and O ring.
10. Use a No. 4 Ezy-Out with ¾ inch [20 mm] cut off to remove the low-reverse cup plug (Fig. 18-28).
11. Take off the dipstick stop and parking-lock bracket.

12. Turn the transaxle so the final-drive C ring is visible. Position the C ring so the open side is facing up. Use the special tool as shown (Fig. 18-29) to push the C ring down and partly off the output shaft. Turn the output shaft 180° so you can use needle-nose pliers to pull the C ring off (Fig. 18-29). Now, the output shaft can be removed.

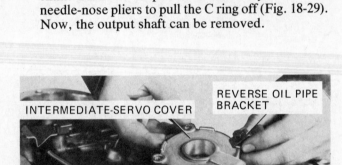

INTERMEDIATE-SERVO COVER

REVERSE OIL PIPE BRACKET

INTERMEDIATE-SERVO GASKET

Fig. 18-26 Removing intermediate-servo cover. (*Chevrolet Motor Division of General Motors Corporation*)

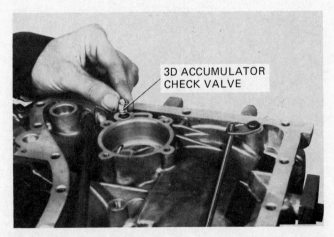

3D ACCUMULATOR CHECK VALVE

Fig. 18-27 Removing third accumulator check valve. (*Chevrolet Motor Division of General Motors Corporation*)

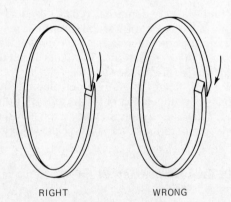

RIGHT   WRONG

Fig. 18-25 Right and wrong ways to install oil-seal rings. (*Chevrolet Motor Division of General Motors Corporation*)

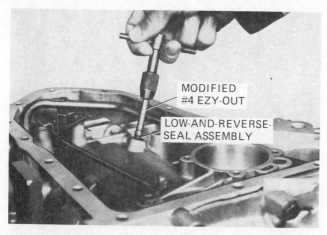

Fig. 18-28 Removing low-and-reverse seal assembly. (*Chevrolet Motor Division of General Motors Corporation*)

✿ 18-17 Removing the valve body, cover, and sprocket-link assembly

1. Remove the valve-body cover. Discard the gasket.
2. Take out the screws and remove the throttle-lever-and-bracket assembly with the throttle-valve cable link.

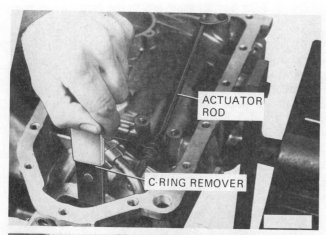

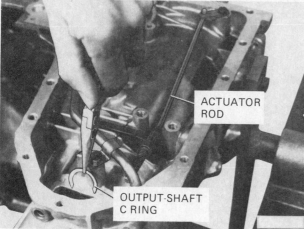

Fig. 18-29 Top, using special tool to partly remove snap ring. Bottom, completing removal of snap ring with pliers. (*Chevrolet Motor Division of General Motors Corporation*)

Fig. 18-30 Pump-cover and control-valve-assembly attaching screws. (*Chevrolet Motor Division of General Motors Corporation*)

3. Remove the pump cover screws except for the one shown in Fig. 18-30. Loosen this screw.
4. Take out the rest of the screws and remove the control valve and pump assembly. Place it, machined surface up, on the bench.
5. Remove the check ball from the spacer plate and take out the pump shaft.
6. Remove the spacer plate and gasket and the five check balls in the case cover (Fig. 18-31).
7. Check the input-shaft-to-case-cover end play. Turn the transaxle so the right axle end is up. Install a special adapter and input-shaft loading tool (Fig. 18-32). Turn the handle all the way down. Rotate the transaxle so the case cover is up. Install a special tool that contacts the input shaft. Then install a dial indicator so it can measure the input-shaft end movement (Fig. 18-33). Press down and lift up

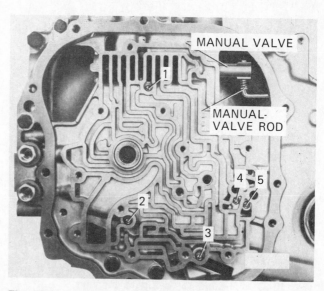

Fig. 18-31 Locations of five check balls. (*Chevrolet Motor Division of General Motors Corporation*)

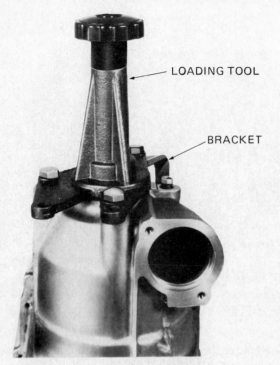

Fig. 18-32 Installing loading tool. (*Chevrolet Motor Division of General Motors Corporation*)

Fig. 18-34 Removing case-cover screws from the converter side. (*Chevrolet Motor Division of General Motors Corporation*)

to measure the end play. If it is not right, a selective snap ring of a different thickness is required.

8. Disconnect the manual-valve rod from the manual valve (Fig. 18-31).
9. From the converter side, remove the screws (Fig. 18-34). Then remove the remaining case-cover screws.
10. Take the case cover and gasket off. Lay the case cover on the bench, accumulator side up. The 1-2 accumulator pin may fall out.
11. Remove the accumulator spring and case center gasket (Fig. 18-35).

12. Lift off the thrust washer from the drive sprocket. Remove the driven-sprocket thrust bearing. Then remove the drive and driven sprockets with the link assembly (Fig. 17-3). The thrust washers under the sprockets may come off with the sprockets.

❀ **18-18 Removing input-unit parts** Remove the parts from the input-unit as follows.

1. Use a ⅛-inch [3-mm] pin punch to remove the pin attaching the detent lever to the manual shaft.
2. Use a large screwdriver to lift out the manual-shaft-to-case retaining pin. Now, the detent lever, manual shaft, and parking-lock actuator lever can be lifted off.
3. Remove the driven-sprocket support and thrust washer (Fig. 18-36).
4. Note the position of the intermediate-band stop. Remove the band anchor-hole plug and the band.
5. Lift out the direct- and forward-clutch assemblies (Fig. 18-37). Separate the direct and forward clutches.
6. Remove the input internal-gear thrust washer and the gear.
7. Remove the input carrier assembly and thrust washers.
8. Remove the input sun gear and input drum.

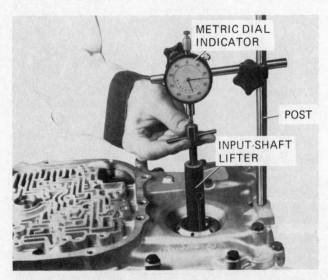

Fig. 18-33 Dial-indicator setup to measure input-shaft end play. (*Chevrolet Motor Division of General Motors Corporation*)

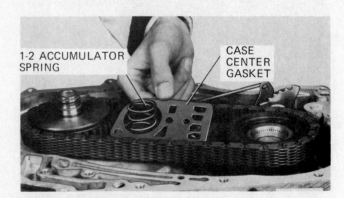

Fig. 18-35 Removing case center gasket and 1-2 accumulator spring. (*Chevrolet Motor Division of General Motors Corporation*)

Fig. 18-36 Removing driven-sprocket support. (*Chevrolet Motor Division of General Motors Corporation*)

❀ 18-19 Removing reaction-unit parts Before proceeding further, check the end play of the reaction sun gear. Install the special tool and dial indicator as shown (Fig. 18-38), using two case-cover bolts to hold it in place. Position the thickness gauge under the extension shoulder as shown. Set the dial indicator to zero. Rotate the selective snap ring and swing the thickness gauge out. Note the amount of dial-indicator needle movement. This is the end play. If it is not right, select a snap ring of a different thickness.

Next, check the low-and-reverse clutch end play. Leave the tool and dial indicator in place as shown in Fig. 18-38. Then use a screwdriver through the parking-pawl-case opening (Fig. 18-39) and lift the reaction internal gear to check the end play. If it is not right, select a washer of the proper thickness to correct the end play. Proceed as follows to remove reaction-unit parts.

1. Remove the reaction sun gear and the large clutch-housing-to-case snap ring. Use a screwdriver to pry the snap ring out.
2. Use the tool as shown in Fig. 18-40 to remove the low-and-reverse clutch housing. Take out the spacer ring that was under the housing.

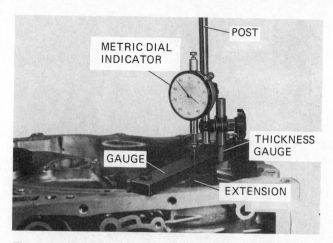

Fig. 18-38 Checking for proper reaction-sun-gear selective snap ring. (*Chevrolet Motor Division of General Motors Corporation*)

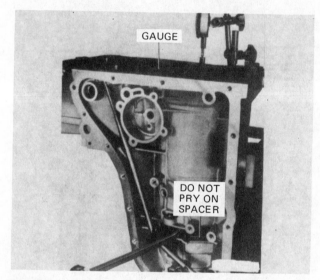

Fig. 18-39 Checking for proper low roller-clutch-race selective thrust washer. (*Chevrolet Motor Division of General Motors Corporation*)

Fig. 18-37 Removing forward- and direct-clutch assemblies. (*Chevrolet Motor Division of General Motors Corporation*)

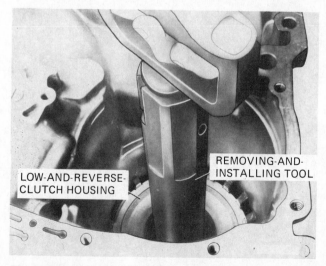

Fig. 18-40 Removing low-and-reverse clutch housing. (*Chevrolet Motor Division of General Motors Corporation*)

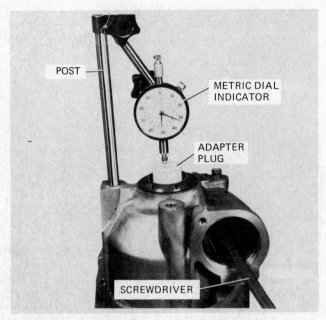

Fig. 18-41 Checking final-drive end play. (*Chevrolet Motor Division of General Motors Corporation*)

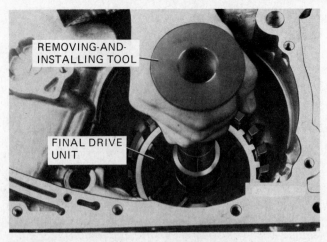

Fig. 18-42 Removing final-drive assembly. (*Chevrolet Motor Division of General Motors Corporation*)

3. Grasp the final-drive sun-gear shaft and lift out the reaction gearset.
4. Lift the roller-clutch and reaction-carrier assembly off the sun-gear shaft.
5. Remove the thrust washer with four tangs from the reaction carrier or the internal gear.
6. Take the clutch plates off the sun-gear shaft.
7. Remove the thrust bearing from the reaction gear.
8. Take the reaction internal gear off the sun-gear shaft.

✿ 18-20 Removing final-drive parts First check the final-drive-to-case end play, as follows. Rotate the

transaxle so the right axle end is up. Install tools as shown in Fig. 18-41. Use a large screwdriver through the governor bore as shown to lift up on the governor drive gear. Note the end play. The selective washer that controls this end play is located between the differential carrier and the differential-carrier-case thrust bearing. Use a selective washer of a different thickness if the end play is not correct.

Remove the tools. Rotate the transaxle so the case-cover side is up. Remove the large snap ring and the spacer above the internal gear. Then use the tool as shown in Fig. 18-42 to remove the final-drive assembly. Figure 18-43 shows the differential and final drive disassembled. Note the locations of the snap rings and thrust washers.

Refer to Fig. 18-43 if further disassembly of the differential is required. The governor drive gear is a press

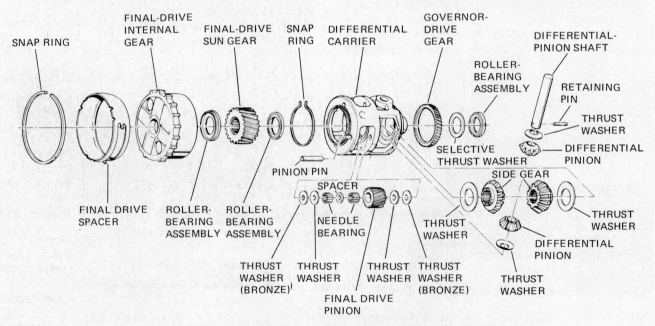

Fig. 18-43 Disassembled differential and final drive. (*Chevrolet Motor Division of General Motors Corporation*)

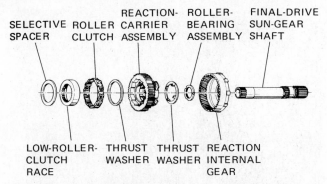

SELECTIVE SPACER • ROLLER CLUTCH • REACTION-CARRIER ASSEMBLY • ROLLER-BEARING ASSEMBLY • FINAL-DRIVE SUN-GEAR SHAFT

LOW-ROLLER-CLUTCH RACE • THRUST WASHER • THRUST WASHER • REACTION INTERNAL GEAR

Fig. 18-44 Disassembled reaction gearset. (*Chevrolet Motor Division of General Motors Corporation*)

fit on the differential carrier, and if it must come off, a gear puller is required. Tap the new gear into place with a soft hammer. To remove the differential pinions, the retaining pin is driven out of the pinion shaft. Then the shaft is removed so the pinions can come out.

The end play of the final-drive pinions can be measured with a thickness gauge between the side of the pinion and the case. The pinions can be replaced

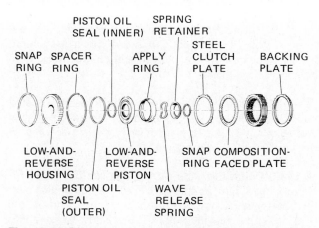

PISTON OIL SEAL (INNER) • SPRING RETAINER • STEEL CLUTCH PLATE • BACKING PLATE

SNAP RING • SPACER RING • APPLY RING

LOW-AND-REVERSE HOUSING • LOW-AND-REVERSE PISTON • SNAP RING • COMPOSITION-FACED PLATE

PISTON OIL SEAL (OUTER) • WAVE RELEASE SPRING

Fig. 18-45 Disassembled low-and-reverse clutch. (*Chevrolet Motor Division of General Motors Corporation*)

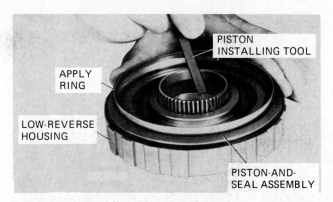

PISTON INSTALLING TOOL • APPLY RING • LOW-REVERSE HOUSING • PISTON-AND-SEAL ASSEMBLY

Fig. 18-47 Installing piston into low-and-reverse clutch housing, using the piston-installing tool. (*Chevrolet Motor Division of General Motors Corporation*)

by removing the pinion pin. Note, in Fig. 18-43, the stack-up of parts in the pinion—two sets of needle bearings with a spacer between, and four thrust washers. For assembly, the needles are held in place with petroleum jelly.

✿ **18-21 Reaction gearset** The reaction gearset is shown disassembled in Fig. 18-44. Note the locations of the selective spacer and the thrust washers. The tanged thrust washer is installed with the tangs fitted into the reaction carrier assembly. If the end play is not correct, a spacer of a different thickness must be installed.

✿ **18-22 Low-and-reverse clutch** The low-and-reverse clutch is shown disassembled in Fig. 18-45. When disassembling the clutch, note the positions of the clutch plates and other parts. The cupped side of the spring retainer faces the apply ring. When installing inner and outer piston seals, face the lips away from the apply-ring side.

To install the piston, use a tool such as that shown in Fig. 18-46. The loop of wire is worked around the seal as shown in Fig. 18-47 to work the lip over the

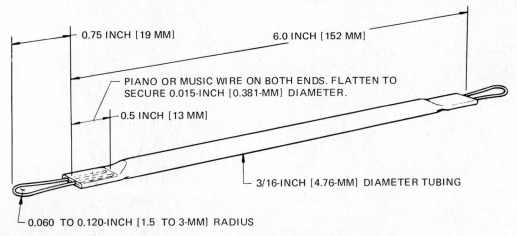

0.75 INCH [19 MM]   6.0 INCH [152 MM]

PIANO OR MUSIC WIRE ON BOTH ENDS. FLATTEN TO SECURE 0.015-INCH [0.381-MM] DIAMETER.

0.5 INCH [13 MM]

3/16-INCH [4.76-MM] DIAMETER TUBING

0.060 TO 0.120-INCH [1.5 TO 3-MM] RADIUS

Fig. 18-46 How to make a piston-installing tool. (*Chevrolet Motor Division of General Motors Corporation*)

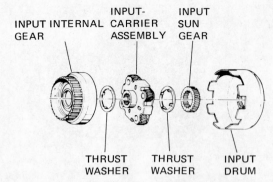

INPUT INTERNAL GEAR
INPUT-CARRIER ASSEMBLY
INPUT SUN GEAR
THRUST WASHER
THRUST WASHER
INPUT DRUM

Fig. 18-48 Disassembled input unit. (*Chevrolet Motor Division of General Motors Corporation*)

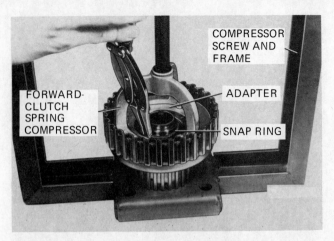

COMPRESSOR SCREW AND FRAME
FORWARD-CLUTCH SPRING COMPRESSOR
ADAPTER
SNAP RING

Fig. 18-50 Using spring compressor and snap-ring pliers to remove snap ring. (*Chevrolet Motor Division of General Motors Corporation*)

shaft or into the housing. On reassembly, use the proper reaction-sun-gear selective snap ring as determined in Fig. 18-38. Also, use the proper thickness of selective thrust washer as determined in Fig. 18-39.

✱ **18-23 Input-unit parts** The input-unit parts are shown separated in Fig. 18-48. Note that the tangs on the thrust washers face in toward the input carrier assembly.

✱ **18-24 Forward-clutch assembly** Figure 18-49 shows the clutch disassembled. To remove the clutch-pack snap ring, use the spring compressor as shown in Fig. 18-50 to compress the clutch pack. Then remove the snap ring with snap-ring pliers. Use the special tool (Figs. 18-46 and 18-47) when installing the piston.

✱ **18-25 Direct-clutch assembly** Figure 18-51 shows the direct clutch disassembled. Note the positions of the piston, apply ring, and release-spring assembly. Use the piston-installing tool as shown in Figs. 18-46 and 18-47 to install the piston.

Reassembling the transaxle

✱ **18-26 Installing forward and direct clutches** Put the forward-clutch assembly on the bench, shaft

up. Install the direct-clutch assembly on the forward-clutch housing. Be sure it is seated.

Stick the input-shaft thrust washer, stepped side out, on the input shaft with petroleum jelly (Fig. 18-52). Then install the direct-and-forward-clutch assembly in the case. Rotate it as you let it down into place. Do not force it. When it is correctly installed, the measurement between the case face and direct-clutch housing should be about $1\frac{11}{16}$ inches [42 mm].

✱ **18-27 Installing the intermediate band** Install the band with the eye end in the case and the lugged end aligned with the apply-pin bore (Fig. 18-53). Install the band anchor-hole plug.

✱ **18-28 Installing the driven-sprocket support** Before installing the support (Fig. 18-36), make sure the bearing is in good condition. If it is not, pull the old bearing with a bearing puller. Press a new bearing into place using the special tools specified.

✱ **18-29 Manual shaft** The manual shaft and the detent-lever assembly are a matched set. If one part requires replacement, replace both with a new set. To

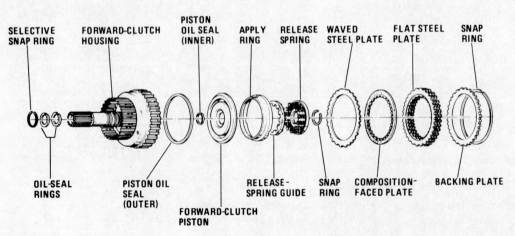

SELECTIVE SNAP RING
FORWARD-CLUTCH HOUSING
PISTON OIL SEAL (INNER)
APPLY RING
RELEASE SPRING
WAVED STEEL PLATE
FLAT STEEL PLATE
SNAP RING

OIL-SEAL RINGS
PISTON OIL SEAL (OUTER)
FORWARD-CLUTCH PISTON
RELEASE-SPRING GUIDE
SNAP RING
COMPOSITION-FACED PLATE
BACKING PLATE

Fig. 18-49 Disassembled forward-clutch assembly. (*Chevrolet Motor Division of General Motors Corporation*)

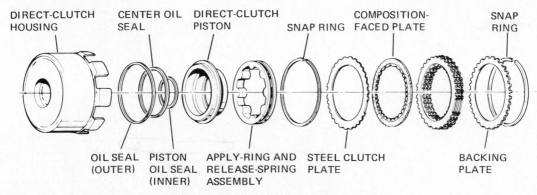

DIRECT-CLUTCH HOUSING   CENTER OIL SEAL   DIRECT-CLUTCH PISTON   SNAP RING   COMPOSITION-FACED PLATE   SNAP RING

OIL SEAL (OUTER)   PISTON OIL SEAL (INNER)   APPLY-RING AND RELEASE-SPRING ASSEMBLY   STEEL CLUTCH PLATE   BACKING PLATE

Fig. 18-51 Disassembled direct-clutch assembly. (*Chevrolet Motor Division of General Motors Corporation*)

install, install the manual shaft to retain the detent-lever assembly. Secure with the detent-lever-to-manual-shaft retaining pin.

✿ **18-30 Drive-link-and-sprocket assembly** Be sure the sprocket thrust washers are under the sprock-

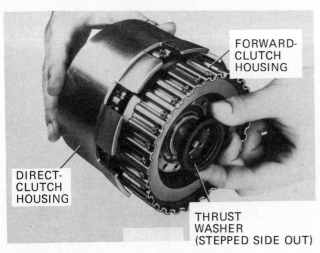

Fig. 18-52 Installing input-shaft thrust washer. (*Chevrolet Motor Division of General Motors Corporation*)

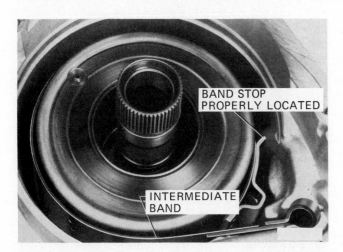

Fig. 18-53 Installing intermediate band. (*Chevrolet Motor Division of General Motors Corporation*)

ets when installing the sprocket and link assembly. Retain them with petroleum jelly. Figure 18-35 shows the assembly installed in the case. The colored guide link, which has numbers, should face the case. Install the driven-sprocket roller bearing with the outer race against the sprocket.

✿ **18-31 Case-cover service** If the vent assembly is damaged, replace it. Install a new assembly by coating the end with Loctite and tap the assembly into place with a rubber hammer.

If the cooler-line connectors are damaged, remove them. Coat the new connectors with a thread sealer. Tighten them to the specified torque.

Check other parts and replace them if necessary. These include the detent-and-roller assembly, the axle oil seal, manual valve, 1-2 accumulator piston, oil-seal ring, and thermostatic valve. Refer to the factory service manual for procedures.

When installing the cover case, use a new gasket. Figure 18-54 shows the locations of the attaching bolts. Figure 18-34 shows the locations of the two screws that are installed from the converter-housing side. Figure 18-31 shows the attachment of the manual valve to the

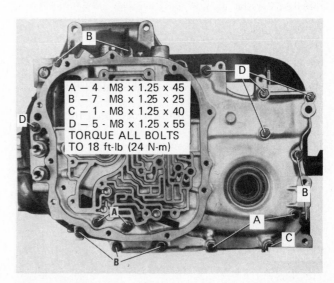

Fig. 18-54 Case-cover bolt locations. (*Chevrolet Motor Division of General Motors Corporation*)

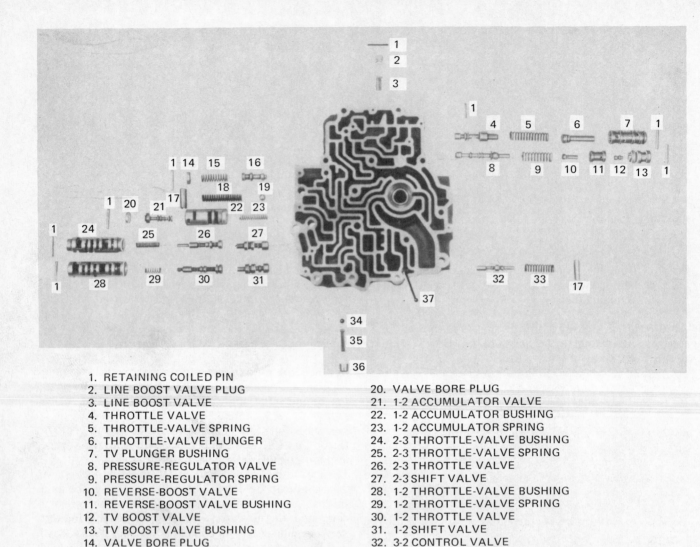

1. RETAINING COILED PIN
2. LINE BOOST VALVE PLUG
3. LINE BOOST VALVE
4. THROTTLE VALVE
5. THROTTLE-VALVE SPRING
6. THROTTLE-VALVE PLUNGER
7. TV PLUNGER BUSHING
8. PRESSURE-REGULATOR VALVE
9. PRESSURE-REGULATOR SPRING
10. REVERSE-BOOST VALVE
11. REVERSE-BOOST VALVE BUSHING
12. TV BOOST VALVE
13. TV BOOST VALVE BUSHING
14. VALVE BORE PLUG
15. SHIFT–TV SPRING
16. SHIFT–TV VALVE
17. SPRING-RETAINING SLEEVE
18. PRESSURE-RELIEF SPRING
19. PRESSURE-RELIEF BALL

20. VALVE BORE PLUG
21. 1-2 ACCUMULATOR VALVE
22. 1-2 ACCUMULATOR BUSHING
23. 1-2 ACCUMULATOR SPRING
24. 2-3 THROTTLE-VALVE BUSHING
25. 2-3 THROTTLE-VALVE SPRING
26. 2-3 THROTTLE VALVE
27. 2-3 SHIFT VALVE
28. 1-2 THROTTLE-VALVE BUSHING
29. 1-2 THROTTLE-VALVE SPRING
30. 1-2 THROTTLE VALVE
31. 1-2 SHIFT VALVE
32. 3-2 CONTROL VALVE
33. 3-2 VALVE SPRING
34. LOW BLOW-OFF BALL
35. LOW BLOW-OFF SPRING AND PLUG ASSEMBLY
36. LOW BLOW-OFF VALVE PLUG
37. CHECK BALL #1

Fig. 18-55 Control-valve body with all parts removed. (*Chevrolet Motor Division of General Motors Corporation*)

valve rod. Install the correct input-shaft selective snap ring as determined by the setup shown in Fig. 18-33.

❈ **18-32 Control-valve-and-pump assembly** The control-valve-and-pump assembly is shown with all valves, plugs, and springs removed in Fig. 18-55. This illustration, and Fig. 18-31, will be your guide to the servicing of the control-valve system. Figure 18-56 shows the spacer plate and the locations of the various fluid passages in the control-valve assembly.

❈ **18-33 Installing the output shaft** To retain the output shaft after installing it, install a new C ring (Fig. 18-29).

❈ **18-34 Reverse pipe and parking bracket** If the parking-lock bracket is damaged, it should be removed

and a new bracket installed. Figure 18-57 shows the location of the bracket, actuator rod, and dipstick stop.

❈ **18-35 Intermediate servo** Figure 18-58 is a disassembled view of the intermediate servo. Figure 18-26 shows the installation of the intermediate servo cover and reverse oil-pipe bracket. How to check the apply pin to make sure it is the correct length is explained in ❈ 18-16.

❈ **18-36 Oil strainer and pan** Install a new strainer and O ring. Then install the oil-pan gasket and pan, torquing the retainer bolts to specifications.

❈ **18-37 Governor and speedometer-drive gear** If the governor-shaft seal ring is damaged, cut it off. Install a new ring, lubricated with petroleum jelly. Put

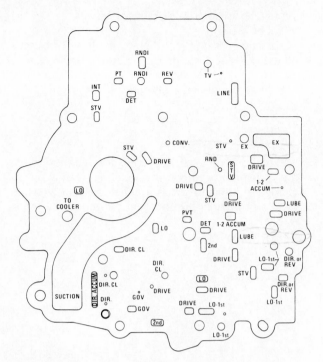

Fig. 18-56 Spacer plate. (*Chevrolet Motor Division of General Motors Corporation*)

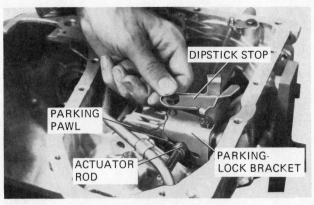

Fig. 18-57 Installing dipstick stop. (*Chevrolet Motor Division of General Motors Corporation*)

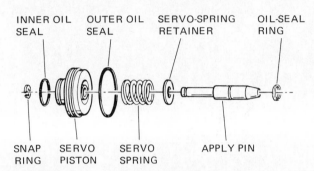

Fig. 18-58 Disassembled intermediate servo. (*Chevrolet Motor Division of General Motors Corporation*)

the speedometer drive gear and thrust washer on the governor assembly. Use a new governor O ring in the governor cover. Install the governor and cover on the transaxle. Install the speedometer driven gear and retainer.

❉ **18-38 Installing transaxle on vehicle**   Install the transaxle on a transmission jack. Install the converter, securing it with a C clamp or a retainer strap. Raise the transaxle under the vehicle and install the transaxle

(❉ 18-14). Adjust the throttle-valve control cable and shifter cable (❉ 18-6 and 18-7). Add fluid (❉ 18-3 and 18-4). Road-test the vehicle to make sure the transaxle performs normally.

Chapter 18 review questions

Select the *one* correct, best, or most probable answer to each question. Then check your answer against the correct answers given at the end of the book.

1. Diagnosis of transaxle troubles should start with checks of:
   a. drivability and stalling
   b. fluid and linkages
   c. clutches and bands
   d. oil and air pressure
2. Mechanic A says the differential uses the same lubricant as the transaxle. Mechanic B says the differential is sealed off from the transaxle. Who is right?
   a. A only
   b. B only
   c. both A and B
   d. neither A nor B
3. Mechanic A says that if the engine lacks power and performs poorly, it can make the transaxle look

bad. Mechanic B says that a poorly operating engine can cause late shifting. Who is right?
   a. A only
   b. B only
   c. both A and B
   d. neither A nor B
4. The fluid level should be checked with the:
   a. engine idling and in D
   b. engine not running and in D
   c. engine hot and running in N
   d. engine hot, idling, and in P
5. No drive in drive range could be caused by the forward clutch not engaging, one-way clutch not holding, or:
   a. low fluid pressure
   b. low fluid level
   c. misadjusted manual linkage
   d. all of the above
6. High fluid pressure could be caused by:
   a. a malfunctioning torque converter

b. a defective accumulator

c. stuck valves

d. clutch not holding

7. Mechanic A says failure to upshift from first can be caused by failure of the intermediate band to apply or by stuck valves. Mechanic B says the failure to upshift from first could be due to a sticking 1-2 shift valve or to failure of the servo to work. Who is right?

a. A only

b. B only

c. both A and B

d. neither A nor B

8. If the transaxle drives in neutral, the trouble could be due to:

a. a misadjusted manual linkage or a stuck forward clutch

b. a misadjusted TV cable or a stuck one-way clutch

c. a stuck regulator or pressure valve

d. all of the above

9. Mechanic A says both bands in the 125 can be adjusted. Mechanic B says neither band can be adjusted, but correction is made by replacing whichever band is not holding. Who is right?

a. A only

b. B only

c. both A and B

d. neither A nor B

10. Mechanic A says that if there is foreign material in the transmission fluid, you must get a new or exchange torque converter. Mechanic B says you can flush out the torque converter and reuse it. Who is right?

a. A only

b. B only

c. both A and B

d. neither A nor B

OTHER AUTOMATIC TRANSMISSIONS

After studying this chapter, you should be able to:

1. Describe the construction and operation of the Turbo Hydra-Matic transmissions used in General Motors full-size front-drive cars.
2. Describe the construction and operation of the Toyota Toyoglide automatic transmission.
3. Describe the construction and operation of the Volkswagen automatic transmission.
4. Describe the construction and operation of the belt-drive transmission.
5. Describe the construction and operation of the toroidal-drive transmission.

19-1 Introduction to other automatic transmissions A variety of automatic transmissions have been produced in recent years. Other chapters in this book cover the most popular automatic transmissions and transaxles. This chapter describes the Turbo Hydra-Matic transmission used on some full-size front-drive Cadillacs and Oldsmobiles, one automatic transmission used by Toyota, and a Volkswagen automatic transmission.

✿ 19-2 Turbo Hydra-Matic for full-size front-drive cars Figure 19-1 shows the Turbo Hydra-Matic automatic transmission for the front-drive Cadillac and Oldsmobile. It is essentially a type 400 Turbo Hydra-Matic automatic transmission which has been cut in two just behind the torque converter so that the planetary-gear system can be placed alongside the torque converter, as shown in Fig. 19-2. A pair of sprockets and a drive chain connect the torque converter with the automatic transmission (Fig. 19-3).

The transmission and torque converter operate essentially in the same manner as the automatic transmission described in Chap. 4. However, the power-flow diagrams look different because of the side-by-side location of the torque converter and gear system. Figures 19-4 to 19-6 show the power flow under different operating conditions. The end views of the front and rear gearsets, shown in the upper left of the figures, indicate the directions the components of the gearsets rotate under the different operating conditions.

✿ 19-3 Toyoglide The Toyoglide automatic transmission, used on Toyota cars, is a three-speed unit with a three-member torque converter and compound planetary-gear system controlled by two multiple-disk clutches, two brake bands, and an overrunning clutch. Figure 19-7 is a sectional view of the transmission, and Fig. 19-8 is a partial cutaway view of the planetary-gear system with the controlling clutches and bands.

Figure 19-9 is a table showing the conditions in the transmission in the different selector-lever positions. Figures 19-10 to 19-13 show the power flow under different operating conditions. Plate 31 illustrates the hydraulic circuit in the Toyoglide when in direct drive.

✿ 19-4 Volkswagen automatic transmission The Volkswagen automatic transmission provides three forward speeds, automatically selected when the transmission is in the drive range, and reverse. It includes one compound planetary gearset, two multiple-disk clutches, two bands, and a one-way, or overrunning, clutch. Figure 19-14 is a cutaway view of the planetary gearset. Figure 19-15 is a cutaway view of the forward clutch. The shifts are controlled by a hydraulic system similar to those used in other automatic transmissions.

Figure 19-16 shows the six selector-lever positions. P for park, R for reverse, and N for neutral are the same as with other transmissions. 3 is the same as D, or drive, in other automatic transmissions and is for all normal driving conditions. 2 is the same as D2 or S in

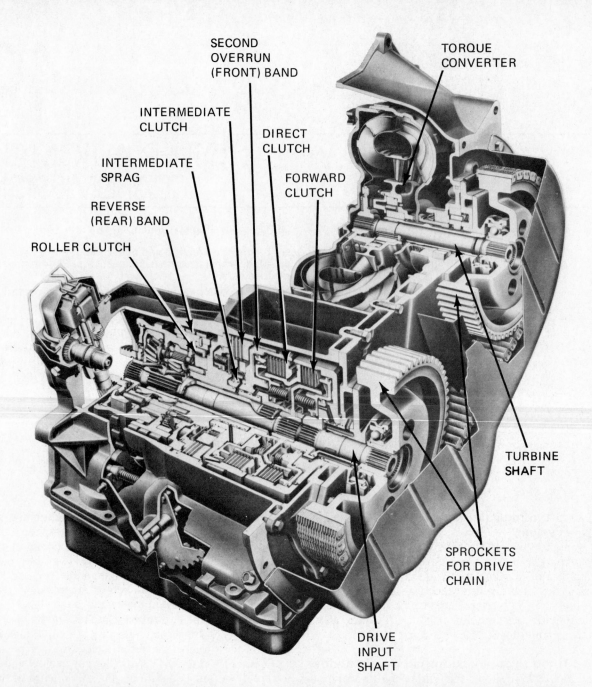

SELECTOR POSITION	PUMP PRESSURE	FORWARD CLUTCH	DIRECT CLUTCH	2ND OVERRUN BAND	INT. CLUTCH	INT. SPRAG	ROLLER CLUTCH	REV. BAND
PARK—NEUT.	60-150	OFF	OFF	OFF	OFF	OFF	OFF	OFF
DRIVE 1	60-150	ON	OFF	OFF	OFF	OFF	ON	OFF
LEFT 2	60-150	ON	OFF	OFF	ON	ON	OFF	OFF
3	60-150	ON	ON	OFF	ON	OFF	OFF	OFF
DRIVE 1	150	ON	OFF	OFF	OFF	OFF	ON	OFF
RIGHT 2	150	ON	OFF	ON	ON	ON	OFF	OFF
LO 1	150	ON	OFF	OFF	OFF	OFF	ON	ON
2	150	ON	OFF	ON	ON	ON	OFF	OFF
REV.	95 - 230	OFF	ON	OFF	OFF	OFF	OFF	ON

Fig. 19-1 Turbo Hydra-Matic automatic transmission for a full-size front-drive automobile, partly cut away so that internal construction can be seen. The table shows the internal conditions for different selector positions. (*Cadillac Motor Car Division of General Motors Corporation*)

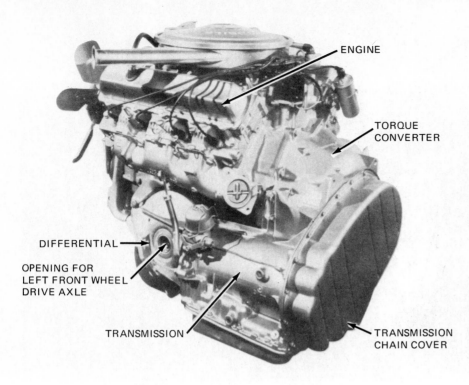

ENGINE

TORQUE CONVERTER

DIFFERENTIAL

OPENING FOR LEFT FRONT WHEEL DRIVE AXLE

TRANSMISSION

TRANSMISSION CHAIN COVER

Fig. 19-2 Engine and transmission assembly for a full-size front-wheel-drive car, as seen from the left rear. (*Oldsmobile Division of General Motors Corporation*)

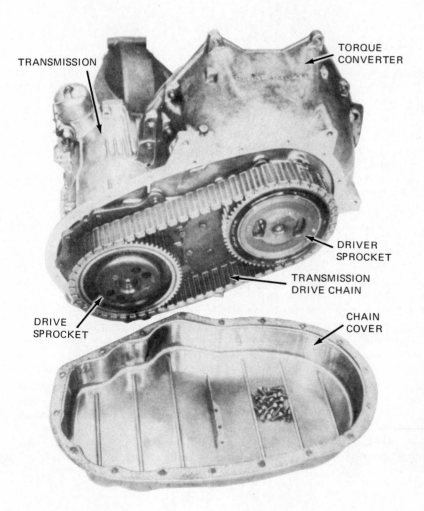

TRANSMISSION

TORQUE CONVERTER

DRIVER SPROCKET

TRANSMISSION DRIVE CHAIN

DRIVE SPROCKET

CHAIN COVER

Fig. 19-3 Transmission assembly for a full-size front-wheel-drive car with the transmission-chain cover removed so that the chain and sprockets can be seen. (*Oldsmobile Division of General Motors Corporation*)

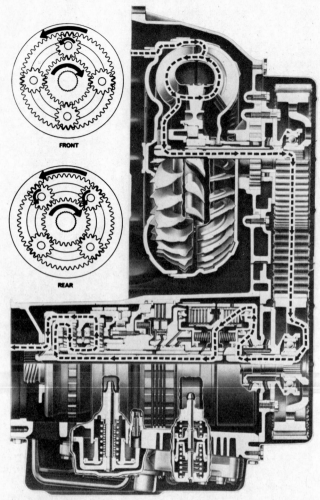

Fig. 19-4 Power flow in drive range, first gear, in the Turbo Hydra-Matic automatic transmission for full-size front-drive cars. (*Cadillac Motor Car Division of General Motors Corporation*)

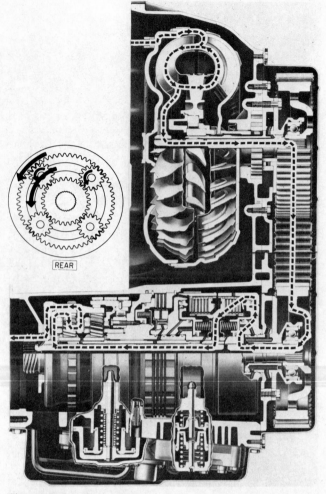

Fig. 19-5 Power flow in drive range, second gear, in the Turbo Hydra-Matic automatic transmission for full-size front-drive cars. (*Cadillac Motor Car Division of General Motors Corporation*)

other automatic transmissions and is for intermediate driving conditions. 1 is the same as L, or low, in other automatic transmissions and is used to pull heavy loads or for engine braking when going down a long hill.

Figure 19-17 is a simplified sectional view of the Volkswagen automatic transmission with all essential parts named.

✿ 19-5 Continuously variable transmissions Today the typical automatic transmission has a three-element torque converter, a compound-planetary gearset, and various combinations of bands and clutches. These are used to provide the three or four forward gear ratios, reverse, and neutral. However, this type of automatic transmission provides only a certain number of gear ratios. Even with torque-converter action increasing the "range" of each gear ratio, the transmission fails to allow the engine to operate at peak efficiency, regardless of vehicle speed, at all times. Theoretically, a continuously variable transmission which has, in effect, an infinite number of gear ratios can do this.

An early type of continuously variable transmission, called *friction drive,* was used in the Cartercar from 1906 to 1909. Other early car makers also used friction drive. The basic design was of a large metal disk that rotated in contact with a smaller wheel. To change the drive ratio, the smaller wheel was moved across the face of the disk.

However, as bigger engines were used and cars became heavier, friction drive was abandoned. Gear-type transmissions became almost universal in production cars. Then, after World War II, the manually shifted gear-type transmission began to loose popularity to the self-shifting type of automatic transmission that we know today.

Recent downsizing of the automobile has sparked renewed interest in the continuously variable transmission. One reason is that the typical automatic transmission does not get as good fuel mileage as the comparable manual transmission. Also, the typical automatic transmission is heavier, and with many small engines, too much power is required to drive the transmission. Therefore, to retain operation without manual shifting, reduce vehicle weight, and improve the mechanical efficiency of the drive train, continuously variable transmissions are being reexamined.

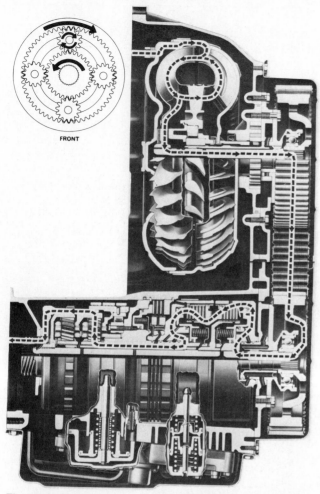

Fig. 19-6 Power flow in reverse in the Turbo Hydra-Matic automatic transmission for full-size front drive cars. (*Cadillac Motor Car Division of General Motors Corporation*)

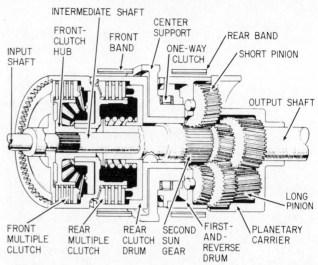

INTERMEDIATE SHAFT
FRONT-CLUTCH HUB
FRONT BAND
CENTER SUPPORT
ONE-WAY CLUTCH
REAR BAND
SHORT PINION
INPUT SHAFT
OUTPUT SHAFT
LONG PINION
FRONT MULTIPLE CLUTCH
REAR MULTIPLE CLUTCH
REAR CLUTCH DRUM
SECOND SUN GEAR
FIRST-AND-REVERSE DRUM
PLANETARY CARRIER

Fig. 19-8 Cutaway view of the planetary-gear system and controls in the Toyoglide. (*Toyota Motor Sales Company, Ltd.*)

Two types of continuously variable transmissions, belt drive (✸ 19-6) and traction drive (✸ 19-7), are discussed below.

✸ 19-6 Belt drive transmissions For many years, the Dutch car maker DAF has marketed a Variomatic transmission which uses belt drive. Figure 19-18 shows the construction of the belt drive automatic transmission. Earlier versions had two belts, one for each wheel. Basically, this type of transmission is the same as in some motorcycles and snowmobiles. It uses a *centrifugal clutch* to drive the geared differential. Gear reduction is provided by a single pair of variable-diameter pulleys. Diameter of the pulleys is controlled by a combination of engine intake-manifold vacuum and centrifugal force.

The centrifugal clutch (also known as a "belt torque converter" or "Salsbury clutch") is basically two variable-width pulleys connected by a drive belt. One pul-

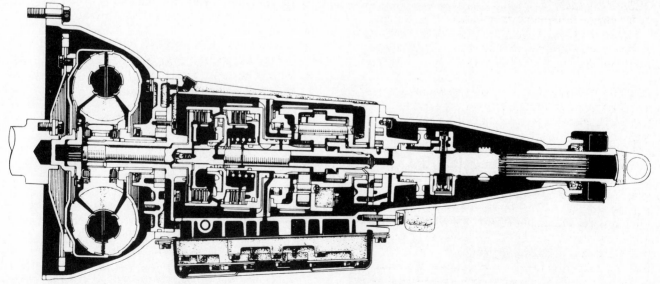

Fig. 19-7 Sectional view of the Toyoglide. (*Toyota Motor Sales Company, Ltd.*)

SELECTOR-LEVER POSITIONS	GEAR	FRONT CLUTCH	REAR CLUTCH	FRONT BAND	REAR BAND	ONE-WAY CLUTCH	GEAR RATIO
P	Neutral				Applied		
R	Reverse		Engaged		Applied		1.920
N	Neutral						
D or 2	First	Engaged				Locked	2.400
D or 2	Second	Engaged		Applied		Overrunning	1.479
D	Third	Engaged	Engaged			Overrunning	1.000
L	First	Engaged			Applied		2.400

Fig. 19-9 Table of the conditions in the Toyoglide transmission with the selector lever in different positions. (*Toyota Motor Sales Company, Ltd.*)

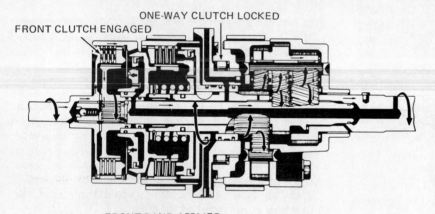

Fig. 19-10 Power flow in the Toyoglide in first gear, drive range. (*Toyota Motor Sales Company, Ltd.*)

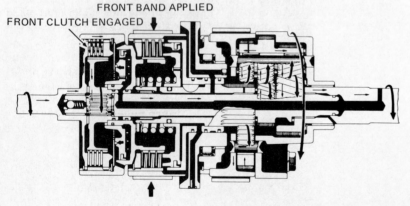

Fig. 19-11 Power flow in the Toyoglide in second gear, drive range. (*Toyota Motor Sales Company, Ltd.*)

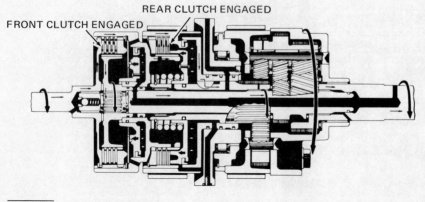

Fig. 19-12 Power flow in the Toyoglide in third gear, drive range. (*Toyota Motor Sales Company, Ltd.*)

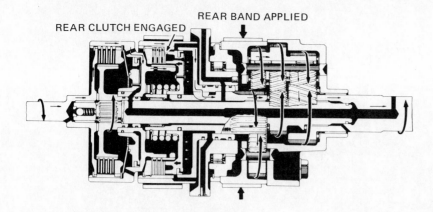

REAR CLUTCH ENGAGED   REAR BAND APPLIED

Fig. 19-13 Power flow in the Toyoglide in reverse gear. (*Toyota Motor Sales Company, Ltd.*)

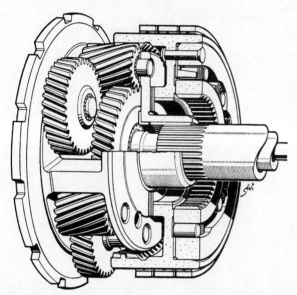

Fig. 19-14 Cutaway view of the compound planetary gearset used in the Volkswagen automatic transmission. (*Volkswagen of America, Inc.*)

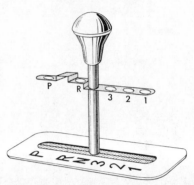

Fig. 19-16 Selector-lever positions for the Volkswagen automatic transmission. (*Volkswagen of America, Inc.*)

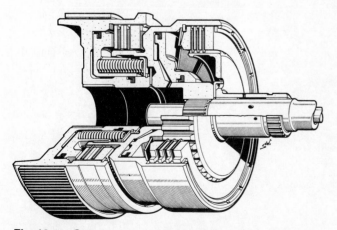

Fig. 19-15 Cutaway view of the forward clutch used in the Volkswagen automatic transmission. (*Volkswagen of America, Inc.*)

ley is the *drive pulley.* It is connected to the engine crankshaft. The other pulley, the *driven pulley,* is connected to the differential gearing.

At idle speed, a spring holds the sides of the drive pulley so far apart that they do not grip the belt. Therefore the belt does not turn. This is shown in Fig. 19-19*a*.

As engine speed increases, the centrifugal force overcomes the spring and pushes the sides of the drive pulley closer together until they grip the belt. Now the belt begins to turn the driven pulley (Fig. 19-19*b*). This condition provides the maximum torque to the driven pulley.

With further increases in engine speed, the increased centrifugal force causes the sides of the drive pulley to move even closer together. At the same time, the sides of the driven pulley are moving apart an equal distance. This changes the gear ratio through the centrifugal clutch, as shown in Fig. 19-19c. Now the gear ratio through the pulleys decreases the torque and increases the speed of the driven pulley.

Figure 19-20 shows the belt drive transmission in a design by Fiat that is computer-controlled. In this system, there is no mechanical connection between the accelerator and the carburetor. As the accelerator pedal is depressed, the computer regulates the throttle opening. Therefore, the engine is always running at the speed providing the lowest fuel consumption for the required power output. At the same time, the computer selects the proper gear ratio and pulley diameters.

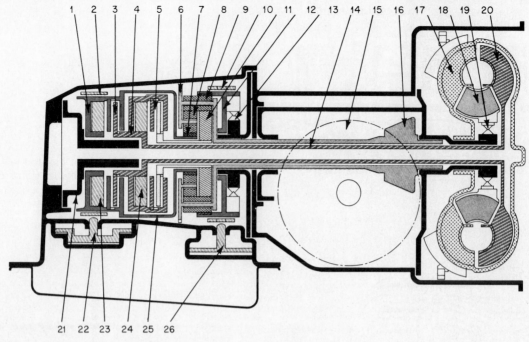

1. DIRECT-AND-REVERSE-CLUTCH DRUM
2. SECOND-GEAR BAND
3. DIRECT-AND-REVERSE CLUTCH
4. FORWARD-CLUTCH DRUM
5. FORWARD CLUTCH
6. PLANET CARRIER
7. SMALL SUN GEAR
8. LARGE PLANET PINION
9. SMALL PLANET PINION
10. FIRST-GEAR-AND-REVERSE BAND
11. LARGE SUN GEAR
12. ANNULUS GEAR
13. FIRST-GEAR ONE-WAY CLUTCH
14. TURBINE SHAFT
15. DIFFERENTIAL RING GEAR
16. DRIVE PINION
17. IMPELLER
18. STATOR
19. STATOR CLUTCH
20. TURBINE
21. OIL-PUMP HOUSING
22. PISTON FOR SECOND-GEAR BAND
23. PISTON FOR DIRECT-AND-REVERSE-GEAR CLUTCH
24. PISTON FOR FORWARD CLUTCH
25. DRIVING SHELL
26. PISTON FOR FIRST-AND-REVERSE-GEAR BAND

Fig. 19-17 Schematic layout of the essential parts in the Volkswagen automatic transmission. (*Volkswagen of America, Inc.*)

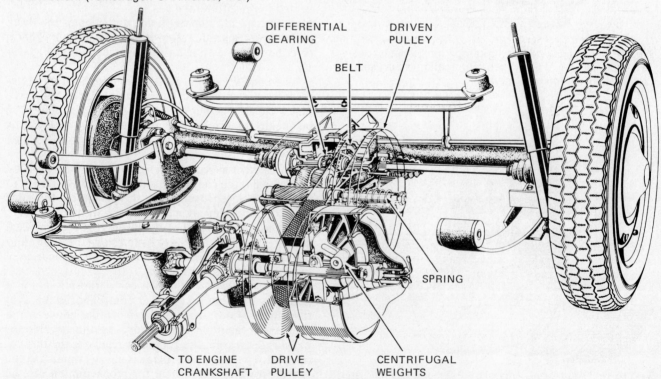

Fig. 19-18 A belt drive type of continuously variable transmission used in an automobile. (*DAF B.V. of Holland*)

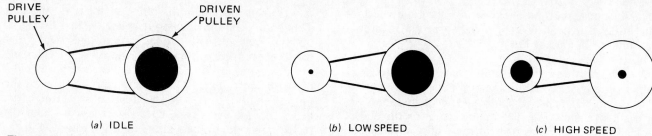

DRIVE PULLEY   DRIVEN PULLEY

(a) IDLE

(b) LOW SPEED

(c) HIGH SPEED

Fig. 19-19 Operation of a centrifugal clutch. (*ATW*)

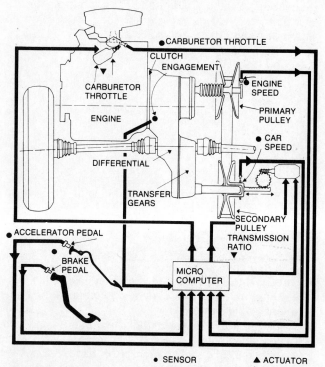

Fig. 19-20 A computer-controlled belt drive transmission. (*Fiat Motors of North America, Inc.*)

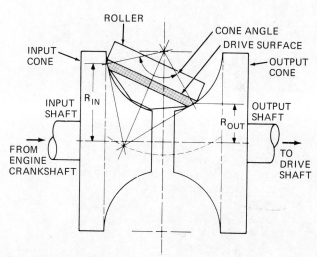

Fig. 19-21 Basic construction of a cone-roller toroidal drive.

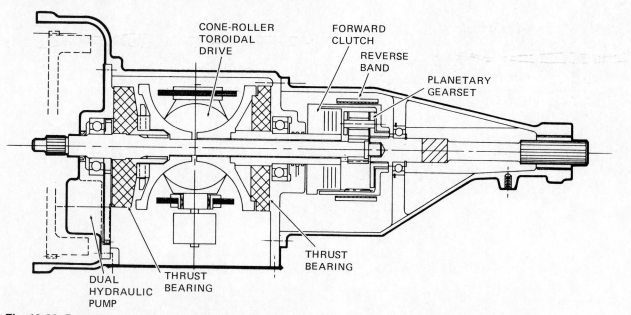

Fig. 19-22 Prototype of a toroidal drive for a compact car. The transmission is continuously variable from 40:1 to 0.4:1. (*Excelermatic, Inc.*)

Small electric motors, or *actuators,* vary the width of the pulleys instead of centrifugal force and manifold vacuum as in Fig. 19-18.

❀ 19-7 Traction-drive transmissions Figure 19-21 shows the basic construction of a cone-roller toroidal drive (CRTD). This is another type of continuously variable transmission. The input cone is on the end of a shaft which is connected to the engine crankshaft. The output cone is attached to a shaft which connects to the drive shaft or final drive. The only connection between the two shafts is through the movable roller. By tilting the roller from right to left (in Fig. 19-21), a continuously variable gear ratio is provided.

Various types of traction drives have been developed. The design objective in all of them is to transmit power from one smooth rolling element to another. This should be done without belts, chains, gears, or pulleys. Instead, balls, cones, or toroidal-shaped elements are moved to change the ratio while the elements are transmitting power. For example, in the toroidal drive shown in Fig. 19-21, tilting the roller toward the output shaft provides a speed increase through the drive. As tilt decreases, speed decreases. When the roller is parallel to the shafts, the drive ratio is 1:1.

Figure 19-22 shows a toroidal-drive transmission that has been tested in a car. Note the use of a planetary gearset and a band and clutch. These controls are required to provide the operating modes needed in an automotive automatic transmission.

Chapter 19 review questions

Select the *one* correct, best, or most probable answer to each question. Then check your answers against the correct answers given at the end of the book.

1. In the Turbo Hydra-Matic transmission shown in Fig. 19-1, power is transferred from the torque converter to the transmission by:
   a. a chain
   b. a toothed belt
   c. gears
   d. a fluid coupling
2. The biggest difference between the transmission shown in Fig. 19-1 and the General Motors type 350 is the:
   a. operation of the bands
   b. method of operating the clutches
   c. size of the roller clutch
   d. location of the torque converter
3. When the transmission shown in Fig. 19-1 fails to move forward in any forward selector position, a possible cause is:
   a. high pump pressure
   b. a defective direct clutch
   c. a defective forward clutch
   d. a defective stator clutch
4. The Toyota Toyoglide transmission has:
   a. three multiple-disk clutches, two bands, and one overrunning clutch
   b. two multiple-disk clutches, two bands, and one overrunning clutch
   c. one multiple-disk clutch, two bands, and two overrunning clutches
   d. none of the above
5. The type of compound planetary gearset shown in Fig. 19-14 is:
   a. used only by Volkswagen
   b. used in many types of automatic transmission
   c. used only in automatic overdrive transmissions
   d. none of the above

GLOSSARY

This glossary of automobile words and phrases provides a ready reference for the automotive technician. The definitions may differ somewhat from those given in a standard dictionary. They are not intended to be all-inclusive but to cover only what specifically applies to the automotive service field.

Acceleration An increase in velocity or speed.

Accelerator A foot-operated pedal linked to the throttle valve in the carburetor; used to control the flow of fuel to the engine.

Accessories Devices not considered essential to the operation of a vehicle, such as the radio, car heater, and electric window lifts.

Accumulator A device used in automatic transmissions to cushion the shock of clutch and servo actions.

Additive A substance added to gasoline or oil to improve some property of the gasoline or oil.

Adjust To bring the parts of a component or system into a specified relationship, or to a specified dimension or pressure.

Adjustments Necessary or desired changes in clearances, fit, or settings.

Alignment The act of lining up; also, the state of being in a true line.

Annular groove On a spool valve, the space between two lands, or raised areas.

Annulus See "Internal gear."

Antifriction bearing Name given to almost any type of ball, roller, or tapered-roller bearing.

Apply pressure Hydraulic pressure, controlled by the control valves, which acts to apply a band.

Arbor press A small, hand-operated shop press used when only a light force is required against a bearing, shaft, or other part.

Assembly A component part, itself made up of assembled pieces which form a self-contained, independently mounted unit. For example, in the automobile, the transmission is an assembly.

Automatic overdrive transmission An automatic transmission that overdrives the output shaft in high gear.

Automatic transmission A transmission in which gear ratios are changed automatically, eliminating the necessity of hand-shifting gears.

Automatic-transmission fluid (ATF) Any of several types of special oil used in automatic transmissions.

Axis The center line of a rotating part, a symmetrical part, or a circular bore.

Axle A theoretical or actual crossbar which supports a vehicle and on which one or more wheels turn.

Axle ratio The ratio between the rotational speed (rpm) of the drive shaft and that of the driven wheel; gear reduction through the differential, determined by dividing the number of teeth on the ring gear by the number of teeth on the drive pinion.

Axle shaft In a vehicle drive axle, a shaft which transmits power from the differential side gear to the wheel hub.

Backlash In gearing, the clearance between the meshing teeth of two gears.

Ball-and-trunnion universal joint A non-constant-velocity universal joint which combines the universal joint and slip joint; it uses a drivable housing connected to a shaft with a ball head through two other balls mounted on trunnions.

Ball bearing An antifriction bearing with an inner race and an outer race, and one or more rows of balls between them.

Ball check valve A valve consisting of a ball and a seat. Fluid can pass in one direction only; flow in the other direction is checked as the ball seats tightly on the seat.

Band In an automatic transmission, a hydraulically controlled brake band installed around a metal clutch drum; used to stop or permit drum rotation.

Bearing A part that transmits a load to a support and in so doing absorbs the friction of moving parts.

Bearing oil clearance The space purposely provided between a shaft and a bearing through which lubricating oil can flow.

Bevel gear A gear shaped like the lower part of a cone; used to transmit motion through an angle.

bhp Abbreviation for brake horsepower.

Body On a vehicle, the assembly of sheet metal and plastic panels and sections, together with windows, doors, seats, and other parts, that provides enclosures for the passengers, engine, and luggage compartments.

Boil tank A large tank of boiling parts-cleaning solution, used for cleaning cylinder blocks, axle housings, and other large metal parts. Also called a *hot tank*.

Bolt A type of fastener having a head on one end and threads on the other; usually used with a nut.

Boost valve A valve which, when acted upon by oil pressure, adds to the force of the spring in the pressure-regulator valve.

Borderline lubrication A type of poor lubrication resulting from greasy friction; moving parts are coated with a very thin film of lubricant.

Bore An engine cylinder, or any cylindrical hole. Also used to describe the process of enlarging or accurately refinishing a hole, as "to bore an engine cylinder." The bore size is the diameter of the hole.

Brake An energy-conversion device used to slow, stop, or hold a vehicle or mechanism. A device which changes the kinetic energy of motion into useless and wasted heat energy.

Brake horsepower (bhp) Power delivered by the engine and available for driving the vehicle; bhp = torque × rpm/5252.

Burr A feather edge of metal left on a part being cut with a file or other cutting tool.

Bushing A one-piece sleeve placed in a bore to serve as a bearing surface.

Bypass A separate passage which permits a liquid, gas, or electric current to take a path other than that normally used.

Calibrate To check or correct the initial setting of a test instrument.

Caliper A tool that can be set to measure the thickness of a block, the diameter of a shaft, or the bore of a hole (in the third case, an inside caliper).

Cam A rotating lobe or eccentric which can be used with a cam follower to change rotary motion to reciprocating motion.

Capacity The ability to perform or to hold.

Case hardening The carburizing method used on low-carbon steel or other alloys to make the case or outer layer of the metal harder than its core.

cc Abbreviation for cubic centimeter.

Centrifugal See "Centrifugal force."

Centrifugal clutch A clutch that uses centrifugal force to apply a greater force against the friction disk as the clutch spins faster.

Centrifugal force The force acting on a rotating body which tends to move it outward and away from the center of rotation. The force increases as rotational speed increases.

Chassis The assembly of mechanisms that make up the major operating systems of the vehicle; usually assumed to include everything except the car body.

Check To verify that a component, system, or measurement complies with specifications.

Check valve A valve that opens to permit the passage of air or fluid in one direction only, or operates to prevent (check) some undesirable action.

CID Abbreviation for cubic-inch displacement.

Circuit The complete path of an electric current, including the current source. When the path is continuous, the circuit is closed, and current flows. When the path is broken, the circuit is open, and no current flows. Also used to refer to fluid paths, as in refrigerant and hydraulic systems.

Clearance The space between two moving parts, or between a moving and a stationary part, such as a journal and a bearing. The bearing clearance is filled with lubricating oil when the mechanism is running.

Clutch Device used to transmit rotary motion from one shaft to another while permitting engagement or disengagement of the shafts during rotation of one or both. Normally, the two shafts are in line and rotate about the same axis.

Clutch disk See "Clutch Plate."

Clutch plate In an automatic transmission, a disk with friction surfaces which is splined to a shaft or locked to a drum, used to transmit power when engaged.

cm Abbreviation for centimeter.

Coil spring A spring made of an elastic metal such as steel, formed into a wire and wound into a coil.

Companion flange A mounting flange that fixedly attaches a drive shaft to another drive train component.

Component A part of a whole assembly, system, or unit, which may be identified and serviced separately. For example, a bulb is a component of the lighting system.

Compound planetary gearset A planetary gearset that has two planet-pinion carriers and two ring gears revolving around a single sun gear.

Converter clutch The clutch in the torque converter that can be engaged to lock the pump and turbine together, eliminating any loss through the fluid.

Corrosion Chemical action, usually by an acid, that eats away (decomposes) a metal.

Cotter pin A type of fastener, made from soft steel in the form of a split pin, that can be inserted in a drilled hole. The split ends are spread to lock the pin in position.

Coupling point In a torque converter, the speed at which the oil begins to strike the back faces of the stator vanes; occurs when the turbine and pump speeds reach a ratio of approximately 9:10.

Crankshaft The main rotating member or shaft of the engine, with cranks to which the connecting rods are attached; converts up-and-down, or reciprocating, motion into circular or rotary motion.

Cubic centimeter (cu cm, cm³, or cc) A unit of volume in the metric system; equal to approximately 0.03 fluid ounce.

Cubic-inch displacement (CID) The cylinder volume displaced or swept out by the pistons of an engine as they move from BDC to TDC, measured in cubic inches.

cu cm Abbreviation for cubic centimeter.

Cylinder A circular tubelike opening in an engine cylinder block or casting in which a piston moves up and down.

Deceleration A decrease in velocity or speed. Also, allowing the car or engine to coast to idle speed from a higher speed with the accelerator at or near the idle position.

Degree Part of a circle; 1° is 1/360 of a complete circle.

Detent A small depression in a shaft, rail, or rod into which a pawl or ball drops when the shaft, rail, or rod is moved; this provides a locking effect.

Device A mechanism, tool, or other piece of equipment designed to serve a special purpose or perform a special function.

Dexron II A special type of automatic-transmission fluid (ATF) specified for use in vehicles built by General Motors and Chrysler.

Diagnosis A procedure followed in locating the cause of a malfunction; also, to specifically identify and answer the question *What is wrong?*

Dial indicator A gauge that has a dial face and a needle to register movement; used to measure variations in dimensions and movements too small to be measured accurately by other means.

Diaphragm A thin dividing sheet or partition which separates an area into compartments; used in fuel pumps, modulator valves, vacuum-advance units, and other control devices.

Differential A gear assembly between axle shafts that permits one wheel to turn at a different speed than the other, while transmitting power from the drive shaft to the wheel axles.

Direct drive The condition in a gearset when both the input shaft and the output shaft turn at the same speed, with a ratio of 1:1.

Disassemble To take apart.

Displacement In an engine, the total volume of air-fuel mixture an engine is theoretically capable of drawing into all cylinders during one operating cycle. Also, the volume swept out by the piston in moving from one end of a stroke to the other.

Dowel A metal pin attached to one object which, when inserted into a hole in another object, ensures proper alignment.

Downshift To shift a transmission into a lower gear.

Drivability The general operation of a vehicle, usually rated from good to poor; based on characteristics of concern to the average driver, such as smoothness of idle, even acceleration, ease of starting, quick warm-up and tendency to overheat at idle.

Driveline An assembly of one or more drive shafts usually with universal joints and some type of slip joint, used to transmit torque through varying angles from one shaft to another.

Drive plate A thin disk fastened to the rear end of the crankshaft to connect it to the torque converter.

Drive shaft An assembly of one or two universal joints connected to a shaft or tube; used to transmit power from the transmission to the differential. Also called the *propeller shaft*.

Drive train See "Power train."

Drum The rotating metal drum which is held stationary or allowed to rotate by application or release of the band.

Dry friction The friction between two dry solids.

Duration The length of time during which something exists or lasts.

Dynamic balance The balance of an object when it is in motion (for example, the dynamic balance of a rotating drive shaft).

Dynamometer A device for measuring the power output, or brake horsepower, of an engine. An engine dynamometer measures the power available at the flywheel; a chassis dynamometer measures the power available at the drive wheels.

Eccentric A disk or offset section (of a shaft, for example) used to convert rotary to reciprocating motion. Sometimes called a *cam*.

ECU See "Electronic control unit."

Efficiency The ratio between the power of an effect and the power expended to produce the effect; the ratio between an actual result and the theoretically possible result.

Electromechanical Describing a device whose mechanical movement is dependent upon an electric current.

Electronic control unit A solid-state device that receives information from sensors and is programmed to operate various circuits and systems on the basis of that information.

Electronics Electrical assemblies, circuits, and systems that use electronic devices such as diodes and transistors.

End play The distance that a shaft can move forward and backward in its housing or case.

Energize To activate; to cause movement or action.

Energy The capacity or ability to do work. Usually measured in work units of foot-pounds or meter-kilograms, but also expressed in heat-energy units (Btu's or joules).

Engine A machine that converts heat energy into mechanical energy. A device that burns fuel to produce mechanical power; sometimes referred to as a *power plant*.

External gear See "Spur gear."

Extreme-pressure lubricant A special lubricant for use in hypoid-gear differentials; needed because of the heavy wiping loads imposed on the gear teeth.

Fatigue failure A type of metal failure resulting from repeated stress which finally alters the character of the metal so that it cracks. In engine bearings, frequently caused by excessive idling, or slow engine idling speed.

Feeler gauge See "Thickness gauge."

Filter A device through which air, gases, or liquids are passed to remove impurities.

Final drive The final speed-reduction gearing in the power train.

Flex plate See "Drive plate."

Fluid Any liquid or gas.

Fluid coupling A device in the power train consisting of two rotating members. It transmits power from the engine, through a fluid, to the transmission.

Flywheel A heavy metal wheel that is attached to the crankshaft and rotates with it. It helps smooth out the power surges from the engine power strokes, and also serves as part of the clutch and engine cranking system.

Force Any push or pull exerted on an object; measured in pounds and ounces, or newtons (N) in the metric system.

Four-speed A transmission having four forward-gear ratios.

Four-wheel drive On a vehicle, the presence of driving axles at both front and rear, so that all four wheels can be driven.

Frame The assembly of metal structural parts and channel sections that supports the car engine and body and is supported by the wheels.

Freewheeling clutch See "Sprag clutch."

Friction The resistance to motion between two bodies in contact with each other.

Friction bearing A bearing in which there is sliding contact between the moving surfaces. Sleeve bearings, such as those used in connecting rods, are friction bearings.

Friction horsepower The power used up by an engine in overcoming its own internal friction; usually increases as engine speed increases.

Front-wheel drive A kind of drive in which a vehicle has its drive wheels located on the front axle.

Full throttle Wide-open throttle position, with the accelerator pedal pressed all the way down to the floorboard.

Fusion Melting; conversion from the solid to the liquid state.

Gallery A passageway inside a wall or casting. The main oil gallery within the block supplies lubrication to all parts of the engine.

Gap The air space between two parts or electrodes.

Gasket A layer of material, usually made of cork, paper, plastic, composition, or metal, or a combination of more than one of these, placed between two parts to make a tight seal.

Gasket cement A liquid adhesive material, or sealer, used to install gaskets.

Geared speed The theoretical vehicle speed, based on engine rpm, transmission gear ratio, drive-axle ratio, and tire size.

Gear lubricant A type of grease or oil blended especially to lubricate gears.

Gear ratio The number of revolutions of a driving gear required to turn a driven gear through one complete revolution. For a pair of gears, the ratio is found by dividing the number of teeth on the driven gear by the number of teeth on the driving gear.

Gears Toothed wheels that transmit power between shafts.

Gearshift A linkage-type mechanism by which the gears in an automobile transmission are engaged and disengaged.

Gear-type pump A pump in which a pair of rotating gears mesh to force oil (or some other liquid) from between the teeth to the pump outlet.

Goggles Special glasses worn over the eyes to protect them from flying chips, dirt, dust and spraying refrigerant.

Governor A device that controls, or governs, another device, usually on the basis of speed or load; for example, the governor used in certain automatic transmissions to control gear shifting in relation to car speed.

Governor pressure Pressure, at one end of the shift valve, which is controlled by governor action. One of the two signals that determine when the transmission will shift.

Greasy friction The friction between two solids coated with a thin film of oil.

Grommet A device, usually made of hard rubber or a similar material, used to encircle or support a component.

Helical gear A gear in which the teeth are cut at an angle to the centerline of the gear.

Heli-Coil See "Threaded insert."

Horsepower A measure of mechanical power, or the rate at which work is done. One horsepower equals 33,000 ft-lb (foot-pounds) of work per minute; it is the power necessary to raise 33,000 pounds a distance of 1 foot in 1 minute.

Hydraulic press A piece of shop equipment that develops a heavy force by use of a hydraulic piston-and-jack assembly.

Hydraulic pressure Pressure exerted through the medium of a liquid.

Hydraulics The use of a liquid under pressure to transfer force or motion, or to increase an applied force.

Hydraulic valve A valve in a hydraulic system that operates on, or controls, the hydraulic pressure in the system. Also, any valve that is operated or controlled by hydraulic pressure.

Hydrodynamic A hydraulic device, such as the torque converter, in which the force of the fluid in motion is used to multiply torque.

Hydrostatic A hydraulic device, such as a valve in the automatic-transmission valve body, which operates because of the incompressibility of the fluid working on it.

Idle Engine speed when the accelerator pedal is fully released and there is no load on the engine.

Idler gear The middle gear in a three-gear set which does not change the gear ratio but only the direction of rotation of the driven gear.

Impeller A rotating finned disk; used in centrifugal pumps, such as water pumps, and in torque converters.

Indicator A device used to make some condition known by use of a light or a dial and pointer.

Inertia The property of an object that causes it to resist any change in its speed or the direction of its travel.

Inspect To examine a part or system for surface condition or function to answer the question *Is something wrong?*

Install To set up for use on a vehicle any part, accessory, option, or kit.

Integral Built into, as part of the whole.

Interchangeability The manufacture of similar parts to close tolerances so that any one of them can be substituted for another in a device and the part will fit and operate properly; the basis of mass production.

Internal gear A gear with teeth pointing inward, toward the hollow center of the gear.

Journal The part of a rotating shaft which turns in a bearing.

Key A wedgelike metal piece, usually rectangular or semicircular, inserted in a groove to transmit torque while holding two parts in the same relative position. Also, the small strip of metal with coded peaks and grooves used to operate a lock, such as that for the ignition switch.

kg/cm₂ Abbreviation for kilograms per square centimeter, a metric engineering term for the measurement of pressure; 1 kilogram per square centimeter equals 4.22 pounds per square inch.

Kickdown In automatic transmissions, a system that produces a downshift when the accelerator is pushed down to the floorboard.

Kilogram (kg) In the metric system, a unit of weight and mass; approximately equal to 2.2 pounds.

Kilometer (km) In the metric system, a unit of linear measure; equal to 0.621 miles.

Kilowatt (kW) 1000 watts; a unit of power, equal to about 1.34 horsepower.

Kinetic energy The energy of motion; the energy stored in a moving body through its momentum; for example, the kinetic energy stored in a rotating flywheel.

Knock A heavy metallic sound which varies with speed; usually caused by a loose or worn bearing.

kPa Abbreviation for kilopascals, the metric measurement of pressure; 1 kilopascal equals 0.145 pounds per square inch.

kW Abbreviation for kilowatt.

Lash The amount of free motion in a gear train, between gears, or in a mechanical assembly, such as the lash in a valve train.

Linear measurement A measurement taken in a straight line; for example, the measurement of shaft end play.

Linkage An assembly of rods, or links, used to transmit motion.

Liter (L) In the metric system, a measure of volume; approximately equal to 0.26 gallon (U.S.), or about 61 cubic inches (33.8 fluid ounces, or 1 quart 1.8 ounces). Used as a metric measure of engine-cylinder displacement.

Locknut A second nut turned down on a holding nut to prevent loosening.

Lockup torque converter A torque converter in which the pump can be mechanically locked to the turbine, eliminating any loss through the fluid.

Lockwasher A type of washer which, when placed under the head of a bolt or nut, prevents the bolt or nut from working loose.

Lubricant Any material, usually a petroleum product such as grease or oil, that is placed between two moving parts to reduce friction.

Lugging Low-speed, full-throttle engine operation in which the engine is heavily loaded and overworked; usually caused by failure of the driver to shift to a lower gear when necessary.

Machining The process of using a machine to remove metal from a metal part.

Magna-flux A process in which an electromagnet and a special magnetic power are used to detect surface and subsurface cracks in iron and steel which otherwise might not be seen.

Make A distinctive name applied to a group of vehicles produced by one manufacturer; may be further subdivided into car lines, body types, etc.

Malfunction Improper or incorrect operation.

Manual lever The lever which moves the manual valve as the shift lever moves.

Manual low Position of the units in an automatic transmission when the driver moves the shift lever to the low or first-gear position on the quadrant.

Manual transmission A transmission that the driver must shift by hand, or manually.

Manual valve A spool valve in the valve body that is manually positioned by the driver through linkage.

Manufacturer Any person, firm, or corporation engaged in the production or assembly of motor vehicles or other products.

Mass production The manufacture of interchangeable parts and similar products in large quantities.

Matter Anything that has weight and occupies space.

Measuring The act of determining the size, capacity, or quantity of an object.

Mechanical advantage In a machine, the ratio of the output force to the input force applied to it.

Mechanical efficiency In an engine, the ratio between brake horsepower and indicated horsepower.

Mechanism A system of interrelated parts that make up a working assembly.

Member Any essential part of a machine or assembly.

Meshing The mating, or engaging, of the teeth of two gears.

Meter (m) A unit of linear measure in the metric system, equal to 39.37 inches. Also, the name given to any test instrument that measures a property of a substance passing through it, as an ammeter measures electric current. Also, any device that measures and controls the flow of a substance passing through it, as a carburetor jet meters fuel flow.

Micrometer A precision measuring device used to measure small bores, diameters, and thicknesses. Also called a *mike*.

Mike Slang term for micrometer.

Millimeter (mm) In the metric system, a unit of linear measure, approximately equal to 0.039 inches.

mm Abbreviation for millimeter.

Mode Term used to designate a particular set of operating characteristics.

Model year The production period for new motor vehicles or new engines, designated by the calendar year in which the period ends.

Modification An alteration; a change from the original.

Modulator A pressure-regulated governing device; used, for example, in automatic transmissions.

Motor vehicle A vehicle propelled by a means other than muscle power, usually mounted on rubber tires, which does not run on rails or tracks.

mph Abbreviation for miles per hour, a measure of speed.

Multiple-disk clutch A clutch with more than one friction disk; usually there are several driving disks and several driven disks, alternately placed.

Multiple-plate clutch See "Multiple-disk clutch."

Neck A portion of a shaft that has a smaller diameter than the rest of the shaft.

Needle bearing An antifriction bearing of the roller type, in which the rollers are very small in diameter (needle-sized).

Needle valve A small, tapered, needle-pointed valve which can move into or out of a valve seat to close or open the passage through the seat. Used to control the carburetor float-bowl fuel level.

Neoprene A synthetic rubber that is not affected by the various chemicals that are harmful to natural rubber.

Neutral In a transmission, the setting in which all gears are disengaged and the output shaft is disconnected from the drive wheels.

Neutral-start switch A switch wired into the ignition switch to prevent engine cranking unless the transmission shift lever is in neutral or the clutch pedal is depressed.

Nut A removable fastener used with a bolt to lock pieces together; made by threading a hole through the center of a piece of metal which has been shaped to a standard size.

Odometer The meter that indicates the total distance a vehicle has traveled, in miles or kilometers; usually located in the speedometer.

OEM Abbreviation for original-equipment manufacturer.

Oil A liquid lubricant usually made from crude oil and used to provide lubrication between moving parts.

Oil cooler A small radiator that lowers the temperature of oil flowing through it.

Oil seal A seal placed around a rotating shaft or other moving part to prevent leakage of oil.

Oil strainer A wire-mesh screen placed at the inlet end of the oil-pump pickup tube. It prevents dirt and other large particles from entering the oil pump.

One-way clutch See "Sprag clutch."

Orifice A small, calibrated hole in a line carrying a liquid or gas.

O ring A type of sealing ring, usually made of rubber or rubberlike material. In use, the O ring is compressed into a groove to provide the sealing action.

Oscillating Moving back and forth, as a swinging pendulum.

Output shaft The main shaft of the transmission; the shaft that delivers torque from the transmission to the drive shaft.

Overdrive A device or transmission gear arrangement which causes the output shaft to overdrive or rotate faster than the input shaft.

Overflow Spilling of the excess of a substance; also, to run or spill over the sides of a container, usually because of overfilling.

Overhaul To completely disassemble a unit, clean and inspect all parts, reassemble it with the original or new parts, and make all adjustments necessary for proper operation.

Overheat To heat excessively; also, to become excessively hot.

Overrunning clutch See "Sprag clutch."

Oxidation Burning or combustion; the combining of a material with oxygen. Rusting is slow oxidation, and combustion is rapid oxidation.

Parallel The quality in which two items are the same distance from each other at all points; usually applied to lines and, in automotive work, to machined surfaces.

Parking pawl An arm on a lever that can be moved to lock the output shaft to the transmission case so the car cannot be moved.

Part A basic mechanical element or piece, that normally cannot be further disassembled, of an assembly, component, system, or unit. Also applied to any separate entry in a parts catalog, or one that has a "part number."

Particle A very small piece of metal, dirt, or other impurity which may be contained in the air, fuel, or lubricating oil used in an engine or transmission.

Passage A small hole or gallery in an assembly or casting through which air, coolant, fuel, or oil flows.

Passenger car Any four-wheeled motor vehicle manufactured primarily for use on streets and highways and carrying 10 or fewer passengers.

Peen To mushroom, or spread, the end of a pin or rivet.

Petroleum The crude oil from which gasoline, lubricating oil, and other such products are refined.

Pinion gear The smaller of two meshing gears.

Piston A movable part, fitted in a cylinder, which can receive or transmit motion as a result of pressure changes in a fluid.

Piston displacement The cylinder volume displaced by the piston as it moves from the bottom to the top of the cylinder during one complete stroke.

Pitch The number of threads per inch on any threaded part.

Pivot A pin or shaft upon which another part rests or turns.

Planetary gearset See "Planetary-gear system."

Planetary-gear system A gear set consisting of a central sun gear surrounded by two or more planet pinions, which are, in turn, meshed with a ring (or internal) gear; used in overdrives and automatic transmissions.

Planet carrier In a planetary-gear system, the carrier or bracket that contains the shafts upon which the planet pinions turn.

Planet gears See "Planet pinions."

Planet pinions In a planetary-gear system, the gears that mesh with, and revolve about, the sun gear; they also mesh with the ring (or internal) gear.

Plastic gasket compound A plastic paste which can be squeezed out of a tube to make a gasket in any shape.

Plunger A sliding reciprocating piece driven by an auxiliary power source, having the motion of a ram or piston.

Pour point The lowest temperature at which an oil will flow.

Power The rate at which work is done. A common power unit is the horsepower, which is the power necesary to raise 33,000 pounds a distance of 1 foot in 1 minute.

Power absorber The part of the dynamometer which can place varying loads on the engine and power train.

Power plant The engine or power source of a vehicle.

Power takeoff An attachment for connecting the engine to devices or other machinery for which power is required.

Power tool A tool whose power source is not muscle power; a tool powered by air or electricity.

Power train The mechanisms that carry the power from the engine crankshaft to the drive wheels; these include the clutch, transmission, driveline, differential, and axles.

Preload In bearings, the amount of load placed on a bearing before actual operating loads are imposed. Proper preloading requires bearing adjustment and ensures alignment and minimum looseness in the system.

Press fit A fit between two parts so tight that one part has to be pressed into the other, usually with a shop press.

Pressure Force per unit area, or force divided by area. Usually measured in pounds per square inch (psi) or kilopascals (kPa).

Pressure-regulator valve A device that operates to prevent excessive pressure from developing. In the hydraulic system of automatic transmissions, a valve that opens to release oil from a line when the oil pressure reaches a specified maximum limit.

Pressure relief valve See "Pressure-regulator valve."

Pressurize To apply more than atmospheric pressure to a gas or liquid.

Propeller shaft See "Drive shaft."

psi Abbreviation for *pounds per square inch,* a measurement of pressure.

psig Abbreviation for *pounds per square inch of gauge pressure.*

Pull The result of an unbalanced condition. For example, uneven braking at the front brakes or unequal front-wheel alignment will cause a car to swerve (pull) to one side when the brakes are applied.

Puller Generally, a shop tool used to separate two closely fitted parts without damage. Often contains a screw, or several screws, which can be turned to apply a gradual force.

Pump A device that transfers gas or liquid from one place to another. In a torque converter, often called the *impeller.*

Quadrant A term sometimes used to identify the shift-lever selector mounted on the steering column.

Races The metal rings on which ball or roller bearings rotate.

Radius The distance from the center to the outside edge of a circle.

Ratio The relationship in size or quantity of two or more objects. A gear ratio is derived by dividing the number of teeth on the driven gear by the number of teeth on the drive gear.

Reaction shaft See "Stator shaft."

Reamer A round metal-cutting tool with a series of sharp cutting edges; enlarges a hole when turned inside it.

Reassembly Putting back together the parts of a device.

Reciprocating motion Motion of an object between two limiting positions; motion in a straight line either back and forth or up and down.

Release pressure Hydraulic pressure, controlled by the control valves, which acts to release the band.

Relief valve A valve that opens when a preset pressure is reached. This relieves or prevents excessive pressures.

Remove and reinstall (R and R) To perform a series of servicing procedures on an original part or assembly; includes removal, inspection, lubrication, all necessary adjustments, and reinstallation.

Replace To remove a used part or assembly and install a new part or assembly in its place; includes cleaning, lubricating, and adjusting as required.

Retaining ring A removable fastener used as a shoulder to retain and position a round bearing in a hole.

Ring gear See "Internal gear."

Roller clutch A type of one-way clutch that uses rollers instead of sprags.

Rotary Describing the motion of a part that continually rotates or turns.

Rotary flow The path a drop of oil takes as the oil rotates with the coupling between members.

Rotor oil pump A type of oil pump in which a pair of rotors, one inside the other, produce the pressure required to circulate oil to engine parts.

rpm Abbreviation for *revolutions per minute,* a measure of rotational speed.

RTV sealer *R*oom-*T*emperature *V*ulcanizing gasket material, which cures at room temperature; a plastic paste squeezed from a tube to form a gasket of any shape.

Runout wobble Failure to run true about a center-line.

SAE Abbreviation for Society of Automotive Engineers; used to indicate a grade or weight of oil measured according to Society of Automotive Engineers standards.

Safety Freedom from injury or danger.

Safety stand A pinned or locked type of stand placed under a car to support its weight after the car has been raised with a lift or floor jack. Also called a car stand or jack stand.

Schematic A pictorial representation, most often in the form of a line drawing; a systematic positioning of components, showing their relationship either to others or to an overall function.

Scored Scratched or grooved, as a cylinder wall may be *scored* by abrasive particles moved up and down by the piston rings.

Scraper A device used in engine service to scrape carbon from the engine block, pistons, or other parts.

Screens Pieces of fine-mesh metal fabric; used to prevent solid particles from circulating through any liquid or vapor system and damaging vital moving parts.

Screw A metal fastener with threads that can be turned into a threaded hole. There are many different types and sizes of screws.

Scuffing A type of wear in which there is a transfer of material between parts moving against each other; shows up as pits or grooves in the mating surfaces.

Seal A material, shaped around a shaft, used to close off the operating compartment of the shaft, preventing oil leakage.

Sealer A thick, tacky compound, usually spread with a brush, which may be used as a gasket or sealant to seal small openings or surface irregularities.

Seat The surface upon which another part rests, as a valve seat. Also, to wear into a good fit; for example, new piston rings *seat* after a few miles of driving.

Sediment The accumulation of matter which settles to the bottom of a liquid.

Selector lever The lever on the steering column or on the floor which the driver moves to select the desired driving range, such as park, reverse, neutral, 3, 2, or 1.

Self-locking screw A screw that locks itself in place without the use of a separate nut or lockwasher.

Self-tapping screw A screw that cuts its own threads as it is turned into an unthreaded hole.

Serviceable Describing parts or systems that can be repaired and maintained to continue in operation.

Service manual A book published annually by each vehicle manufacturer, listing the specifications and service procedures for each make and model of vehicle. Also called a *shop manual*.

Servo A device in a hydraulic system that converts hydraulic pressure to mechanical movement; consists of a piston which moves in a cylinder as hydraulic pressure acts on it.

Setscrew A type of metal fastener that holds a collar or gear on a shaft when its point is turned down into the shaft.

Shift lever The lever used to change gears in a transmission.

Shift valve In an automatic transmission, a valve that moves to produce the shifts from one gear ratio to another.

Shim A slotted strip of metal used as a spacer to adjust the front-end alignment on many cars; also used to make small corrections in the position of body sheet metal and other parts.

Shim stock Sheets of metal of accurately known thicknesses which can be cut into strips and used to measure or correct clearances.

Shop layout The locations of aisles, work areas, machine tools, etc., in a shop.

Shrink fit A tight fit of one part into another, achieved by heating or cooling one part and then assembling it to the other part. A heated part will shrink on cooling to provide the tight fit; a cooled part will expand on warming to provide the tight fit.

Side clearance The clearance between the sides of moving parts when the sides do not serve as load-carrying surfaces.

Slip joint In the power train, a variable-length connection that permits the drive shaft to change its effective length.

Snap ring A metal fastener, available in two types: The external snap ring fits into a groove in a shaft; the internal snap ring fits into a groove in a housing. Snap rings often must be installed and removed with special snap-ring pliers.

Solvent A petroleum product of low volatility used in the cleaning of engine and vehicle parts.

Solvent tank In the shop, a tank of cleaning fluid in which most parts are brushed and washed clean.

Specifications (1) Information provided by the manufacturer that describes each automotive system and its components, operation, and clearances; (2) the service procedures that must be followed for a system and its components, operation, and clearances; (3) the service procedures that must be followed for a system to operate properly.

Specs Short for *specifications*.

Speed The rate of motion; for vehicles, measured in miles per hour or kilometers per hour.

Speedometer An instrument that indicates vehicle speed; usually driven from the transmission.

Splines Slots or grooves cut in a shaft or bore. Splines on a shaft are matched to splines in a bore, to ensure that two parts turn together.

Spool valve A rod with indented sections; used to control oil flow in automatic transmissions.

Sprag clutch A one-way clutch; it can transmit power in one direction but not in the other.

Spring A device that changes shape under stress or compression but returns to its original shape when the stress or compression is removed; the component of the automotive suspension system that absorbs road shocks by flexing and twisting.

Sprung weight That part of the car which is supported on springs (including the engine, frame, and body).

Spur gear A gear in which the teeth are parallel to the centerline of the gear.

Squeak A high-pitched noise of short duration.

Squeal A continuous high-pitched noise.

Stall The condition in which an engine quits running, at idle or while driving.

Static balance The balance of an object while it is not moving.

Static friction The friction between two bodies at rest.

Stator The stationary member of a machine, such as an electric motor or generator, in or about which a rotor revolves; in an electronic ignition system, a small magnet embedded in plastic (or a light-emitting diode) which when used with a reluctor replaces contact points. Also, the third member, in addition to the turbine and pump, in a torque converter.

Stator shaft A hollow, stationary shaft which the stator spins around or locks against.

Steam cleaner A machine used for cleaning large parts with a spray of steam, often mixed with soap.

Steering-and-ignition lock A device that locks the ignition switch in the off position and locks the steering wheel so that it cannot be turned.

Steering-column shift An arrangement in which the transmission shift lever is mounted on the steering column.

Stud A headless bolt that is threaded on both ends.

Stud extractor A special tool used to remove a broken stud or bolt.

Substance Any matter or material; may be a solid, a liquid, or a gas.

Sun gear In a planetary-gear system, the center gear that meshes with the planet pinions.

Surface grinder A grinder used to resurface flat surfaces, such as cylinder heads.

Synchronize To make two or more events or operations occur at the same time or at the same speed.

Synthetic oil An artificial oil that is manufactured, not a natural mineral oil made from petroleum.

System A combination or grouping of two or more parts or components into a whole which in operation performs some function that cannot be done by the separate parts.

Tachometer A device for measuring the speed of an engine in revolutions per minute (rpm).

Tap A tool used for cutting threads in a hole.

Taper A gradual reduction in the width of a shaft or hole; in a cylinder, uneven wear, more pronounced at one end than at the other.

Technology The applications of science.

Temperature The measure of heat intensity in degrees. Temperature is not a measure of heat quantity.

Thickness gauge Strips of metal made to an exact thickness, used to measure clearances between parts.

Thread chaser A device, similar to a die, that is used to clean threads.

Thread class A designation indicating the closeness of fit between a pair of threaded parts, such as a nut and a bolt.

Threaded insert A threaded coil that is used to restore the original thread size to a hole with damaged threads. The hole is drilled oversize and tapped, and the insert is threaded into the tapped hole.

Thread series A designation indicating the pitch, or number of threads per inch, on a threaded part.

Throttle pressure Pressure at one end of the shift valve which changes as engine intake-manifold vacuum changes. One of the two signals that determine when the transmission will shift.

Throttle valve A round disk valve in the throttle body of the carburetor; can be turned to admit more or less air-fuel mixture, thereby controlling engine speed in a carbureted spark-ignition engine.

Thrust bearing In the engine, the main bearing, which has thrust faces to prevent excessive end play, or forward and backward movement of the crankshaft.

Torque Turning or twisting effort; usually measured in pound-feet or newton-meters. Also, a turning force such as that required to tighten a connection.

Torque converter In an automatic transmission, a fluid coupling which incorporates a stator to permit a torque increase.

TorqueFlite An automatic transmission used on Chrysler-manufactured cars. It has three forward speeds and reverse.

Torque wrench A wrench that indicates the amount of torque or turning force being applied with the wrench.

Torsional vibration Rotary vibration that causes a twist-untwist action on a rotating shaft, so that a part of the shaft repeatedly moves ahead of, or lags behind, the remainder of the shaft; for example, the action of a crankshaft responding to the cylinder firing impulses.

Tractive effort The force available at the road surface in contact with the driving wheels of a truck. It is determined by engine torque, transmission ratio, axle ratio, tire size, and frictional losses in the driveline. *Rim pull* is also known as tractive effort.

Transaxle A power-transmission device that combines the functions of the transmission and the drive axle (differential) into a single assembly; used in front-wheel-drive cars with front-mounted engines, and in rear-wheel-drive cars with rear-mounted engines.

Transducer Any device which converts an input signal of one form into an output signal of a different form. For example, the automobile horn converts an electric current into sound.

Transfer case A gearbox mounted in back of the transmission which can be used to place the vehicle in four-wheel drive, and which may have a high- and a low-speed range.

Transmission An assembly that transmits power from the engine to the driving axle; it provides different forward-gear ratios, neutral, and reverse, through which engine power is transmitted to the differential.

Transmission-controlled spark (TCS) system A Ford exhaust-emission control system, similar to the General Motors transmission-controlled spark system; allows distributor vacuum advance in high gear only.

Transmission-fluid cooler A small, sometimes finned tube or tank, mounted either separately or as part of the engine radiator, which cools the transmission fluid.

Trouble diagnosis The detective work necessary to find the cause of a trouble.

Truck Any motor vehicle primarily designed for the transportation of property which carries the load on its own wheels.

Truck tractor Any motor vehicle designed primarily for pulling truck trailers and constructed so as to carry part of the weight and load of a semi-trailer.

Turbine In a torque converter, the output section that receives a stream of fluid from the pump (or input section) and converts the kinetic energy of the fluid into mechanical energy.

Turbo Hydra-Matic A three-speed automatic transmission built by General Motors and used on many models of cars and light trucks.

Turbulence The state of being violently disturbed. In the engine, the rapid swirling motion imparted to the air-fuel mixture entering a cylinder.

Turbulent flow Unwanted oil swirl in the center section of the fluid coupling member, caused by the oil striking the turbine vanes with great force.

Unit An assembly or device that can perform its function only if it is not further divided into its component parts.

Universal joint In the driveline, a connecting joint that can transmit torque between shafts while the drive angle varies.

Unsprung weight The weight of that part of the car which is not supported on springs; for example, the wheels and tires.

Upshift To shift a transmission into a higher gear.

Vacuum A pressure less than atmospheric pressure; a negative pressure. Vacuum can be measured in pounds per square inch, but is usually measured in inches or millimeters of mercury (Hg); a reading of 30 inches of mercury [762 mm Hg] would indicate a perfect vacuum.

Vacuum gauge In automotive-engine service, a device that measures intake-manifold vacuum and thereby indicates actions of engine components.

Vacuum modulator In automatic transmissions, a device that modulates, or changes, the main-line hydraulic pressure to meet changing engine loads.

Valve Any device that can be opened or closed to allow or stop the flow of a liquid or gas. There are many different types.

Valve body A casting located at the bottom of the oil pan which contains most of the valves for the hydraulic control system of an automatic transmission.

Vane A flat, extended surface that is moved around an axis by or in a fluid; part of the internal revolving portion of an air pump.

Vaporization A change of state from liquid to vapor or gas by evaporation or boiling; a general term including both evaporation and boiling. In the carburetor, breaking gasoline into fine particles and mixing it with incoming air.

V-block A metal block with an accurately machined V-shaped groove; used to support an armature or shaft while it is checked for roundness.

Vehicle See "Motor vehicle."

Vehicle identification number (VIN) The number assigned to each vehicle by its manufacturer, primarily for registration and identification purposes.

Vent An opening through which air can leave an enclosed chamber.

Ventilation The circulating of fresh air through any space, to replace impure air; the basis of crankcase ventilation systems.

Vibration A rapid back-and-forth motion; an oscillation.

VIN Abbreviation for *vehicle identification number*.

Viscosity The resistance to flow exhibited by a liquid. A thick oil has greater viscosity than a thin oil.

Viscous Thick; tending to resist flowing.

Viscous friction The friction between layers of a liquid.

Volatility A measure of the ease with which a liquid vaporizes; has a direct relationship to the flammability of a fuel.

Vortex flow The spiraling or whirlpool path within a fluid coupling that a drop of oil takes due to centrifugal force and the curvature of the coupling.

Weight, sprung See "Sprung weight."

Weight, unsprung See "Unsprung weight."

Welding The process of joining pieces of metal by fusing them together with heat.

Wet-disk clutch A clutch in which the friction disk (or disks) is (are) operated in a bath of oil.

Wheelbase The distance between the centerlines of the front and rear axles. For trucks with tandem rear axles, the rear center line is considered to be midway between the two rear axles.

Wire thickness gauge A set of round wires of known diameters; often used to check clearances between electric contacts, such as distributor points and spark-plug electrodes.

Work The changing of the position of an object against an opposing force; measured in foot-pounds or meters-newtons. The product of a force and the distance through which it acts.

Worm A type of gear in which the teeth resemble threads; used on the lower end of the steering shaft.

Wot Abbreviation for *wide-open throttle*.

Yoke In a universal joint, the drivable torque-and-motion input and output member, attached to a shaft or tube.

ANSWERS TO REVIEW QUESTIONS

The answers to the chapter review questions are given here. If you want to figure your grade on any quiz, divide the number of questions in the quiz into 100. This gives you the value of each question. For instance, suppose there are 10 questions: 10 goes into a hundred 10 times. Each correct answer, therefore, gives you 10 points. If you answered 8 correct out of the 10 then your grade would be 80 (80 × 10).

If you are not satisfied with the grade you make on a test, restudy the chapter and retake the test. This review will help you remember the important facts.

Remember, when you take a course in school, you can pass and graduate even though you make a grade of less than 100. But in the automotive shop, you must score 100 percent all the time. If you make 1 error out of 100 service jobs, for example, your average would be 99. In school that is a fine average. But in the automotive shop that one job you erred on could cause such serious trouble (a ruined engine or a wrecked car) that it would outweigh all the good jobs you performed. Therefore, always proceed carefully in performing any service job and make sure you know exactly what you are supposed to do and how you are to do it.

CHAPTER 1

1. (d) 2. (d) 3. (b) 4. (c) 5. (b)
6. (d) 7. (d) 8. (a) 9. (c) 10. (b)
11. (c) 12. (b) 13. (c) 14. (c) 15. (c)
16. (a) 17. (b) 18. (d) 19. (a) 20. (c).

CHAPTER 2

1. (b) 2. (a) 3. (a) 4. (d) 5. (d)
6. (a) 7. (a) 8. (c) 9. (c) 10. (b)
11. (c) 12. (b) 13. (c) 14. (b) 15. (d)
16. (a) 17. (c) 18. (a) 19. (b) 20. (b).

CHAPTER 3

1. (c) 2. (c) 3. (a) 4. (b) 5. (c)
6. (a) 7. (b) 8. (c) 9. (a) 10. (d).

CHAPTER 4

1. (c) 2. (b) 3. (a) 4. (d) 5. (c)
6. (a) 7. (a) 8. (b) 9. (b) 10. (c)
11. (d) 12. (b) 13. (b) 14. (b) 15. (a)
16. (b) 17. (d) 18. (a) 19. (b) 20. (c).

CHAPTER 5

1. (b) 2. (d) 3. (c) 4. (b) 5. (c)
6. (c) 7. (a) 8. (c) 9. (b) 10. (b).

CHAPTER 6

1. (b) 2. (a) 3. (b) 4. (d) 5. (b)
6. (c) 7. (b) 8. (a) 9. (a) 10. (c).

CHAPTER 7

1. (d) 2. (a) 3. (d) 4. (b) 5. (b)
6. (d) 7. (b) 8. (b) 9. (a) 10. (c)
11. (d) 12. (c) 13. (b) 14. (c) 15. (b)
16. (c) 17. (c) 18. (a) 19. (b) 20. (d).

CHAPTER 8

1. (c) 2. (c) 3. (d) 4. (c) 5. (a)
6. (c) 7. (b) 8. (a) 9. (b) 10. (d).

CHAPTER 9

1. (b) 2. (c) 3. (b) 4. (b) 5. (d)
6. (d) 7. (a) 8. (c) 9. (b) 10. (d).

CHAPTER 10

1. (c) 2. (d) 3. (a) 4. (c) 5. (b)
6. (a) 7. (b) 8. (b) 9. (a) 10. (b).

CHAPTER 11

1. (c) 2. (d) 3. (b) 4. (a) 5. (d)
6. (b) 7. (a) 8. (c) 9. (c) 10. (b).

CHAPTER 12

1. (c) 2. (d) 3. (b) 4. (a) 5. (a)
6. (b) 7. (b) 8. (a) 9. (d) 10. (b)
11. (c) 12. (a) 13. (a) 14. (d) 15. (a)
16. (b) 17. (b) 18. (c) 19. (a) 20. (c).

CHAPTER 13

1. (b) 2. (b) 3. (a) 4. (d) 5. (a)
6. (b) 7. (c) 8. (a) 9. (b) 10. (d)
11. (c) 12. (a) 13. (b) 14. (b) 15. (d)
16. (c) 17. (a) 18. (b) 19. (d) 20. (b).

CHAPTER 14

1. (a) 2. (c) 3. (b) 4. (b) 5. (c)
6. (d) 7. (d) 8. (b) 9. (c) 10. (d)
11. (a) 12. (d) 13. (c) 14. (c) 15. (c).

CHAPTER 15

1. (c) 2. (c) 3. (d) 4. (a) 5. (c)
6. (c) 7. (d) 8. (c) 9. (c) 10. (b).

CHAPTER 16

1. (c) 2. (b) 3. (c) 4. (d) 5. (b)
6. (a) 7. (a) 8. (c) 9. (c) 10. (c).

CHAPTER 17

1. (c) 2. (c) 3. (d) 4. (b) 5. (a)
6. (b) 7. (d) 8. (c) 9. (a) 10. (c).

CHAPTER 18

1. (b) 2. (a) 3. (c) 4. (d) 5. (d)
6. (c) 7. (c) 8. (a) 9. (d) 10. (a).

CHAPTER 19

1. (a) 2. (d) 3. (c) 4. (b) 5. (b).